W9-DHF-820

KAORI O'CONNOR

CREATIVE DRESSING

THE UNIQUE COLLECTION OF TOP DESIGNER LOOKS THAT YOU CAN MAKE YOURSELF

Routledge & Kegan Paul

London and Henley

First published in 1980
by Routledge & Kegan Paul Ltd
39 Store Street,
London WC1E 7DD and
Broadway House, Newtown Road,
Henley-on-Thames, Oxon RG9 1EN
Photoset in Palatino, printed and bound
in Great Britain by
Redwood Burn Limited
Trowbridge & Esher

British Library Cataloguing in Publication Data

O'Connor, Kaori
Creative dressing.
1. Dressmaking
2. Costume design – Technique
I. Title
646.4 TT515

ISBN 0 7100 0680 2

Photography by Michel Haddi
Assisted by Peter Thorton
Design Consultant for Fabrics: Frederick Stillman
Hair by Lundy Morrone at Toni & Guy
Make-up by Sue Mann at The Model Agency
Styling by Kaori O'Connor with Terry Quigley

Acknowledgments

I would like to thank all those who have contributed to *Creative Dressing*: Rosalind Woolfson and the European Commission for the Promotion of Silk, Margaret Kinsey and Courtelle/Courtaulds, Jean Wiseman and Iris Bishop of Knitmaster, Robert Beath and Rosalind Ashton for Courtaulds, Michael Roberts and Bally Shoes, Robin Foster, Margaret Lambert and Brenda McAndrew of Sirdar, Diana Martin of Tootals, April Duxbury and Models One Elite, Jeanne Godfrey and Asprey, Simon Wilson and Butler & Wilson, Jennifer Loss and Margaret Kavanaugh of Charles Jourdan, Aristoc, Isobel Kerr and the International Institute for Cotton, Paula McNulty and the International Linen Board, Donald Firth and Suzanne Chivers of the International Wool Secretariat, Jo Grummitt and Sara Ives of ICI Fibres, Mr J. Cannon and British Enkalon, Patricia Owen and Monsanto, Janice Markham and Hoechst, Walter Damnes and Louisa Lambe of Supotco, Richard Wormersley and Zoë Hunt, Michael and Jackie Pruskin, Mrs Turk and Beale & Inman, Robert White & Son, Lady Bealey and Arabesque, G. Wood Fabrics, David Shilling, McMurray Picture Framers and Gilders, Detail, Tailpieces. Also the Hilton Hotel, the Savoy Hotel, Dukes Hotel, the Royal Horseguards Hotel, Nicky Legreq and the Park lane Hotel, all London.

Kato Yamashiro, George O'Connor, Martha Yamashiro, Judy O'Connor, Peter Hopkins, Michel Haddi, Sue Mann at The Model Agency, Lundy Morrone at Toni & Guy, Peter Thorton, Gail Kelly, Owen Ulph, Noel Rees, John Carroll, David Stonestreet, Terry Quigley, Jo Hart, David Babb, Gus Wood, Clive Wylie, Ann Frost, and Martin Hopkins.

Frederick Stillman, Maggie Dyke, Susan Guess, Laraine McCarthy, Susan Collier and Sarah Campbell, David Reeson, Hazel Chapman, Gordon Luke Clarke, Bruce Oldfield, Victor Herbert, Christine Ruffhead, Les Lansdown, Antony Kwock, Sara Fermi, Heinz Edgar Kiewe, Art Needlework Industries of Oxford, Kaffe Fassett, Jennifer Kiernan, Jaime Ortega, Suzanne Barnacle, Joan Proudfoot, Jane Sidey, Mary Wear and Maureen Baker.

Our models: Lenore, Wendy Morgan, Joanna Bartholomew, Sabina, Daisy, Tricia, Janine Freeman, Christina P, Polly Eltes, Esther de Deo, Amanda Bibbey, Linnea, Zara, Gayle Inglis and Sarah Bee of Models One Elite, Gail and David Clay of Gavin Robinson, Christopher Clarke and Helmut of Nevs, Raoul of Bobtons, all London.

Portrait of the Prince of Wales, 1925, by John St Helier Lander, City Art Gallery, Leeds. Portrait of Coco Chanel, Radio Times Hulton Picture Library, picture of Schiaparelli cravat sweater, Victoria and Albert Museum, portrait of Gordon Luke Clarke, *Daily Telegraph*, pictures of Raoul Dufy fabrics and design by Poiret from M. and J. Pruskin, Chenil Galleries, all London.

Additional credits: Hair for Sarong Wrap and Lalique by Caron at Toni & Guy. Make-up for Amy, Shady Lady ensemble and Transcultural Clothes section title by Yashi. Design and Designers section title dress by Victor Herbert. Handknitting section title, all hanks of yarn from Art Needlework Industries. Portrait of H.E. Kiewe by Vernon Brooke Photography of Oxford, portrait of Les Lansdown by Stan McInnes, portrait of Bruce Oldfield by Neil Kirk, portrait of Susan Collier and Sarah Cambell by Rathbone Street Studios, portrait of Michel Haddi by Serge Krouglikoff.

Michel Haddi

Contents

Machine Knitting

Handknitting

Creative Dressing Collection 147

Technological Chic

The Wedding 187

Introduction

The best way to learn about fashion is to look very closely at all the clothes you have banished to the back of your wardrobe. There's the chartreuse sweater – quite pretty, but such a difficult shade to wear. Of course, it *was* the colour of the season. And there's the tailored skirt – very smart still, but Nature has rather taken its course, and the waistband is such *agony* when you sit down. What about the long skirt you made last winter? Oh, you don't think that 'counts'. Here's that frothy frock trimmed with lace – well, it's not really *you*, so you feel a bit silly wearing it now, even though it was the rage back in . . . oh dear, that does date you. By now, you're ready to give way to despair. If only, you think, you had all the money in the world, you could rush out and buy the latest – but stop right there!

If you had spent years, as I have, in designer's studios and the best shops in the world, watching and writing about what people with all the money in the world *do* buy, you'd be very surprised indeed. The clothes are simple and subtle, they are based on good undemanding shapes and imaginative uses of colour, pattern and texture, and above all are *intended to make you feel and look good year after year*. Which is the difference between fashion – your discarded clothes – and style, for looks-of-the-moment and good designs are as different as night and day.

Fashion has become increasingly manufacture-orientated in the last thirty years, and the emphasis is always on things to *buy*, not things to make. But step into a designer's studio, and you'll find yourself a world away from mass production. It's not by chance that a studio looks very like a sewing room at home – a clutter of dress forms, bits of fabric, yarn tumbling out of cupboards, pins everywhere. For *good designs are made, not manufactured* – and that goes for clothes, racing cars, steam engines and anything else you care to name.

There are lots of publications that teach you how to sew or knit, but none that give you the incentive to carry on once you've mastered the techniques. Of course, there are commercial patterns, but they tend either to reflect extreme fashion trends or to rely on outmoded tailoring and fitting techniques that have nothing at all to do with style. And they all stop short of telling you the really important things – what makes a good design, how good designs are created, and how you can use them to develop a style of your own.

That's where CREATIVE DRESSING *starts*, and it goes on to give you a designer's view of style from many different angles. This book is full of the kind of clothes you can buy in the world's top shops – but as you'll see, it's much more exciting to create them yourself. I assume you already know how to sew or knit. No very complicated techniques are called for, and there's no nonsense about conventional sizing – measurements are given in inches and centimetres instead. Adapt the patterns as much as you like – they're guide-lines as well as good designs.

You don't need money or a perfect figure to look good – all you need is imagination and style. So here's to *your* good looks! And may you enjoy CREATIVE DRESSING as much as all of us have enjoyed creating it for you.

Kaori O'Connor

Readers' comments and enquiries are welcome. Write to Kaori O'Connor, *Creative Dressing* c/o the publishers.

Designing with Fabric

Frederick Stillman is a Design Consultant and Senior Lecturer in Textiles and Fashion

It's impossible to make clothes without using cloth! That's what I always tell my students. But obvious as it is, very few people really understand just what that means. Textile designers and printers often design without thinking of the end-product – the clothes that will be made. And equally, many clothes designers design in the abstract, with no thought at all for the fabrics that their sketches will be translated into, let alone the ways in which different fabrics will affect the design. Design must be *total*. The shape, the fabric, the surface textures and patterns, and colours must all work *together*.

If you want to be your own designer, I would suggest you start by collecting remnants and small pieces of different sorts of fabrics – even an eighth of a yard will do – and really get to know them. Learn how they drape, how they fall, how they stretch, how they feel – learn to recognize them with your eyes closed, working by touch alone. Put fabrics of the same weight and type together, mix colours and textures, then stand away from them, have a look, and start all over again. And whenever you're in a fabric shop, spend a bit of time putting one piece of fabric over another, experimenting in the same way. Practice makes perfect.

Shape can't live without colour. You must learn about colours in the same way that you learn about fabrics, breaking each primary colour into all its different shades, then seeing what effect different shades have on each other. You'll know you've got it right when a colour or combination of shades reaches out and grabs you – forces you to *do* something with it, won't *let* you put it down. Then go on to explore surface patterns. Not all shapes lend themselves to patterns. A tight, svelte sheath dress looks marvellous in black or any plain colour, but a large overall pattern would ruin everything. The sheath dress would look like a tattooed lady! Generally, patterned surfaces need to flow – they need movement and room to breathe. Fluid shapes based on drapery are particularly suitable for patterns – circular skirts, for example, have a pretty way of rippling the fabrics *through* the patterns. It's just an illusion of course, but it looks like the wind moving through high grasses. You can destroy many prints by cutting into them – think what a dart or seam does to a rose!

From a designer's point of view, the best fabrics are those that won't be destroyed by cutting. That's one reason why the paisley has survived for so long – it looks the same from every direction, and it's small enough to take the interruption of seams and darts that would kill a more definite design. But paisleys are often very boring – as are some prints that are easy to work with. Don't settle for that. Keep looking for those rare fabrics that combine ease with imagination.

I can't say enough about the value of *practice*. Experiment and experience are the best education you can have. Once you've become a fair dressmaker – one who understands how a paper pattern works – you should take a basic shape that suits you and use it to create designs of your own. Make the sleeves longer or shorter, or take them off altogether. Change the neckline, put a slash in the side seam, add pockets or take them away. Once you see what you can do with a really basic shape, there'll be no holding you back.

When you design for yourself, you can finish your clothes properly, add all the little details like coloured or patterned linings that you don't really get any more, even in *couture*. But more important – you can create a truly individual look, a style of your own.

In mass market production, everything has to be uniform – the cut, the fabric, the colours. As mass manufacturing has taken over, the areas of individual choice have become smaller. It's so depressing to go into the larger shops and see row after row of blue dresses or brown skirts all exactly alike, year after year after year. We all deserve better – and if we do it ourselves, we shall certainly have it.

I would like all of you to see this book as a guide to the possible, full of guidelines rather than hard and fast rules. Don't feel obliged to make up the designs as presented. Adapt and alter them as much as you like, and adapt the instruction sequence to suit the making up methods that work best for you.

Your confidence will increase with experience, and so will your style. Good shapes, fabrics, patterns and colours, all working together, are the basis of good design and good looks. And there are so many wonderful looks, so many beautiful possibilities, you'll find you can go on forever.

Frederick Stillman

Note All dress patterns are shown on a grid of one square = 5 cm (1 15/16 in). These should be transferred to squared pattern paper available at most stores. All measurements are given in inches and centimetres. Use only one set of measurements; do not mix the two.

Transcultural Clothes

Introduction

Everyone wants to look their best, but does that have anything to do with fashion? Designer Yves Saint Laurent doesn't think so. *'I hate fashion'* he says. 'It's a business. And that's a trap.' Fashion has become an industry of change instead of a profession of excellence, a business dedicated to making its own products obsolete every season. All the little tyrannies of the trade – the endless circle of 'new' silhouettes, 'now' colours, the 'right' heel types and hemlines – aim to convince you that 'new' is the same thing as *good*. It isn't.

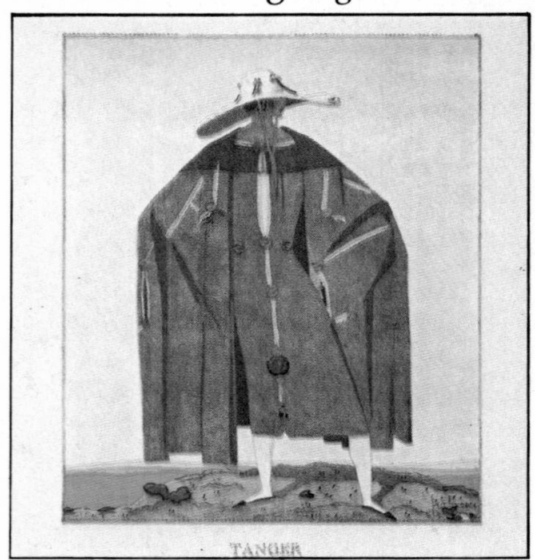

TANGER Design by Paul Poiret, *Gazette du Bon Ton*, February 1920

Imagine yourself without a sewing machine, broad fabrics, paper patterns, a tape measure, or even a sharp pair of scissors. These are the conditions in which folkloric and peasant clothes evolved, and not surprisingly they are the epitome of economic design. The mechanics of construction couldn't be simpler. The clothes for the most part are composed of square and oblong pieces, put together like a jigsaw puzzle with fabric lengths from narrow hand looms. There are few circles, fewer curves and tapered seams – no waste of movement or materials. There is no conventional sizing either – the clothes are meant to fit different bodies equally well, regardless of age and often of sex. It all sounds very primitive, but on closer inspection traditional clothes are everything that most modern fashions are not – practical, versatile, comfortable, durable and flattering. In short – masterpieces of style.

Style always starts with a good *shape*. And good shapes, above all else, are functional and practical. The clothes that achieve this best are the clothes of those who have had to do the most with the least: the traditional, everyday clothes of workers and peasants.

But what has function to do with fashion? Only the fact that the best-known names in fashion design have established their reputations by turning their back on change, and concentrating on style. And where do they seek inspiration? Coco Chanel's famous coats were based on the Breton fisherman's blouse, her chemise dresses on the shirts of Russian *moujiks*. Issey Miyake admits 'I get many ideas from simple clothes worn by *muscle workers* who have nothing to do with fashion, like rickshaw pullers', and Saint Laurent's collections are a veritable *couture* Grand Tour of other cultures. These are the sensational shows in which tziganes and tsaritsas, khadines from the harem and kaftan-clad houris from the kasbah slink down the catwalk arm in arm with Pierrots, Montparnasse painters and swirl-skirted peasants wrapped in cabbage rose shawls. Why do they do it? Saint Laurent speaks for them all. 'The folklore clothing of a peasant is *amazing*! The cut of traditional clothes is perfect simplicity. It's what has enabled them to go through the centuries without changing.'

Traditional shapes have been refined, improved and perfected over time in much the same way that fine horses are bred and great cuisines are created. If something is perfect, why change it for the sake of change alone? Why not just make the most of it? Which is not to say that traditional clothes are stagnant, or that they provide no opportunity for creativity. Things move more slowly in traditional societies, but the impulse for change and improvement is always there, based on the urge for display, changing tastes and the desire to be attractive – which everyone in the world shares. But since the shapes are *already* functional and change slowly, attention is focused on the surfaces – on colour, pattern and texture. If the shape is good, and you know how to use surfaces well, you really have a chance to develop subtlety and style. Here is a selection of some of the best traditional designs under the sun. See what you can learn from them – and enjoy what they can do for you.

The Japanese Kimono

The Japanese kimono, one of the best-known national costumes of the East, is a fine example of traditional design. The kimono has been worn since the early Nara period (AD 645–724) and has continued unchanged except for minor variations in the length of hem and sleeves for over a thousand years. It is cut in the same way for men, women and children, can be made up in summer or winter fabrics, and can be worn singly or in layers. There are no superfluous details on a kimono, not even pockets – the deep sleeves, open at the side, are the traditional place to carry small goods, handkerchiefs and fans. The stunningly simple logic of the kimono is summed up in the *ittan* – a tight cylinder of fabric 14 inches long and as thick as a rolled newspaper, that unrolled, cut, folded and stitched makes up into a kimono as perfectly as a piece of *origami* paper folds up to make a paper crane.

The kimono is distinguished by its symmetrical shape and large area of uninterrupted surface ideal for decoration. The Japanese wore no jewellery, and as their formalized hair and make-up styles offered little scope for individuality, all attention was concentrated on embellishing the kimonos. There were never any rigid rules about decoration, and the fact that the shapes were identical and unchanging stimulated the development of highly individual surfaces. Appreciation of fine fabrics and fabric design was general in all levels of society. Rich brocades, often woven to order, were the favourite fabrics of the Court, but much creative activity centred on embroidery, dyeing, tie-dyeing, hand-painting, *appliqué*, stencilling and block printing, which offered more scope for dramatic designs. In general, a kimono whose surface was covered by a rigid and repetitive overall pattern was considered undistinguished regardless of the quality of the cloth. The highly regarded design was one in which movement and contrast were suggested by colour, by pattern, and by the interplay of the two. Several techniques might be used on a single kimono, turning it into a riot of pattern on pattern, often in striking asymmetric placements.

An elaborate code of seasonal and symbolic motifs developed, and it was considered a matter of form to wear certain patterns and colours at the appropriate time of year. Maple leaf patterns were worn from late September through October, cherry blossom motifs in April, iris motifs in late May and June. Pine and bamboo, the symbols of long life and happiness, were considered appropriate for wear during the New Year festival as they were thought to bring good luck in the coming year.

Colour was as important as pattern, and the selection of good colour harmonies was considered as demanding as the composition of a good *haiku* poem. The sleeves and neck of formal kimonos for women were trimmed with twelve slender bands of colour, meant to suggest the *juni-hitoe*, twelve-layer-kimono, worn at Court in Heian times (AD 794–1185). The bands of colour were composed around certain themes, and the 'pine' colour sequence began with a dark reddish brown band, then a light one, then nine bands shading from yellow to dark green, with a final band of vermilion. If the tone of one of these bands was deemed to be not quite right, too pale or too dark, the whole outfit was considered to be a failure. The colours of linings and inner hems, parts of the kimono not normally visible, were chosen with great care – the hint of colour glimpsed as a sleeve fluttered or a hem swept to one side was another subtle indication of the wearer's taste.

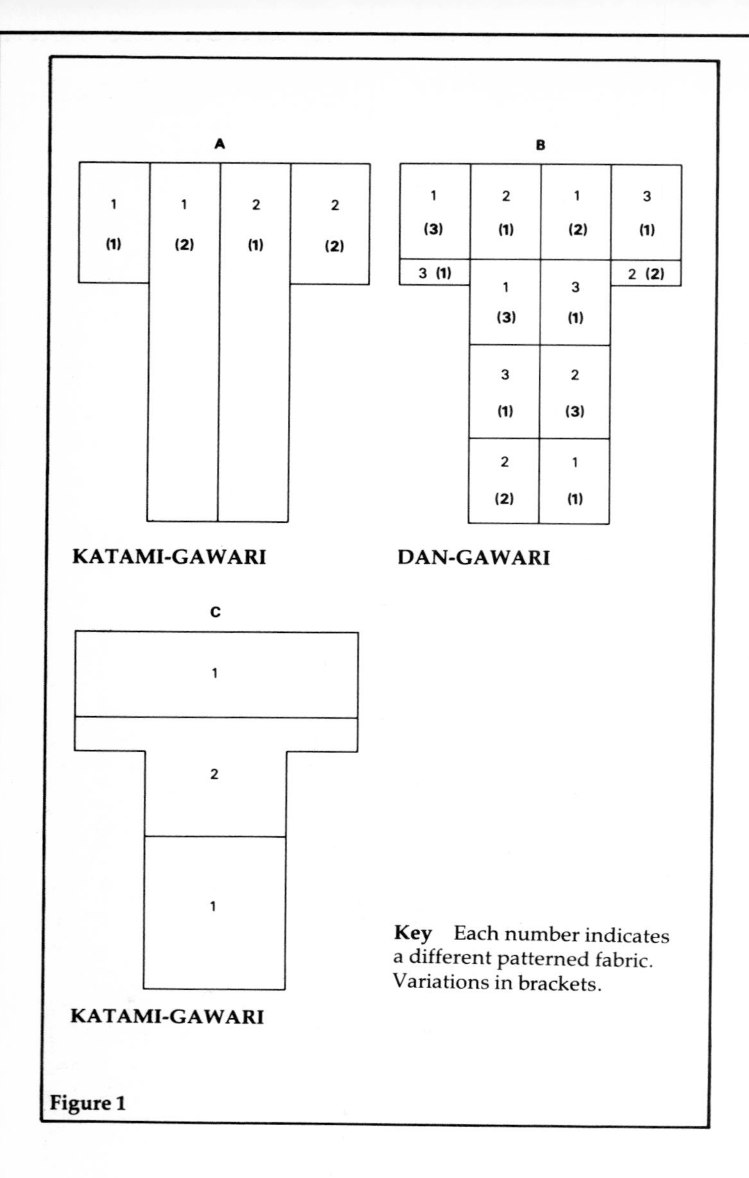

Figure 1

The more extreme of these refinements were not a part of peasant life, nor were the rich brocades, tie-dyeing and hand-painting that made aristocratic kimonos such sumptuous works of art. But the people who could not afford expensive techniques and materials made the most of what they had and perfected the art of fabric combination in the creation of *katami-gawari*, half-body kimonos, and *dan-gawari*, step kimonos. Different areas of the kimonos would be made up in different patterned fabrics, and the classic designs shown in Figure 1 can be extended into many variations. You may prefer the more conventional fabric placements used in the two kimonos here. The important thing is to experiment with this delightful and undemanding art form for which the kimono is the perfect medium.

The kimono is familiar to everyone as a lounging robe, something to wear at home, but you can do much more with it. As you'll see in the next section, a hand-painted kimono makes a lovely evening dress, an embroidered kimono a superb evening coat. The kimono shape is extremely adaptable, but a number of problems are posed by the *obi*, the wide, stiff brocade sash that is difficult to tie yourself, uncomfortable to wear if you are not used to it, and tends in any case to spoil the effect of the surface decorations. Don't worry about tradition – leave the *obi* off. The wide *obi* only came into use during the Genroku era (AD 1688–1703). As an alternative, adopt the soft late Kamakura style. Wear a single silky kimono, slip another kimono over it, tie the waist with a slender sash, then slip the top kimono back to rest on the shoulders, *en décolleté*. To get the most out of traditional designs, you should learn to wear them in an untraditional way. But there *is* one exception where the kimono is concerned. If you are wearing it closed, *always* be sure to wrap the left front *over* the right front.

Japanese Kimono Instructions

MATERIALS **Fabric required:** 4½ yd (4.1 m) of 36 in (90 cm) wide fabric. **Recommended fabrics:** suitable for fabrics that are soft and light in weight such as cotton lawn and fine poplin, polyester and polyester-cotton blends, silk, viscose, fine wool challis and wool-cotton blends such as Viyella. The use of fabric with directional patterns such as stripes may involve an increase in the amount of fabric required. **Made here:** light kimono in a combination of three prints on cotton poplin. Dark kimono made in a combination of two Collier Campbell prints on pure wool challis.

MEASUREMENTS One size, to fit **bust** 32–36 in (81–92 cm); **hips** 34–38 in (86–97 cm). **Back length** of finished garment: 55 in (140 cm). Hem and sleeve lengths can be adjusted; remember to alter fabric requirements accordingly.

NOTE All seam allowances are ½ in (1.2 cm) unless otherwise stated.

PREPARATION **(1)** Cut out pattern pieces following cutting layout. **ASSEMBLE FRONT BODY (1)** Take 1 front panel piece and neaten by turning in ¼ in (6 mm) to wrong side at point A. **(2)** With right sides together, place this point A to point A on kimono front body piece and stitch from A to B at hem. **(3)** Neaten by zigzag and press flat towards front edge. **(4)** Repeat for second front panel. **(5)** Neaten front edges E to hem on both panels by turning in: first turn ⅛ in (3 mm), second turn ⅜ in (1 cm). Slip stitch or machine finish. **JOIN SIDE SEAMS (1)** With right sides together, fold at shoulder point C bringing hems level. Stitch side seams from hem to points D. **ATTACH COLLAR (1)** Fold collar piece in half lengthwise, right side outside. **(2)** Press to establish

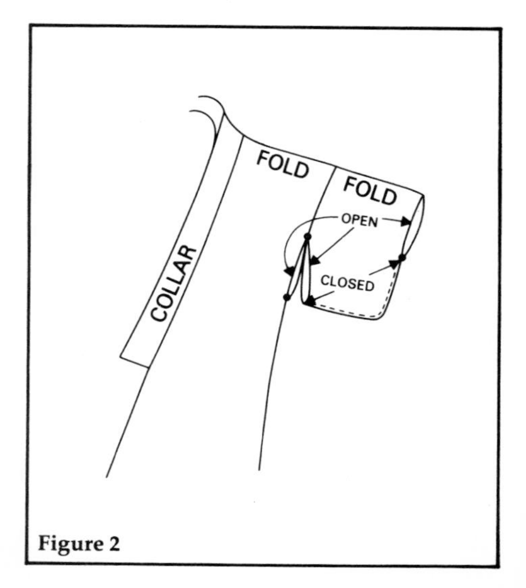

Figure 2

CUTTING LAYOUT

(not drawn to scale)
Fabric width:
36 in (90 cm)

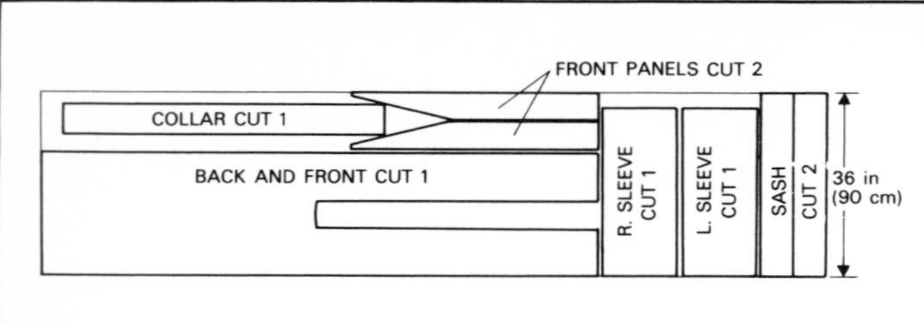

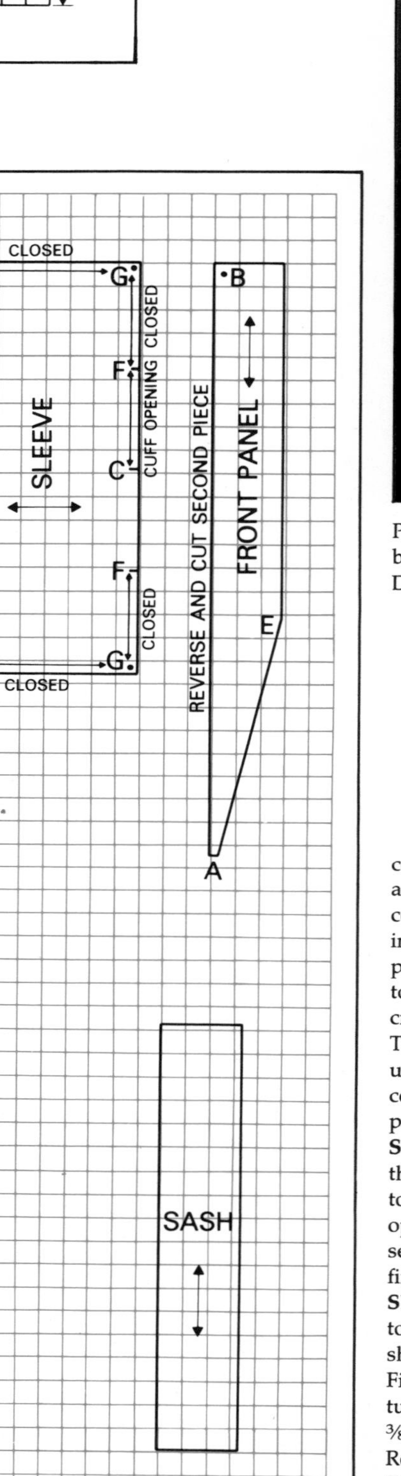

Poncho Top and Soft Skirt in two Collier Campbell prints. Jewellery by Reema Pachachi from Detail. See text on page 14.

centre crease edge. **(3)** Now open out the collar and, with right sides together, pin 1 layer of collar fabric to the back neck of the body, matching up centre point X on the collar and centre point X on back neck of body. **(4)** Stitch from X to point E on the front panel, leaving a ½ in (1.2 cm) turning. **(5)** Repeat for second panel. **(6)** Turn in collar ends ½ in (1.2 cm), then turn under ½ in (1.2 cm) along the inner edge of collar and slip stitch or machine stitch to body part, enclosing all raw edges. **(7)** Press. **ASSEMBLE SLEEVES (1)** Neaten edges by zigzag then fold sleeve piece on line C with right sides together, and stitch F-G-H. **(2)** Neaten cuff opening by turning up: first turn ⅛ in (3 mm), second turn ⅜ in (1 cm). Slip stitch or machine finish. **(3)** Repeat for second sleeve. **ATTACH SLEEVES TO BODY (1)** With right sides together, place sleeve centre C to kimono shoulder centre C. Stitch to J front and back; see Figure 2. **(2)** Neaten open vent below H to H by turning in; first turn ⅛ in (3 mm), second turn ⅜ in (1 cm). Slip stitch or machine finish. **(3)** Repeat for second sleeve. **COMPLETE HEM (1)** Turn up hem: first turn ⅜ in (1 cm), second turn 1½ in (3.8 cm) and slip stitch or machine finish. **(2)** Press. **TO FINISH (1)** Join remaining lengths of fabric to make a sash approximately 3 × 70 in (7.5 × 80 cm).

The Poncho

The rich artistry that the Japanese have lavished on the kimono's surface may slightly obscure the fact that the best thing about traditional clothes is their perfect simplicity of shape. The simplest of all sewn garments is the *poncho* – a Spanish word for a garment found all over the world, in all periods of history. It consists of a piece of skin or cloth with a hole in the centre for the head, fastened at each side. And it is the second oldest design in the world – the oldest being the model without the side fastenings, in which early man first sallied forth from his cave. It's a shape that's a million years old – and a shape doesn't last *that* long unless it's good.

Frederick Stillman's poncho top shows just why this shape has lasted so long and so well. The top suits an exceptionally wide range of figures in exactly the same way that the best designer clothes do. It softens prominent busts and rounds flat ones, sits well on curved or straight shoulders, and the way the fabric curves over the ends of the shoulders flatters the upper arms more than cap sleeves do. The construction couldn't be easier and a minimum of fabric is required. But although it's economical in both senses, the top looks elegant and expensive.

The poncho is also wonderfully functional. Think of the things a top should do. On the simplest level it has to cover you – but it also has to add something to the total look. Half the appeal of good clothes is the cloth they're made of and the poncho is an ideal way to show off a pretty fabric as there are no darts or front fastenings to interfere with the pattern or texture. You can make it up in combinations of fabrics – the body in a plain fabric and the facings in a print, or the other way round. And if you make it up in a plain fabric only, you'll find it lends itself to many sorts of accessories. A top also has to be worn with other things. The poncho suits skirts and trousers equally well, can be tucked in or worn as a tunic, can be made in summer or winter weight fabrics, and provides all the cover you need if you intend to wear it with a cardigan or jacket. And, after all this time, you know it will never go out of date.

Poncho Top Instructions

DETAILS Shown here with matching soft skirt and Balinese trousers.

MATERIALS **Fabric required:** 1½ yd (1.40 m) of 36 in (90 cm) wide fabric. **Recommended fabrics:** suitable for most dress-weight fabrics such as cotton lawn and poplin, polyester, viscose, silk and fine wool challis. Avoid hard or bulky fabrics that do not drape well. **Made here:** in printed silk Surah.

MEASUREMENTS One size, to fit **bust** 33–38 in (84–96.5 cm). **Back length** of finished garment, 24 in (61 cm). Length and width can be altered to suit personal requirements; adjust amount of fabric required accordingly.

STYLING The neck facing can be cut in a contrast fabric, but remember not to mix fabrics of different weights and fibres. Patch pockets can be added. The top can be worn tucked in, as a blouse, or left outside, as a tunic, with skirt or trousers.

NOTE All seam allowances are ¾ in (2 cm) for hem and sides, 3/16 in (5 mm) for neck facing slash.

PREPARATION **(1)** Cut out pattern pieces following cutting layout. **(2)** After cutting, neaten all edges by zigzag, overlocking, or by turning in the edges ⅛ in (3 mm) and machining flat. **ATTACH NECK FACING (1)** With right sides together, lay the neatened neck facing to the neck opening on the body part and stitch with a small seam. **(2)** Clip seam at the lower end of neck opening. **(3)** Turn through to the wrong side and press. **JOIN SIDE SEAMS (1)** Fold body pieces with right sides together and stitch side seams between A and B **only,** leaving open side vents. **HEM SIDE VENTS AND ARMHOLES (1)** Turn in hem at side vents and armholes, first turning ¼ in (6 mm), second turning ½ in (1.2 cm) and topstitch with a ⅜ in (1 cm) seam. **TO FINISH (1)** Turn up and finish bottom hem. **(2)** Press. **(3)** Join 2 sash pieces to make a sash approximately 60 × 3 in (152 × 8 cm) or as required.

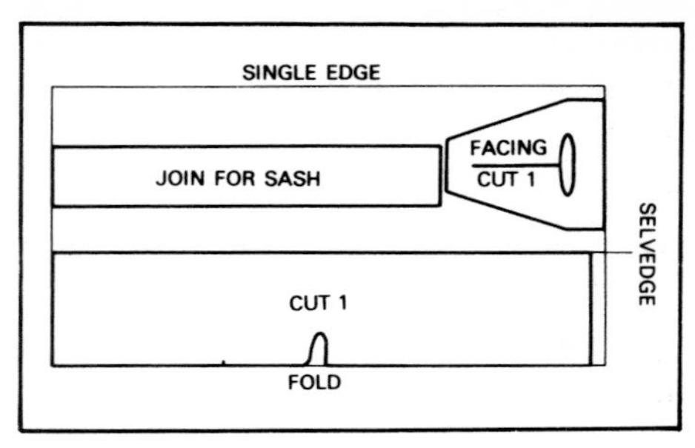

CUTTING LAYOUT (not drawn to scale)
Fabric width: 36 in (90 cm)

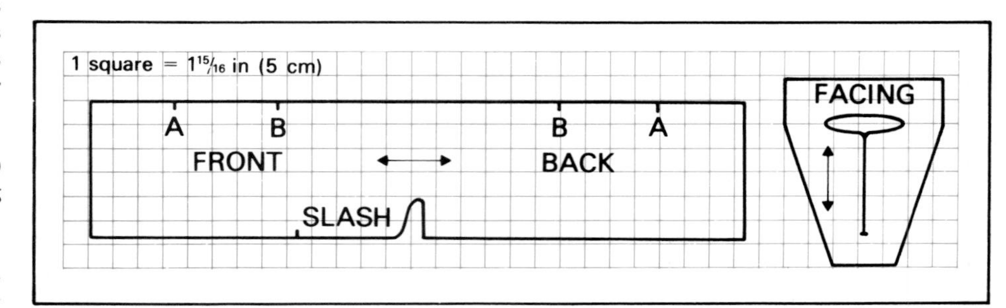

Soft Skirt

To show how adaptable the poncho is, we've shown it with a skirt and trousers, two mainstays of the modern wardrobe. The skirt, like the poncho, is found around the world, and not always as an exclusively feminine garment. Its simplest form is a length of cloth wrapped around the body, then tucked back into itself at the waist, like a sarong or the *lavalavas* worn in the South Pacific. Logically a skirt needn't have side seams or even a separate waistband, let alone a fitted one. However, most sarongs have a tendency to come undone, so closed-side skirts are much more practical – but what sort of skirt is best?

Rigid waistbands are uncomfortable, drawstring skirts tend to bunch at the waist, tailored skirts with darts don't move *with* the body, full gathered skirts are cumbersome and waste fabric. This skirt was developed along the lines suggested by traditional clothes – it is functional, practical and beautifully simple. It has just enough fullness to allow complete freedom of movement, enough economy of line to require little fabric. It suits a wide range of figure types and sizes beautifully, and you can shorten or lengthen it as much as you like without ruining the proportions. It makes the best possible use of elastic – which is as under-used in conventional clothing as zippers are over-used. And the elastic is not enclosed in a fabric casing, so you can change it in less time than it takes to change your nail polish. If you prefer a straight silhouette, try the side slit skirt in the next section. It's constructed on the same principles, and proves beyond the shadow of a doubt that you don't have to skimp to look svelte. You can make the skirt in plain or printed fabrics, and in summer or winter weights. And if you make it up in the same fabric as the poncho top as we have here, you'll have a superbly stylish two-piece dress that can take you anywhere.

Soft Skirt Instructions

DETAILS Shown with matching poncho top, shoes by Charles Jourdan and tights by Aristoc. Hat by Madcaps from Henri Bendel, New York.

MATERIALS **Fabric required:** 1⅝ yd (1.50 m) of 36 in (90 cm) wide fabric. **Recommended fabrics:** suitable for most dress-weight fabrics including cotton lawn and poplin, polyester, viscose, silk and wool challis. Avoid heavy, thick and stiff fabrics that will not gather up under the elastic waistband. **Trimming:** 1 piece of 2 in (5 cm) wide elastic 22–27 in (56–68.5 cm) in length for waistband. Determine final length by stretching the elastic around your waist until you hit on a comfortable fit. Allow for the fact that joining the ends (seam allowance ½ in or 1.2 cm) will reduce the length by a total of 1 in (2.5 cm). **Made here:** in printed silk Surah.

MEASUREMENTS One size, to fit **waist** 24–28 in (61–71 cm), **hips** 36–39 in (91.5–96.5 cm). **Length** 27½ in (70 cm) from bottom of elastic waistband to finished hem. Length can be adjusted to suit personal requirements; alter amount of fabric required if necessary.

STYLING For a waistband finish that does not show the elastic, substitute a length of narrow elastic (1 in or 2.5 cm wide) for the wider elastic. Stitch the elastic to the top of the skirt following instructions below, then fold elastic over to the inside of the skirt.

NOTE Seam allowances for hem and side seams are ¾ in (2 cm); for waistline edge, ⅜ in (1 cm).

PREPARATION (1) Cut out pattern pieces following cutting layout. **(2)** After cutting, neaten the side seam allowances by zigzag, overlocking, or by turning in the edges ⅛ in (3 mm) and machining flat. **NEATEN WAISTLINE (1)** Turn the top edge of the waistline in to the right side ⅜ in (1 cm) and machine stitch edge on back and front parts. **JOIN SIDE SEAMS (1)** Lay the front and back pieces together with right sides inside and join side seams, leaving open the pocket mouths A–B. **(2)** Press side seams open, including the pocket mouths. **ASSEMBLE POCKET BAGS (1)** Lay 2 of the pocket bag pieces together with right sides outside and stitch a ⅛ in (3 mm) seam around pocket edge from exact point D to exact point E. **(2)** Nip the seam allowance to E. **(3)** Turn pocket through and press. **(4)** Stitch around the pocket bag from point D to point E, ¼ in (6 mm) from the edge, locking in the raw edges. **(5)** Neaten the edges of the pocket mouth seam allowances. **(6)** Repeat for second pocket bag. **ATTACH POCKETS TO SIDE SEAMS (1)** Right sides together, take 1 finished pocket bag and pin into position on the side seam allowance, matching up D/A and E/B. **(2)** Stitch the pocket bag mouth seam allowance F to the skirt side seam allowance F. **(3)** Repeat for G/G, as shown in Figures 1 and 2. **(4)** Repeat for second pocket. **(5)** Turn the skirt to the right side and tack the top and bottom corners of the pocket mouths firmly, by hand or machine. **MAKE WAISTBAND (1)** Join ends of elastic to make the waistband and neaten the ends by machining them down flat. **ATTACH WAISTBAND TO SKIRT (1)** Turn the skirt and waistband to their right sides. **(2)** Keeping the waistband flat, slip it over the top edge of the skirt. The bottom of the waistband should come 1 in (2.5 cm) below the finished top edge of the skirt. **(3)** Pin the skirt and elastic waistband together, making sure that the fullness is evenly distributed. **(4)** When the pieces have been pinned together, place machine foot in position on the elastic ⅜ in (1 cm) in from edge on the elastic side. **(5)** Now stretch elastic gently and stitch waistband and skirt together, keeping seam ⅜ in (1 cm) from bottom edge of waistband. Use the rib of the elastic as a guide. **TO FINISH (1)** Turn up hem, first turning ¼ in (6 mm), second turning ½ in (1.2 cm) and press. **(2)** Working from right side of fabric, stitch hem ⅜ in (1 cm) up from turned up edge. **(3)** Press.

CUTTING LAYOUT
(not drawn to scale)
Fabric width:
36 in (90 cm)

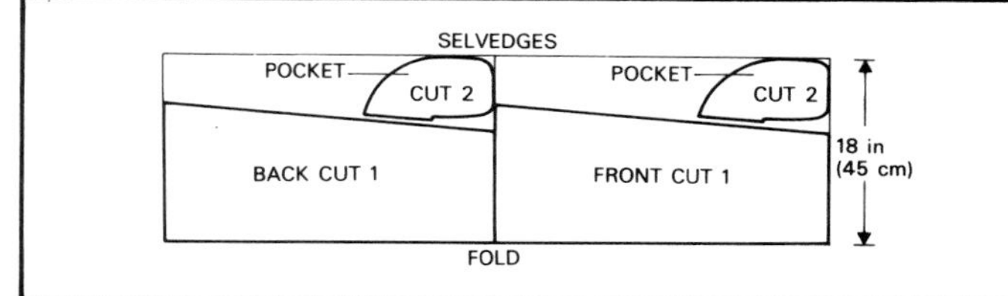

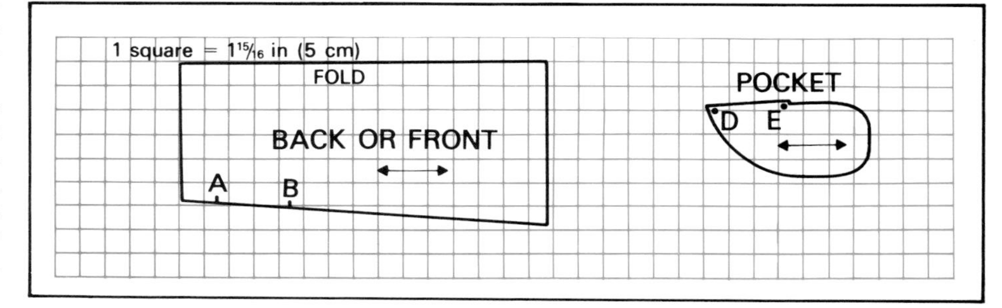

Balinese Trousers

These trousers from the Indonesian island of Bali are a masterpiece of ingenuity. Trousers in general present two problems, the first of which is fit. Loose drawstring trousers present no size problems, but they tend to add bulk around the waist and don't adapt well to wearing with shirts and sun tops. Although fitted trousers come in many styles, trousers that fit properly are very rare indeed. Good jeans are the exception, but they have all the virtues of traditional clothes at their best. The problem lies in the fact that two people with the same waist and hip measurements will never be shaped in exactly the same way. Each body settles into a pair of trousers differently, and with very fitted trousers there's often no place for the body to go. Also, the crotch height in fitted trousers is not adjustable, and if the height is not right, which it often isn't, severe discomfort is the inevitable result. The second problem is cover. In hot climates where the free circulation of air is essential to comfort, conventional trousers are often too restricting. And when trousers are not worn for religious reasons, as they are in Muslim countries, the coverage offered by conventional trousers is sometimes more complete than is really necessary. Shorts are an alternative, but not a realistic one for many figures and age groups.

And what better solution could there be than these wraparound trousers with open sides. They lie flat to the body, suit every style of top, fit a wide range of shapes and lend themselves to different lengths. They make wonderful resort wear and fabulous evening trousers – make them in black satin, top them off with a poncho in white satin, and there you have a superbly chic party piece. You tie the trousers on where they feel best, so you can adjust the crotch height and the degree of wrapover, wear them tied in a floppy knot at the ankle or leave them loose to show a flash of leg. It couldn't be simpler – all you have to do is get into them, and this is how you do it. With right sides inside, fold the trousers down the centre along the inner leg seams to form a U with the crotch at the bottom. Pass the folded trousers between your legs and raise the waistbands to waist level. Open out the back section, bring the ties round to the front and tie. Open out the front section over the first ties, bring the front ties round to the back, and tie in a knot or a bow.

Balinese Trousers Instructions

DETAILS Short trousers shown with shirt by Pua Hawaii and sandals from Marks & Spencer. Printed long trousers shown with matching poncho top. Plain long trousers shown with shoes by Bellesco from Bally.

MATERIALS **Fabric required for short trousers:** 2 yd (1.85 m) of 36 in (90 cm) wide fabric; **for long trousers:** 2¾ yd (2.55 m) of 36 in (90 cm) wide fabric. **Recommended fabrics:** suitable for a wide range of light to medium weight day and evening fabrics such as fine towelling, cotton poplin, voile and lawn, viscose, fine wool, polyester and polyester crepe de Chine, fine flannel, suede fabric and silk. **Made here:** short trousers in Tootal's Osmanda wool and cotton flannel; printed long trousers in silk Surah; plain long trousers in cotton muslin.

MEASUREMENTS One size, to fit **waist** 23–28 in (56.5–71 cm), **hips** 33–38 in (84–96.5 cm). **Finished length** of short trousers, 28 in (71 cm); of long trousers, 41 in (104 cm). Length can be adjusted to suit personal requirements; alter amount of fabric required if necessary.

PATTERN VARIATIONS **For tall sizes** requiring a deeper crotch fitting, simply lower the crotch line on the pattern, retaining the original crotch width and shape. **For smaller sizes** the waist circumference and wrapover can be reduced by inserting extra waist darts back and front, adjusting the waistband facing accordingly.

STYLING The long version can be worn tied at the ankles or left loose, and the short version can be hemmed with turn-up cuffs. Patch pockets can be added to front, back or both.

NOTE Seam allowances: for crotch seam ⅜ in (1 cm); for side hems 1 in (2.5 cm); for hems 1½ in (3.8 cm).

PREPARATION **(1)** Cut out pattern pieces following cutting layout. **(2)** Neaten the side hems by turning in the seam allowance 2 turns to the wrong side, first turning ¼ in (6 mm), second turning ¾ in (2 cm). **(3)** Topstitch with straight or zigzag stitch. **(4)** Neaten leg hems in the same way, first turning ¼ in (6 mm), second turning 1¼ in (3 cm). Topstitch with straight or zigzag stitch. **(5)** Sew all 4 darts and press. **ASSEMBLE POCKETS (OPTIONAL)** **(1)** Cut the patch pocket pieces from the crotch surplus, turn in the edges and press. **(2)** Machine finish the opening edges and press. **(3)** Place the completed patch pockets in position on the leg pieces. **(4)** Topstitch with firm finishing tacks at corners. **JOIN LEG PIECES** **(1)** Lay the 2 leg pieces together with right sides outside and stitch crotch seams A-B, A-B. **(2)** Trim seam allowance to half, turn through to

the wrong side and stitch again, locking in all raw edges. **ASSEMBLE WAIST TIES (1)** Take 1 tie piece and, with right sides together, fold in half lengthwise and stitch down one side and across one end with a ¼ in (6 mm) seam. **(2)** Trim corner thickness, turn through and press. **(3)** Repeat for remaining 3 waist ties. **ATTACH WAISTBAND FACINGS (1)** With right sides together, lay 1 waistband facing to waistline edge of 1 body piece, matching up points C. **(2)** Stitch together with a ⅜ in (1 cm) seam. **(3)** Press facing to inside of body piece. **(4)** Repeat for second facing and body piece. **(5)** Insert the raw end of 1 waist tie ½ in (1.2 cm) inside the waistband facing at X on all 4 openings. **(6)** Turn the facing ends in (⅜ in or 1 cm) on to the ties at X and machine stitch to secure ties firmly. **(7)** Turn up the lower edge of the facings in the same way, and machine stitch down to body parts. **TO FINISH (1)** Press.

CUTTING LAYOUT
(not drawn to scale)
Fabric width:
36 in (90 cm)

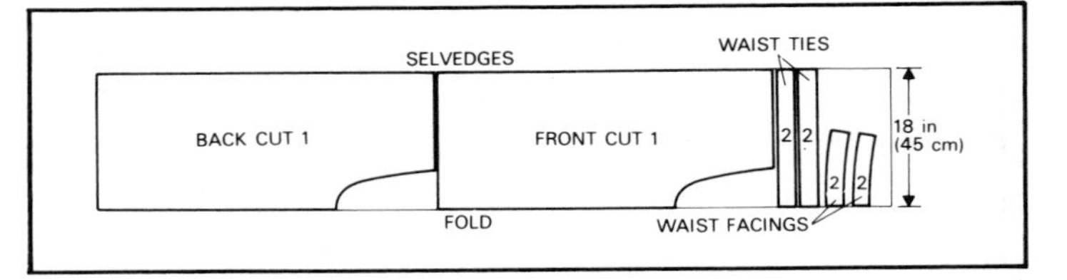

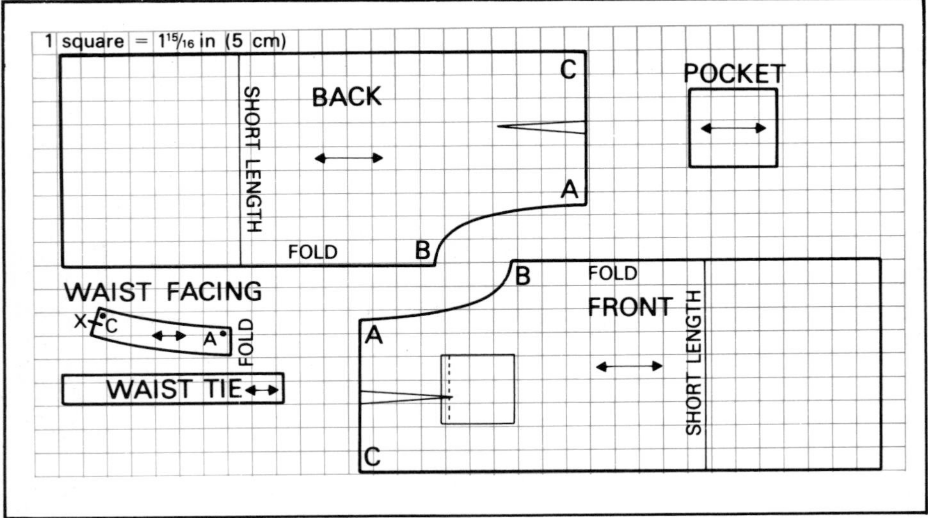

Indochinese Jacket

The poncho, skirt and Balinese trousers are excellent examples of the way in which traditional clothes can achieve a smart, sleek look without any conventional tailoring at all. Why complicate matters needlessly? Fussy, over-fitted fashions can never look or feel as good as shapes that follow the logic of the body.

This design, a version of the loose silk and cotton jackets found throughout Indochina, makes exactly the same point. Sharply tailored classic blazers don't suit everyone, nor do boxy tweed jackets – but a pretty suit is an asset that no woman should have to do without. 'Softly tailored' has come to mean clumsy, baggy versions of masculine shapes carried out in floppy fabrics, but nothing could be more graceful than this example of what soft tailoring for women should be. The lines are all relaxed and gentle, helping to flatter large or over-square shoulders, and although the jacket has a definite shape, you can move with perfect ease. Easy, roomy shapes like this always make large people look smaller, and although it's shown here with a long skirt, it looks just as good with a short one.

Conventional tailored jackets are best made up in a plain fabric or a non-patterned tweed as the cut tends to distort the fabric. This jacket has enough shape and style to look good in one colour, but it works just as well in two – the body in one colour, the lapels in a contrast or toning shade. You can combine a print and a plain fabric, or use several prints, as shown here. The body of the second jacket is made in an all-over print with lapels in a stripe. The skirt is made in a wavy border print that repeats a band of the pattern used for the jacket body, and the poncho blouse is made from the striped fabric cut on the bias.

The jacket can be made up in day or evening, summer or winter fabrics, and you can wear it with the cowl neck blouse or poncho. They add up to a winning suit – and all without a single dart.

Indochinese Jacket Instructions

DETAILS Printed jacket shown with long soft skirt and poncho top, earrings by Butler & Wilson and shoes by Bellesco from Bally. Photographed at the Hilton Hotel, London. Plain jacket shown with long soft skirt and cowl neck blouse, hair comb by Tailpieces and jewellery by Butler & Wilson.

MATERIALS **Fabric required:** 2⅝ yd (2.40 m) of 36 in (90 cm) wide fabric. **Recommended fabrics:** suitable for most soft, dress-weight fabrics including silk, viscose, very fine wools and polyester crepe de Chine. Cottons can be used if soft and reasonably crease-resistant. Plain fabrics are recommended for ease of making, but patterned fabrics can be very effective if chosen with care, see Styling below. **Made here:** plain jacket and skirt in Tootal's Casablanca spun viscose. Printed jacket and skirt in a combination of three Collier Campbell prints on pure wool challis.

MEASUREMENTS One size, to fit **bust** 33–38 in (84–96.5 cm). **Back length** of finished garment, 27½ in (70 cm). Length can be adjusted to suit personal requirements; alter amount of fabric required if necessary.

STYLING Fabrics with all-over patterns that can be cut up or down the grain without distorting the design are easiest to work with. Directional fabrics such as stripes and border prints are more demanding, and may oblige you to increase the amount of fabric required in order to match up the motifs.

NOTE All seam allowances are ⅜ in (1 cm) unless otherwise stated.

PREPARATION **(1)** Lay the pattern pieces on the right side of the fabric and cut following cutting layout. **(2)** After cutting, neaten gussets, sleeve seams, sleeve heads and side seams by zigzag, overlocking or by turning in the edges ⅛ in (3 mm) and machining flat. **ATTACH GUSSETS TO SLEEVES (1)** Lay 1 sleeve piece and 1 gusset piece together, matching points C to C and B to B. **(2)** Stitch together and press seams open. **ATTACH SLEEVE TO BODY (1)** Right sides together. Starting from the shoulder point, stitch front of sleeve to front of body, matching points B to gusset seam B/B, A to A, and down to vent mark. **(2)** Place gusset point BX to BX on the sleeve, stitch right through to cuff and press seam open. **(3)** Starting from the shoulder point, stitch down back to vent mark, matching B to gusset seam BX/BX and A to gusset seam A/A. Press seam open. **(4)** Repeat for second sleeve and gusset. **ATTACH COLLAR FACINGS (1)** With right sides together, place point G on a single layer of the collar piece to point G at centre of back neck on the body piece. **(2)** Tack the collar to the neck and down each front to the hemline, matching up points H to H. **(3)** Stitch collar into position. **(4)** Press seam flat towards collar piece. **FINISH SIDE VENTS (1)** Press back the side vent seam allowances. **(2)** Starting from the hem, topstitch up one side of the side vent, across the side seam ¼ in (6 mm) above the top of the slit, and down the other side to hem. Seam allowance: ¼ in (6 mm). **(3)** Repeat for second side vent. **HEM JACKET (1)** Turn up the hem of jacket all round, including collar piece; first turning ⅛ in (3 mm), second turning ¼ in (6 mm) and topstitch. **COMPLETE COLLAR (1)** Starting at point G at centre back neck, turn under the collar facings ⅜ in (1 cm) and tack to the stitching line, locking in the raw edges. Continue down each front to the hemline. **(2)** Slip stitch the facing to the back of the original seam line. **(3)** Leave the bottom edges of the collar and facings open. **TO FINISH (1)** Hem sleeve cuffs as for jacket. **(2)** Press.

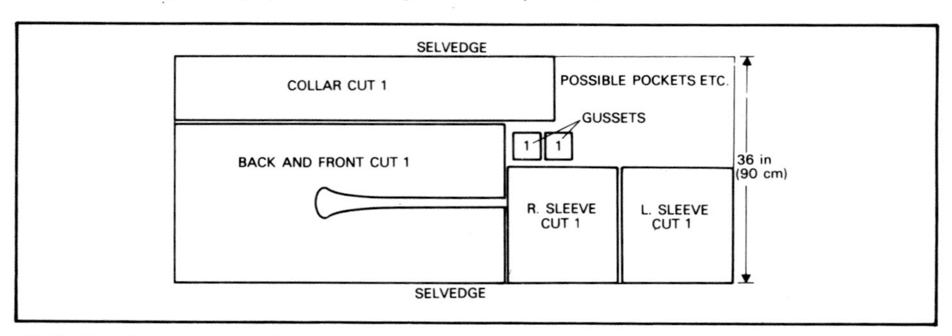

CUTTING LAYOUT (not drawn to scale)
Fabric width: 36 in (90 cm)
Surplus fabric can be used for patch pockets.

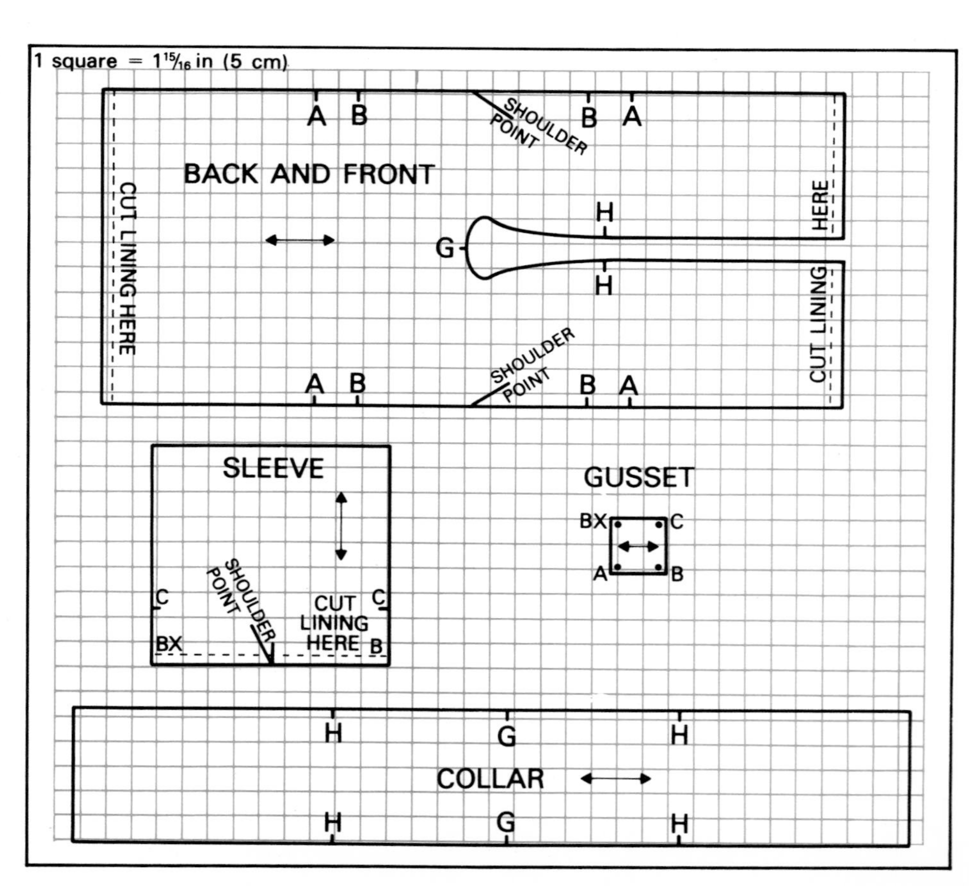

Long Soft Skirt Instructions

MATERIALS **Fabric required:** 2½ yd (2.30 m) of 36 in (90 cm) wide fabric. **Recommended fabrics:** suitable for most dress-weight fabrics including cotton lawn and poplin, polyester, viscose, silk, wool challis and light wool-cotton blends such as Viyella. Avoid heavy, thick and stiff fabrics that will not gather up under the elastic waistband. **Trimming** 1 piece of 2 in (5 cm) wide elastic 22–27 in (56–68.5 cm) in length for waistband. Determine final length by stretching the elastic around your waist until you hit on a comfortable fit. Allow for the fact that joining the ends (seam allowance ½ in or 1.2 cm) will reduce the length by a total of 1 in (2.5 cm). **Made here:** in Tootal's Casablanca spun viscose and in a Collier Campbell print on pure wool challis.

MEASUREMENTS One size, to fit **waist** 24–28 in (61–71 cm), **hips** 36–39 in (91.5–96.5 cm). **Length:** 39½ in (100 cm) from bottom of waistband to finished hem; can be adjusted to suit personal requirements. Alter amount of fabric required if necessary.

STYLING For a waistband finish that does not show the elastic, substitute a length of narrow elastic (1 in or 2.5 cm wide) for the wider elastic. Stitch the elastic to the top of the skirt following instructions below, then fold elastic over to the inside of skirt.

NOTE Seam allowances for hem and side seams are ¾ in (2 cm); for waistline edge, ⅜ in (1 cm).

PREPARATION **(1)** Cut out pattern pieces following cutting layout. **(2)** After cutting, neaten the side seam allowances by zigzag, overlocking or by turning in the edges ⅛ in (3 mm) and machining flat. **NEATEN WAISTLINE (1)** Turn the top edge of the waistline in to the right side ⅜ in (1 cm) and machine stitch edge on front and back parts.

JOIN SIDE SEAMS (1) Lay the front and back pieces together with right sides inside and join side seams, leaving open the pocket mouths A–B. **(2)** Press side seams open, including the pocket mouths. **ASSEMBLE POCKET BAGS (1)** Lay 2 of the pocketbag pieces together with right sides outside and stitch a ⅛ in (3 mm) seam around pocket edge from exact point D to exact point E. **(2)** Nip the seam allowance to E. **(3)** Turn pocket through and press. **(4)** Stitch around the pocket bag from point D to point E, ¼ in (6 mm) from the edge, locking in the raw edges. **(5)** Neaten the edges of the pocket mouth seam allowances. **(6)** Repeat for second pocket bag. **ATTACH POCKETS TO SIDE SEAMS (1)** Right sides together, take 1 finished pocket bag and pin into position on the side seam allowance, matching up D/A and E/B. **(2)** Stitch the pocket bag mouth seam allowance F to the skirt side seam allowance F. **(3)** Repeat for G/G, as shown in Figures 1 and 2. **(4)** Repeat for second pocket. **(5)** Turn the skirt to the right

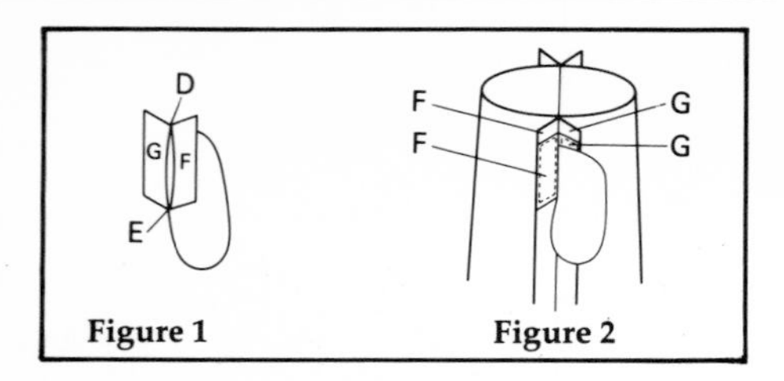

CUTTING LAYOUT
(not drawn to scale)
Fabric width:
36 in (90 cm)

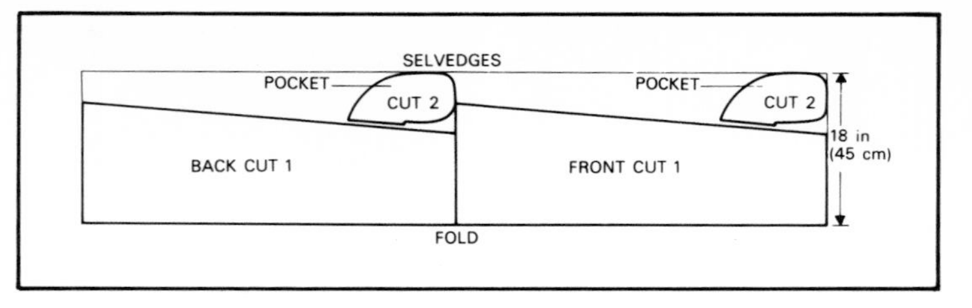

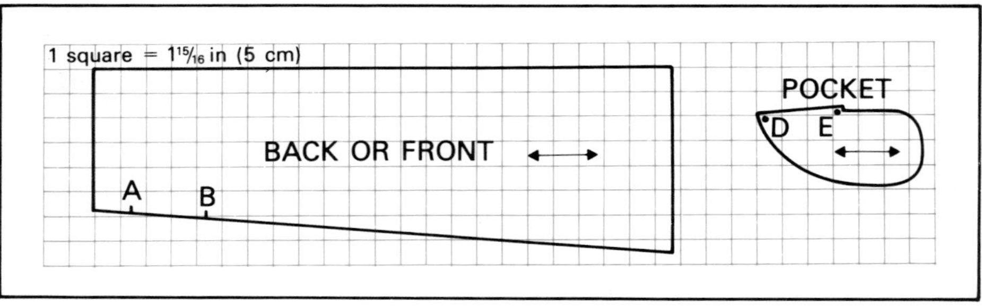

side and tack the top and bottom corners of the pockets mouths firmly, by hand or machine. **MAKE WAISTBAND (1)** Join ends of elastic to make the waistband, and neaten the ends by machining them down flat. **ATTACH WAISTBAND TO SKIRT (1)** Turn the skirt and waistband to their right sides. **(2)** Keeping the waistband flat, slip it over the top edge of the skirt. The bottom of the waistband should come 1 in (2.5 cm) below the finished top edge of the skirt. **(3)** Pin the skirt and elastic waistband together, making sure that the fullness is evenly distributed. **(4)** When the pieces have been pinned together, place machine foot in position on the elastic ⅜ in (1 cm) in from edge on the elastic side. **(5)** Now stretch elastic gently and stitch waistband and skirt together, keeping seam ⅜ in (1 cm) from bottom edge of waistband. **TO FINISH (1)** Turn up hem, first turning ¼ in (6 mm), second turning ½ in (1.2 cm) and press. **(2)** Working from right side of fabric, stitch hem ⅜ in (1 cm) up from turned up hem edge. **(3)** Press.

Cowl Neck Top Instructions

MATERIALS **Fabric required:** 1¾ yd (1.60 m) of 26 in (66 cm) wide fabric. **Recommended fabrics:** any very soft fabrics that drape well, such as polyester jersey or polyester crepe de Chine. Avoid stiff or bulky fabrics that do not drape well. Also avoid printed fabrics with designs that will be distorted by the gathering of the cowl neck. **Made here** in polyester jersey.

MEASUREMENTS One size, to fit **bust** 32–36 in (81–91 cm). **Back length:** 22 in (56 cm).

STYLING For a softer finish, the topstitching can be done with a zigzag rather than a straight stitch. The top can be lengthened and worn as a tunic over skirt or trousers, with patch pockets as a further option. The cowl neck piece can be made in a different print or colour from the rest of the body, but remember not to mix fabrics of different fibres or weights.

NOTE Seam allowances for body and hem are ½ in (1.2 cm); for neck and armholes, ¼ in (6 mm); for top of cowl neck piece, 1 in (2.5 cm).

PREPARATION (1) Cut out pattern pieces from single layer of fabric following cutting layout. **The cowl neck piece must be cut so that the top fold edge lies to the bias, or softest grain, of the fabric. (2)** Neaten all edges with zigzag. **GATHER UP COWL NECK PIECE (1)** Gather up A–B on each side of cowl neck piece to fit A-B on bodice. **(2)** Fold down the facing at top of cowl; do not press or stitch. **ATTACH COWL TO NECK (1)** Stitch cowl to neck opening from C to slightly rounded points at X. **(2)** Complete the neck by turning the back neck seam allowance C-C ³⁄₁₆ in (5 mm) to the wrong side, then topstitch by zigzag from the right side. **COMPLETE ARMHOLES (1)** Turn in the neatened edges and topstitch as for back neck. **JOIN SIDE SEAMS (1)** Lay top with right sides together and stitch from armholes to top of vents. **(2)** Turn up seam allowance on vents and hem ⅜ in (1 cm) and topstitch. **TO FINISH (1)** Make up a sash by zigzagging the edges of a single thickness of fabric approximately 60 x 4 in (153 x 10 cm) or as required.

CUTTING LAYOUT
(not drawn to scale)
Fabric width:
26 in (66 cm)

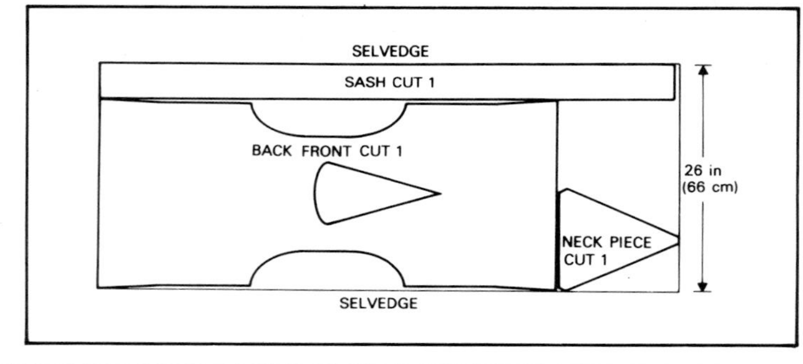

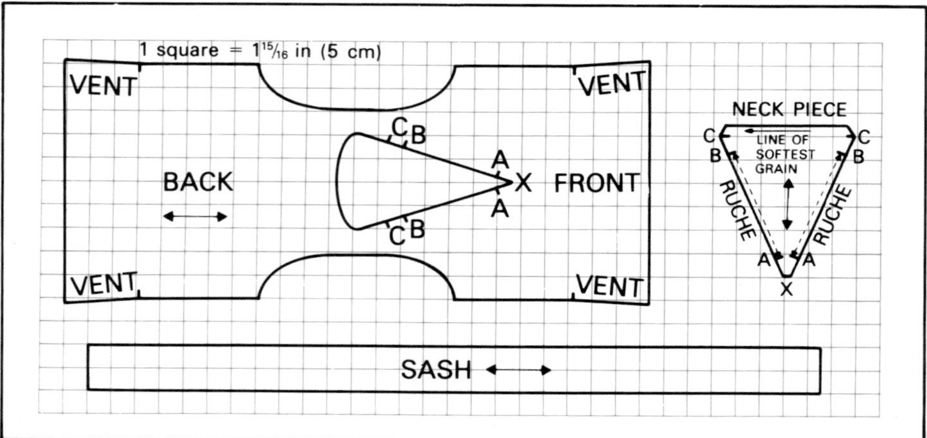

Ottoman Kusak Dressing Gown

Traditional Eastern clothes are nothing new to the West, although their influence on traditional Western clothes is rarely acknowledged. When Constantinople fell in 1454, the great cities of Italy were flooded with refugees whose oriental robes in magnificent Byzantine fabrics started a vogue for dressing *alla Turchesa*. Some fifty years later, the Turkish style of dress was so generally accepted that Lorenzo the Magnificent wore a Turkish turban to his wedding celebrations as a matter of course. Slowly but surely the Turkish influence spread across Europe. English court records of 1510 show Henry VIII leading an evening of festivities in the Palace of Whitehall clad in a golden Turkish gown, girded with a scimitar, his head wrapped in a turban of red velvet and gold.

These Ottoman Turkish gowns, called dolmans or stamboulines, were something of a cross between kaftans and coats. Made in brightly coloured brocades trimmed with embroidery, fur and sometimes jewels, they were worn over baggy trousers and soft leather boots. Some gowns had elaborate ornamental double sleeves which trailed almost to the floor, and all gowns were belted with a wide *kusak* sash in brilliant silk, wound around the body ten to fifteen times.

That this is an exceptionally handsome design for men is shown by Frederick Stillman's majestic *kusak* gown adapted for wear as a dressing gown. The double sleeve has been simplified into deep shaped cuffs, the *kusak* sash slimmed down into a shaped tie that conveys the opulence of the original with none of its bulk. But the basic shape has not been altered, and cannot be improved upon. It achieves to perfection what classic masculine tailoring never quite manages to carry off – the illusion of a figure that is wide through the shoulders, tapering down to a slender waist. Classic tailoring tries to accomplish this by pinching the body until it looks and feels like it's trapped in a corset. The secret of the Ottoman gown lies in the large dolman sleeve set into the body just above the waist. When the sash is tied, the dolman line automatically makes the shoulders look broader and the waist smaller. Where conventional dressing gowns rely on artificial pulling and padding, this design exaggerates the shape, builds it into the garment, and lets it fall into place naturally with plenty of room to spare.

Above all else, a dressing gown is something to relax in, and this design sits on the body so lightly you hardly know it's there. The perfect accessories? A tray of Turkish Delight and a few well-turned quatrains from the Rubaiyat of Omar Khayyam.

Ottoman Kusak Dressing Gown Instructions

DETAILS Shown with wing collar shirt and white tie from Moss Bros, London.

MATERIALS **Fabric required for body:** 4 ⅜ yd (4 m) of 45 in (115 cm) wide fabric. **Fabric required for contrast facing:** 2¼ yd (2.10 m) of 45 in (115 cm) wide fabric. **Interfacing:** 1 yd (95 cm) of soft iron-on interfacing such as Vilene Superdrape. **Recommended fabrics for body:** wool challis and other lightweight worsteds, soft brocade, fine velvet, broadcloth and other soft fabrics suitable for dressing gowns. Avoid thick, stiff and bulky fabrics. **Recommended fabrics for contrast facing:** satin or wool challis. **Made here:** body in printed acrylic broadcloth, facings in polyester satin.

MEASUREMENTS One size, to fit **chest** measurements 40–44 in (101.5–112 cm). **Back length** of finished garment: 46 in (117 cm). **Sleeve length** from neckline over the shoulder to front cuff, 30 in (76 cm).

PATTERN VARIATIONS The hem can be increased to floor length without destroying the balance and proportions of the design.

STYLING The collar has been designed to be worn up or down at centre back.

NOTE All seam allowances are ½ in (1.2 cm) unless otherwise stated.

PREPARATION **(1)** Cut out pattern pieces following cutting layouts. **(2)** Neaten all edges on the body pieces (**not** on the contrast facings) by zigzag, overlocking or by turning in the edges ⅛ in (3 mm) and machining flat. **APPLY INTERFACING (1)** Apply the iron-on interfacing to the **wrong** side of the 2 contrast pocket facings. Cover completely and trim edges. **(2)** Apply interfacing to the wrong side of the contrast collar piece. Cover half only, from centre line to the edge. **(3)** Apply interfacing to the wrong side of the 2 contrast sleeve cuff pieces. Cover completely and trim edges. **(4)** Apply interfacing to the wrong side of the kusak sash. Cover approximately 18 in (45 cm) each side of the centre point, so that a total area of approximately 36 in (90 cm) carries a firm lining of interfacing. **ASSEMBLE AND ATTACH INNER POCKET (1)** Make up inner pocket by turning the top edge in twice, first turning ¼ in (6 mm), second turning ½ in (1.2 cm), and topstitching across ⅜ in (1 cm) from folded edge. **(2)** Turn the remaining 3 edges in ¼ in (6 mm), and topstitch pocket into position on the right body part as shown on pattern. **ASSEMBLE PATCH POCKETS (1)** Take 1 contrast pocket facing, fold to sandwich interfacing, and stitch to top of 1 patch pocket piece. **(2)** Neaten edges together and press seam allowance down towards bottom of pocket. **(3)** Repeat for second pocket. **ATTACH PATCH POCKETS (1)** Turn in the front and bottom edges of 1 prepared patch pocket piece ⅜ in (1 cm). **(2)** Place pocket on body part and topstitch into position along the 2 folded edges, ¼ in (6 mm) from edge. **(3)** Stitch third side of pocket to side seam allowance. **(4)** Trim and finish with firm machine tacks on the front corners. **(5)** Repeat for second pocket. **JOIN SHOULDERS (1)** With right sides together, join both shoulder

CUTTING LAYOUT
(not to scale)
Fabric width:
45 in (115 cm)

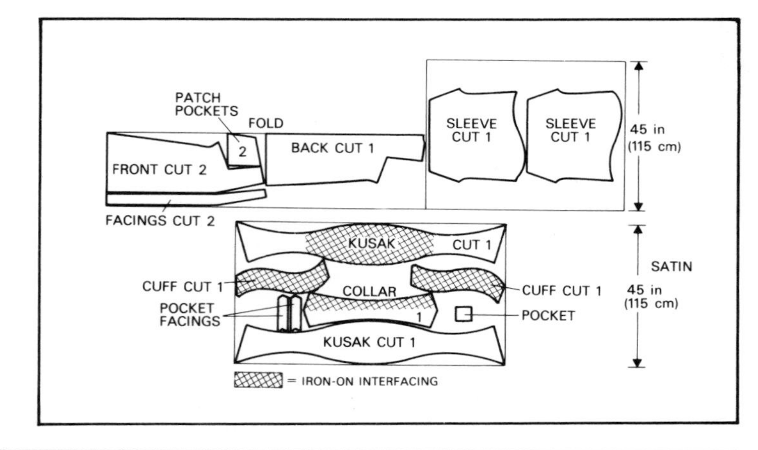

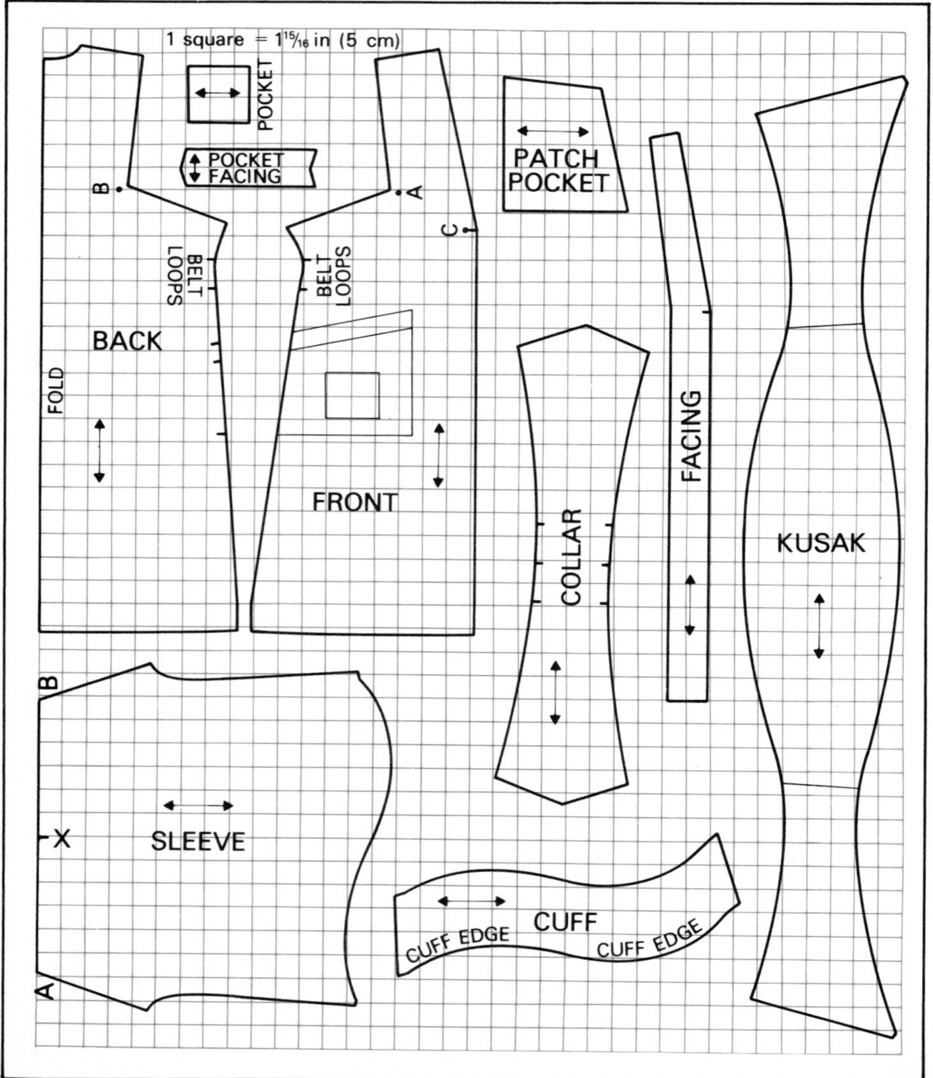

seams and press open. **ATTACH SLEEVES TO BODY (1)** Note that the shaped sleeve cuffs are **longer at the back** than at the front; be sure that your sleeves face in the correct direction. **(2)** With right sides together, pin from shoulder point X on 1 sleeve to point A on the front body, and from shoulder point X on sleeve to point B on the back body. **(3)** Stitch from A to B. **(4)** Turn under ½ in (1 cm) and lap the remaining sleeve seams on to the body parts, A to the side seam front, B to the side seam back. **(5)** Top-stitch from side seam front to side seam back; seam allowance ¼ in (6 mm). **(6)** Repeat for second sleeve. **JOIN SIDE SEAMS (1)** Lay the side and sleeve seams together with right sides inside and stitch from hem through to sleeve ends. **(2)** Press seams open, but leave armhole curve to roll. **ATTACH INNER SLEEVE CUFFS (1)** Join ends of 1 contrast cuff piece and press open. **(2)** With right sides together, pull contrast cuff piece over matching sleeve end, lining up the seams. **(3)** Stitch cuff and sleeve together ¼ in (6 mm) in from the sleeve end. **(4)** Turn through and press gently so the contrast facing and sleeve cuff lie smoothly together. **(5)** Turn in the top edge of the facing ¼ in (6 mm), and stitch facing to inner sleeve by hand or machine. **(6)** Repeat for second cuff and sleeve. **ASSEMBLE AND ATTACH COLLAR (1)** Fold collar lengthwise, right sides together, and stitch across the bottom ends with a ¼ in (6 mm) seam. **(2)** Trim corners, turn to right side and press lightly. **(3)** The collar part with the interfacing now becomes the **under** collar. With right sides together, pin the folded collar to the neck edge from the centre back to point C on each body part. **(4)** Tack the collar into position by hand or machine, using an easy stitch just inside the edge. **(5)** Neaten the raw edges at the back neck by zigzag. **ATTACH FRONT FACINGS (1)** Take 1 facing, neaten and turn in seam allowance at shoulder edge. **(2)** With right sides together, tack it into position from shoulder seam to the hem notch, locking in the collar. **(3)** Stitch from centre back neck to hem; seam allowance ¼ in (6 mm). **(4)** Repeat for second facing. **(5)** Turn in the facings and hand-catch them to fronts. Press lightly. **COMPLETE HEM (1)** Turn up the hem twice, first turning ½ in (1.2 cm), second turning 1¾ in (4.5 cm). **(2)** Slip stitch by hand and press lightly. **ASSEMBLE KUSAK SASH (1)** Lay the 2 sash pieces together with right sides inside and stitch together with a ¼ in (6 mm) seam allowance, leaving an opening along one edge. **(2)** Trim corners and turn sash through to right side. **(3)** Close opening by hand and press lightly. **TO FINISH (1)** Attach side belt loops approximately 3 in (7.5 cm) deep starting from a point 2½ in (6.5 cm) down from sleeve seam.

Indian Shirt Dress

The Islamic influence on dress is seen most dramatically in India. Before the Moghul conquest, Indian traditional dress for men and women consisted of minimal sarongs in plain, striped or plaid cotton – and little else. The Moghuls brought buttons and the Islamic religion with its strict notions of modesty, neither of which had existed in India before. The result was the *shalvar kamize* – loose buttoned tunic and drawstring trousers – that are found through India and still serve as the everyday dress of Pathan men and women on the Northwest Frontier and in Afghanistan.

As fierce and proud as the ancient Spartans, the Pathans disapprove of any unnecessary ostentation in matters of dress. Their *shalvar kamize*, always made in plain colours and never trimmed with embroidery or braid, look positively severe next to the riotously bright and busy versions favoured in the neighbouring Punjab. But within the strict limits set by their code, the Pathans consider the possession of a satisfactory wardrobe a matter of some importance. 'Though man has one appearance', says one of their proverbs, 'his clothes have a hundred varieties.' Throughout the Frontier, every city has its version of the Fabric Bazaar in Peshawar, the capital, where hundreds of open-fronted stalls jostle under cover, some spilling over with pyramids of cottons and Chinese silks, others sheltering tailors and their apprentices who spin out the process of bargaining with interminable trays of sweetmeats and tea. The women wear cotton for day and silk for evening, their costume completed by a long silk or muslin scarf that they wrap around the neck when at home and drape over the head when they go out. Much care is taken in the selection of just the right cloth and colour for, taken in context, the simple lines and subtle colours of the Pathan *shalvar kamize* has an understated elegance all its own.

Like all Muslim clothes, the *kamize* tunic is meant to cover the body and disguise its shape as well. The shirt yoke sets the tunic on the shoulders and the side panels and underarm gussets ease the tunic into a naturally rounded shape that skims the contours of the body without clinging to them. Worn in the Western way without the *shalvar* trousers, the *kamize* makes a superbly stylish and adaptable dress. Its straight lines give an illusion of slimness, the clever shaping permits complete freedom of movement, when you tie it at the waist you'll find that it never rides up, and it gives you a better shape than any Western shirtdress. You can wear it with sashes, scarves, thin or wide belts, hats or plaits, dress it up or down as much as you like. You can make it in prints, plain fabrics or combinations, in day or evening fabrics, and in summer or winter weights. In short, it's a design that can take you around the world. So if you decide to go on our own transcultural tour, here's a bag to take with you – a feather-light travelling bag with wrap-around straps and a completely soft finish. You could pack all three of our shirt dresses in it, and still have plenty of room to spare.

Indian Shirt Dress Instructions

DETAILS Yellow dress shown with shoes by Charles Jourdan, tights by Aristoc, crepe paper turban by Lundy Morrone and props from Supotco. Print dress shown with shoes by Charles Jourdan, earrings and necklace by Tailpieces and tights by Aristoc. Plaid dress shown with hat by David Shilling and shoes from Midas.

MATERIALS **Fabric required:** 3¼ yd (3 m) of 36 in (90 cm) wide fabric. **Trimming:** 4 buttons. **Recommended fabrics:** suitable for a wide range of light to medium weight fabrics such as cotton shirtings, polyester, fine wools, viscose and silk. **Made here:** yellow dress in cotton crepon, print dress in a Collier Campbell print on wool challis, plaid dress in two Madras cottons and cotton poplin.

MEASUREMENTS One size, to fit **bust** 33–37 in (84–94 cm), **hips** 34–39 in (86.5–99 cm). **Back length** of finished dress, 45 in (114 cm); length can be adjusted to suit personal requirements.

PATTERN VARIATIONS The dress can be shortened and worn as a tunic, with or without pockets.

STYLING Raise the side slits eight to twelve inches, and wear the dress as a long tunic over trousers. The design is suitable for prints, plain fabrics and combinations. You can mix several fabrics, as on the plaid model, or use one fabric overall and a second contrast fabric for sleeve, neck, sash and side slit facings. When combining fabrics, make sure that they are of the same weight and fibre type.

NOTE Seam allowances for hem and cuffs are ¾ in (2 cm); for pocket bags and neck circumference, ⅛ in (3 mm). All other seam allowances, ½ in (1.2 cm).

PREPARATION **(1)** Cut out pattern pieces following cutting layout. **(2)** Neaten edges by zigzag or other method suitable to your fabric. **ASSEMBLE GUSSETS (1)** Take one gusset piece and, with right sides together, pin it to the top of the front side panel piece, matching A/A–B/B. **(2)** Stitch to the exact point B, leaving the seam allowance free. **(3)** Lay the matching side and back panels together with right sides inside, and stitch from top of the side vent opening to 2 in (5 cm) below point B. **(4)** Pin gusset corner C to the top of the back side panel point C. **(5)** Stitch to B, then complete the side seam to B. **(6)** Repeat for second gusset. **ATTACH SLEEVES (1)** With right sides together, pin point D on one sleeve to point D on the front side panel, matching up the notches at A. **(2)** Stitch E on sleeves to exact point E on gusset. **(3)** Turn sleeve through and repeat F-C-E on the back side panel. **(4)** Stitch sleeve from point E to cuff. **(5)** Repeat for second sleeve. **ATTACH NECK FACINGS (1)** With right sides together, pin the facings to the neck slash opening. **(2)** Stitch both facings with a ½ in (1.2 cm) seam. **(3)** Finish with firm machine tacks at X. **(4)** Snip the seam allowance on the body parts only to point X; **do not snip the facings. (5)** Lay one facing strip over to the inside **under** the snip and one outside **over** the snip. **(6)** Fold the top facing to the inside, turn under the seam allowance and slip stitch to the back of the original machine stitching. **(7)** Repeat for the under facing. **(8)** Finish the ends of the facings below the neck opening, the top by machine and the inside by hand. **JOIN SHOULDERS (1)** With right sides together, join both shoulder seams and press open. **JOIN SIDES TO BODY (1)** Take one side body and sleeve section and, with right sides together, pin together the shoulder points of sleeve and shoulder seams. **(2)** Stitch from shoulder points to front hem line, leaving open the pocket mouth between the notches. **(3)** Stitch from shoulder points to back hem line. **(4)** Repeat for second side. **ASSEMBLE POCKET BAGS (1)** Take 2 pocket bag pieces and, with wrong sides

CUTTING LAYOUT
(not drawn to scale)
Fabric width:
36 in (90 cm)

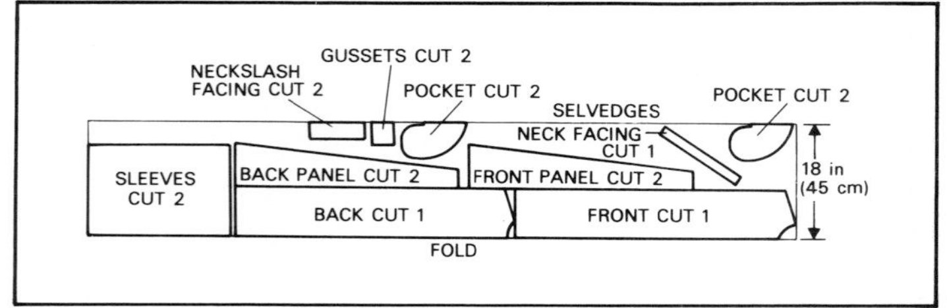

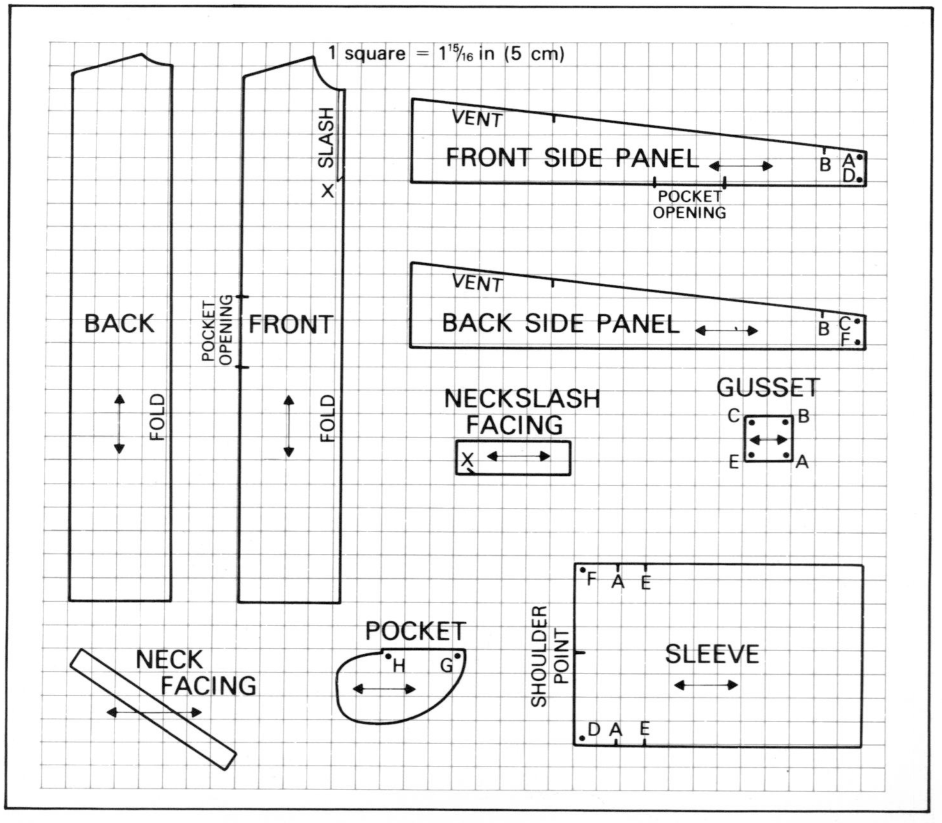

together, stitch a ⅛ in (3 mm) seam around the bag G-H, leaving the pocket mouth open. **(2)** Turn bag through and seam round again, locking in the raw edges. **(3)** Repeat for second pocket. **ATTACH POCKET BAGS (1)** Place the open pocket mouth to the opening on the body part and machine the pocket mouth seam allowance flat to the body part seam allowance. **(2)** Repeat for second pocket. **FINISH NECK (1)** With right sides together, stitch the collar band to the neck edge. **(2)** Turn under and finish the edges in line with the front neck opening. **(3)** Neaten the inside of the neck by turning in the seam allowance. Finish by hand, enclosing all raw edges. **MAKE BUTTONHOLES (1)** Make 1 horizontal buttonhole on the neckband, ⅜ in (1 cm) in from edge. **(2)** Now make 3 vertical buttonholes down the centre of the top neck facing. The top vertical buttonhole should be 1½ in (3.8 cm) down from the neckband seam. Suggested buttonhole size: ¾ in (2 cm). **TO FINISH (1)** Sew on buttons. **(2)** Make up tie belt from surplus fabric.

Travelling Bag Instructions

MATERIALS **Fabric required for bag:** 1 yd (0.95 m) of 44/45 in (115 cm) wide quilted fabric *or* 1¼ yd (1.15 m) of 36 in (90 cm) wide quilted fabric. **Fabric required for lining:** a piece of lining fabric 32 × 30 in (83 × 77 cm). **Trimming:** 1 20 in (51 cm) zip fastener; 1 zip pull ring; 1 8 in nylon zip fastener for inner pocket of lining. **Also:** a piece of stiff cardboard approximately ⅛ in (3 mm) thick for base of bag; 18¼ × 7¼ in (46.5 × 18.5 cm). **Recommended fabrics for bag:** quilted natural or man-made fibre fabrics that can be washed by hand or machine, preferably one that does not require ironing as this may compress the quilt wadding. **Recommended fabrics for lining:** man-made fibre fabric suitable for washing such as acrylic, nylon and polyester. **Made here:** both bags in quilted cuprammonium rayon with polyester wadding and nylon backing by Hoffman California Fabrics.

STYLING The cardboard base should be placed in the bottom of the completed bag to give it stability. The cardboard base can be dispensed with if you prefer a bag with a completely soft finish. If using the cardboard base, put it into a slip cover of lining fabric before inserting it in the bag. Remember to remove cardboard base before washing.

PREPARATION (1) Cut out pattern pieces following cutting layout. **(2)** Mark the strap position, as shown on the pattern, on to the body part with pins or tailors tacks. **ASSEMBLE STRAPS (1)** With right sides together, join the 4 strap pieces with a ¼ in (6 mm) seam to make 1 long strap, then join the ends together to make a circle. **(2)** Press seams open. **(3)** Now press both raw edges of strap in ⅜ in (1 cm) to the wrong side. **(4)** With right side outside, fold and press strap to half width, 1¼ in (3.2 cm), pin the edges together and machine topstitch ⅛ in (3 mm) all round to close the strap opening. **(5)** Machine topstitch the same amount on the folded edge to balance. **(6)** Machine at least 2 rows of topstitching all round to make the strap firm. **(7)** Press. **ATTACH STRAPS TO BODY (1)** Taking care not to twist the strap, place it into position on the right side of the body part and machine topstitch into position with at least 2 rows of stitching; see Figure 1. **(2)** Finish by back-tacking firmly at each of the 4 ends. **ATTACH ZIP FASTENER (1) If your zip is open-ended and longer than the bag opening, do not cut it off. Tuck the end down inside the bag; the lining will enclose and secure it.** **(2)** Turn in the 1 in (2.5 cm) allowance at bag opening. **(3)** Pin or tack the zip into position starting ½ in (1.2 cm) in from the side seam allowance notch, allowing any surplus

CUTTING LAYOUT
(not drawn to scale)
Fabric width:
45 in (115 cm)

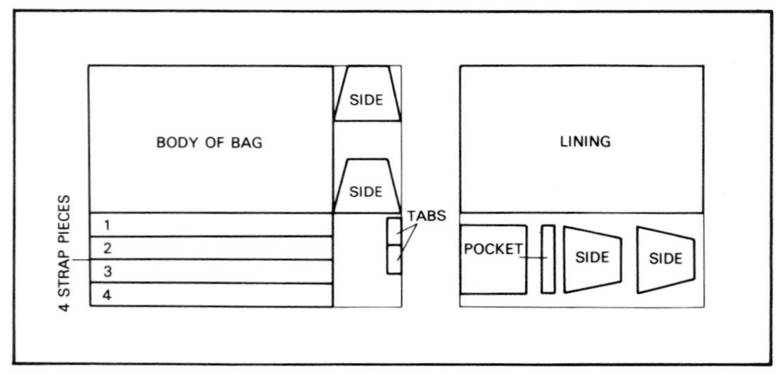

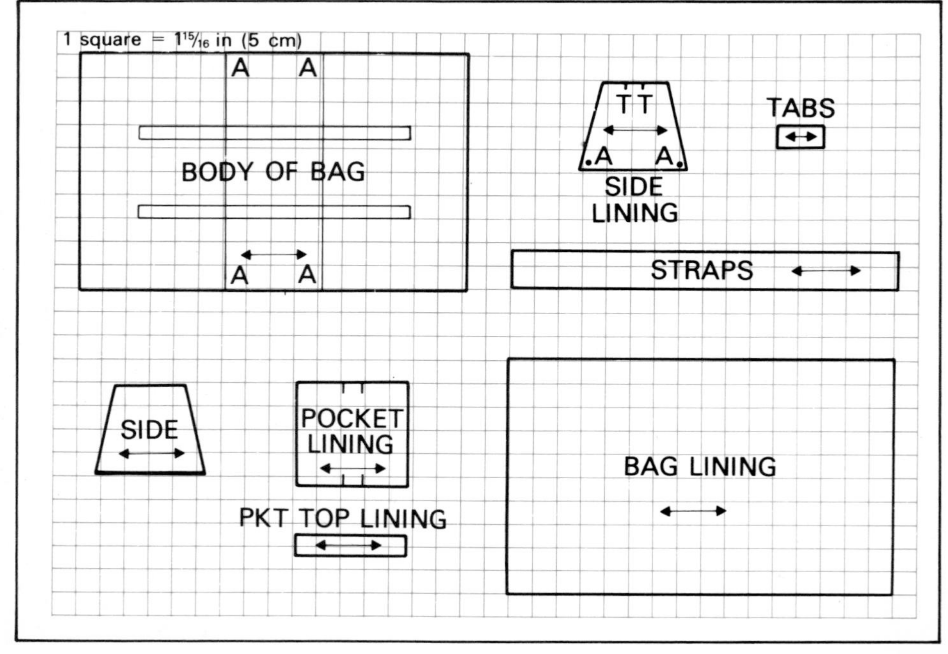

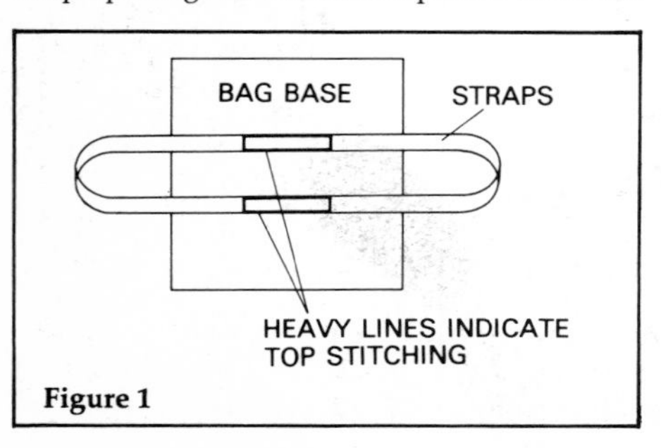

Figure 1

zip length to extend over the end of the bag opening. **(4)** After placing the zip into position on both sides of the opening, secure by topstitching with 2 rows of machine stitching approximately ¼ in (6 mm) from the zip teeth. **ATTACH SIDE PANELS (1)** Note that seam allowance for side panels is ½ in (1.2 cm). **(2)** With right sides together, stitch one panel piece into position at bottom of bag matching A/A to A/A as pattern. Machine stitch firmly to **exact** points; do not stitch past the points on to the seam allowance. **(3)** Machine sides of bag to side panel, nipping corner seam allowance if necessary. **(4)** Repeat for second panel. **MAKE UP TABS (1)** Remove quilt wadding from the tab pieces. **(2)** Fold over one tab piece right side inside, and machine down 2 sides with ¼ in (6 mm) seam. **(3)** Turn tab through and topstitch all round to make tab firm. **(4)** Repeat for second tab. **CLOSE TOPS OF SIDE PANELS ENCLOSING TABS (1)** Pin or tack the tabs between the notches on the right side of the side panels. **(2)** Turn the bag through and, with right sides together and the tabs sandwiched in the centre, machine stitch across the tops of the side zip. If in doubt, finish this seam by hand.

(3) When the side panels have been completed, strengthen all corners by double stitching. **ASSEMBLE LINING (1)** After cutting out, mark the patch pocket position on to the lining. **(2)** Neaten the edges of the patch pocket pieces by zigzag. **(3)** Stitch the zip fastener between the large and small pieces with ⅜ in (1 cm) seam allowance. **(4)** Press over the 4 neatened edges of the completed patch pocket ⅜ in (1 cm), and press up the pleats by placing notch to notch, as on pattern. **(5)** Topstitch the pocket to the lining with 2 rows of machine stitching, first row ⅛ in (3 mm) from edge, second row ¼ in (6 mm) from edge. **COMPLETE LINING (1)** Complete the lining bag by sewing in the side panels exactly as for the body, but leaving open the centre of the top panel seams to facilitate the attaching of lining to bag. **(2)** Reinforce the corner with double stitching. **ATTACH LINING TO BAG (1)** Trim off all loose threads and fit lining to the inside of the bag. **(2)** Turn in the lining seam allowance and pin into position. **(3)** Slip stitch the lining by hand to the back of the zip machine stitching, and finish firmly at zip ends.

Singapore Pyjamas

Paul Poiret was the first modern designer to look East for inspiration, and although circumstances obliged him to restrict his Eastern travels to voyages of the imagination, his designs were always based on careful research in museums and private collections closer to home. During a visit to the Victoria and Albert Museum in London, he chanced upon a collection of turbans that introduced him to the splendours of India, and Poiret was enchanted. 'I admired unwearyingly the diversity of their so logical and so elegant forms' he enthused. 'There was the little close turban of the sepoys, that ends in a panel negligently thrown over the shoulder; and there was the enormous Rajah's turban, mounted like a gigantic pincushion, to receive all the costliest aigrettes and jewels.' Poiret immediately summoned an assistant from Paris, and they spent eight days in the Museum copying the turbans in meticulous detail. A few weeks later, Poiret's Indian turbans took Paris by storm.

India also inspired Poiret's designs for lounging pyjamas and culotte skirts, which he introduced to Europe. These Singapore pyjamas were copied from Poiret's designs by the clever Chinese seamstresses of Singapore in the palmy days of the British Empire. These are the elusive shapes glimpsed through the shimmering heat in the stories of Somerset Maugham: the frocks of the memsahibs fluttering like silken butterflies along Cad's Alley at Raffles Hotel, where planters from upcountry lingered over gin slings.

Circles are neither the simplest nor the most economical shape, but used as effectively as they are here, their function cannot be faulted. The design had to be loose enough to let the body breathe in the sultry climate of the Straits, had to be frankly feminine when looking like a lady was a point of honour, had to be practical enough to allow for climbing in and out of rickshaws and for leaping over the deep drains that then lined every little street in Singapore, and had to hold its shape in an atmosphere that could pucker a seam in half a minute. The series of circles that makes up the pyjamas provides an ideal solution. The soft drape cuddles the body into a series of graceful curves that never lose their shape, the wide sleeves and hem sway back and forth like a lazy fan, the culottes are a model of pretty practicality, and the pyjamas are far more subtle and suggestive than frills could ever be.

Singapore pyjamas look just as good in cold countries as in hot ones. So if you aren't going to wear them in the tropics, here's something to put you in the mood. Gin slings were invented in the Long Bar at Raffles by the barman Ngiam Tong Boon in 1915, and here's the Raffles recipe:

RAFFLES SINGAPORE GIN SLING
2/4 Tanqueray gin
1/4 Peter Heering cherry brandy
drops of Benedictine and Cointreau
1/4 mixed orange, pineapple and fresh lime juice
dash Angostura bitters
Shake well and strain.

Singapore Pyjamas Instructions

DETAILS Grey Singapore pyjamas shown with necklace from Butler & Wilson and shoes by Charles Jourdan. Orange Singapore pyjamas shown with jewellery from Butler & Wilson and shoes by Manolo Blahnik, all London. Cream Singapore pyjamas shown with jewellery from Butler & Wilson and umbrella from People's Park, Singapore.

MATERIALS **Fabric required for top:** 1⅞ yd (1.75 m) of 48 in (122 cm) wide fabric. **Fabric required for culottes:** 4⅛ yd (3.80 m) of 48 in (122 cm) wide fabric. **Trimming for top:** 10 in (25 cm) of narrow stay tape. **Trimming for culottes:** piece of 2 in (5 cm) wide elastic the same length as your waist measurement. **Recommended fabrics:** polyester knit jersey and other fine knit jersey fabrics. Any soft, drapey, non-fray fabric can be used, but allow for the fact that the use of a fabric other than jersey will alter the hang of the drapes. Avoid bulky, loosely-woven fabrics that do not drape well. **Made here:** cream pyjamas made in printed polyester jersey from G. Wood, London. Orange, grey and two-tone pyjamas made in printed polyester jersey by Beeren Fabrics of Holland. All trimmed with English Sewing's wide elastic in toning colours.

MEASUREMENTS One size, to fit **bust** 33–38 in (84–97 cm); **waist** 24–30 in (61–76 cm); **hips** 34–40 in (87–102 cm). **Culotte length** from bottom of waistband to finished hem, 35½ in (90 cm). Length and waist size can be adjusted to suit personal requirements.

STYLING The top can be turned into a dress or tunic by increasing the length and adding two deep side slits and a sash. Side pockets can be added to the culottes, and skirt and sleeve hems can be finished with fancy machine stitching instead of zigzag. For a waistband finish that does not show the elastic, substitute a length of narrow elastic 1 in (2.5 cm) wide for the wide elastic. Stitch to the top of the skirt following instructions for wide elastic, then fold over to the inside of the skirt.

NOTE All seam allowances are ⅜ in (1 cm) unless otherwise stated.

Top

PREPARATION **(1)** Cut out pattern pieces following cutting layout. **(2)** Neaten all edges by zigzag. **(3)** Stitch stay tape to wrong side of back, in the crease edge at back neck, with 1 row of machine stitching. **ASSEMBLE TOP** **(1)** Lay back and front body pieces together with right sides inside and stitch sleeve joins E–A, making sure to fasten firmly on to the ends of the back neck stay at E. **Do not snip corners.** The back and front neck facings are designed to tuck inside when the top is worn, so do not stitch them down. **(2)** Stitch side seams B–F. **(3)** Topstitch side vents and hem F–C–D with ¼ in (6 mm) seam. **TO FINISH** **(1)** Press.

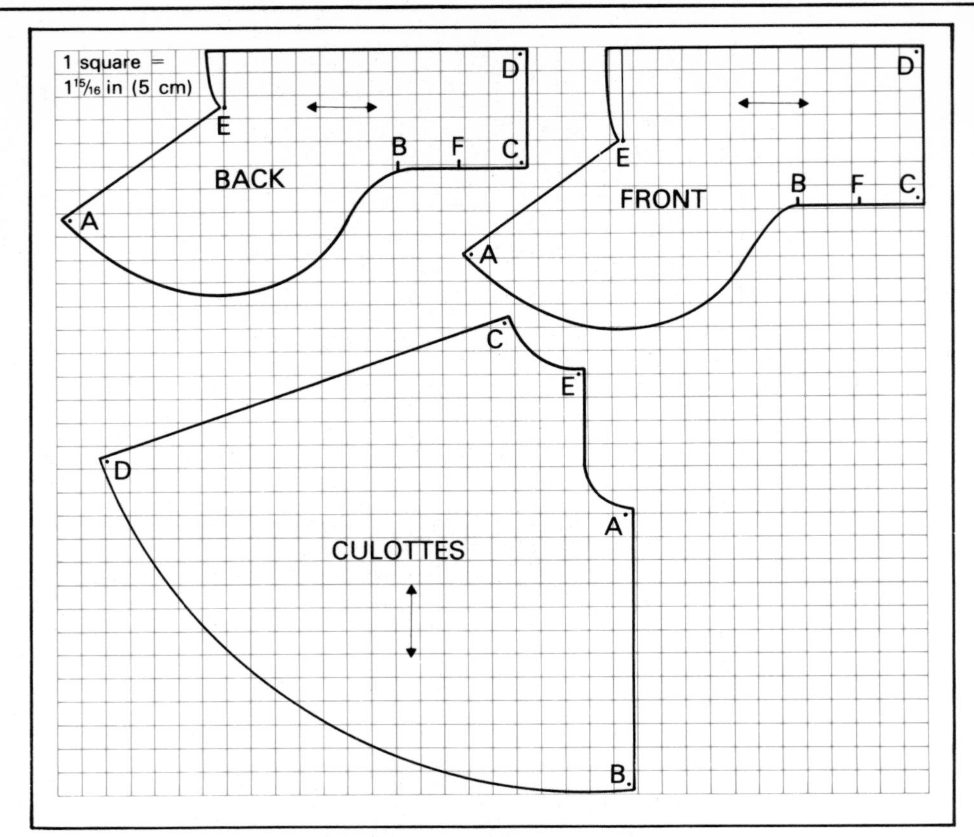

Culottes

PREPARATION **(1)** Open fabric to full width then fold as shown on cutting layout, with right side of fabric inside. **(2)** Cut 2 of each leg pattern piece. **(3)** As these pieces are exactly the same size and shape, place 2 together with right sides of fabric inside, and stitch A–B and C–D to make 1 leg piece. **(4)** Repeat for second leg. **(5)** Neaten seams, hemlines and waist edges C–E by zigzag. **JOIN LEGS** **(1)** With right sides of fabric outside, join the 2 completed leg pieces by stitching the crotch seam E–A–E. **(2)** Now trim seam to half the seam allowance, reverse the culottes and stitch again, locking in the raw edges. **WAISTBAND** **(1)** Join ends of elastic to make a waistband with a circumference of 22 in (56 cm) or measurement required, and neaten join. **(2)** With the waistband elastic on the **outside** of the skirt, stretch the band evenly on to the waist of the culottes and stitch ⅜ in (1 cm) from the edge on the elastic side, using a zigzag stitch. **TO FINISH** **(1)** Press. **(2)** If desired, make a sash from surplus fabric.

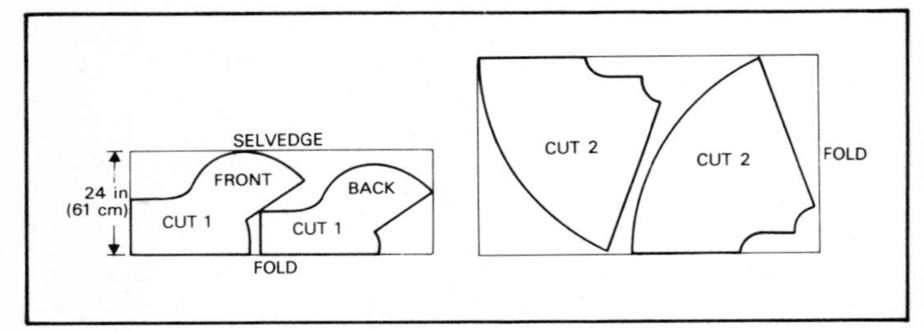

CUTTING LAYOUT
(not drawn to scale)
Fabric width:
48 in (122 cm)

Tartar Khalat Coat

Poiret's feeling for shape amounted to genius, and in his hands the simplest materials assumed grand proportions of style – even in mid-Atlantic. He once improvised a superb cloak on the deck of a cruise liner, using only the scissors he always carried in his pocket and a plaid travelling rug. Word of the incident was telegraphed ahead, and when the liner arrived in New York harbour, the quay was packed with journalists and photographers clamouring for a sight of what had already become the most famous design in the world. But the beautiful simplicity of most of Poiret's shapes was often obscured by his fondness for magnificent fabrics and opulent trimmings reminiscent of the fabled Caliph Harun al-Rashid of Baghdad. And it was this very theme – the romance of Baghdad and the stories of Scheherezade – that inspired the best-known of his oriental designs.

Poiret presented his collection in Paris in 1911, during a gala entertainment that has never been forgotten – the *Fête of the Thousand and Second Night*. Nubian slaves with torches led guests through perfumed gardens to a robing room where each was provided with a costume taken from one of the many colourful periods in the history of Baghdad, exact in every detail. Poiret played the role of the Sultan, and Madame Poiret that of the Sultan's favourite, dressed in a harem costume and locked in a golden cage. Gauze-clad houris danced arabesques on priceless carpets, fortune tellers whispered to the feasters as fireworks filled the sky, and Poiret's mannequins drifted over the lawns in the soft harem skirts, graceful pantaloons, slender tunics and sweeping coats that were to free women from corsets forever.

This design, like many worn that evening, is a *khalat* coat that came from the East with the Tartar hordes who captured Baghdad in AD 1258, and represents the perfect marriage of shape and surface. The *khalat*'s exceptional versatility can be seen in these two treatments. The first, made in black velvet, is worn wrapped at the front and belted with a wide gold belt. The second, made in quilted silky fabric, is worn open – and it's difficult to believe they're the same design. Both styles of dress are traditional, and *khalats* are still worn by Tartars and tribesmen who live along the old Silk Road, in fabrics that recall the time when two days were needed to pass through the Cloth Bazaar of Tabriz. The Kzan Tartars make their *khalats* in brocade, the Turkomen in silver-embroidered velvet, the Uzbeks in quilted cotton or silk, and you can make your *khalat* as subtle or opulent as you please. You can have it in a plain fabric lined and piped in a print, make it in a combination of printed fabrics, in quilted or flat fabrics, untrimmed or embellished like a carpet. It suits all figures, can be worn over any style of undergarments, can be anything *you* want it to be. Which is what creative dressing is all about.

Tartar Coat Instructions

DETAILS Black velvet coat shown with hair comb by Tailpieces, necklace and earrings from Butler & Wilson and belt to order from Rony, London. Photographed at the Park Lane Hotel, London. Quilted print coat shown with *cloisonné* necklace from Liberty. Photographed at the Dukes Hotel, London.

MATERIALS **Fabric required for coat:** 5¾ yd (5.30 m) of 36 in (90 cm) wide fabric. **Fabric required for lining:** 4 yd (3.70 m) of 36 in (90 cm) wide fabric. **Recommended fabrics:** suitable for most quilted and coat-weight fabrics. Patterned fabrics may oblige you to increase the amount of fabric required. **Made here:** black coat made in gold-striped velvet, lined with printed silk. Print coat made in quilted cuprammonium rayon by Hoffman California Fabrics, trimmed with red velvet and lined with silk.

MEASUREMENTS One size, to fit bust 33–38 in (86–96 cm). **Back length of finished garment:** 49 in (124.5 cm).

PREPARATION (1) Cut out pattern pieces following cutting layout. **ASSEMBLE POCKETS (1)** Lining for pockets must be cut slightly off grain to ensure a soft finish. **(2)** Take pocket piece and turn in the front and hem allowances ½ in (1.2 cm). (3) Turn down the top edge allowance 1½ in (3.8 cm). **(4)** Press or tack into position. **(5)** Repeat for second pocket. **ATTACH POCKET LINING (1)** Slip stitch the lining 1 in (2.5 cm) down from top edge and ¼ in (6 mm) inside front and hem edges. **(2)** With a loose tension stitch, topstitch across the top of the pocket mouth ¾ in (2 cm) down from edge. **(3)** If making up in velvet fabric, gently steam press and brush down the pile using another piece of velvet as a brush. **(4)** Neaten remaining raw edge by zigzag. **(5)** Repeat for second pocket. **ATTACH POCKETS TO FRONTS (1)** Attach pockets to fronts as marked on pattern. **(2)** Topstitch down front edge and across bottom of pocket with a loose tension stitch: seam allowance ¼ in (6 mm). **(3)** Hand tack pocket corner from the **inside** of the coat using a small square of lining as a stay. **JOIN**

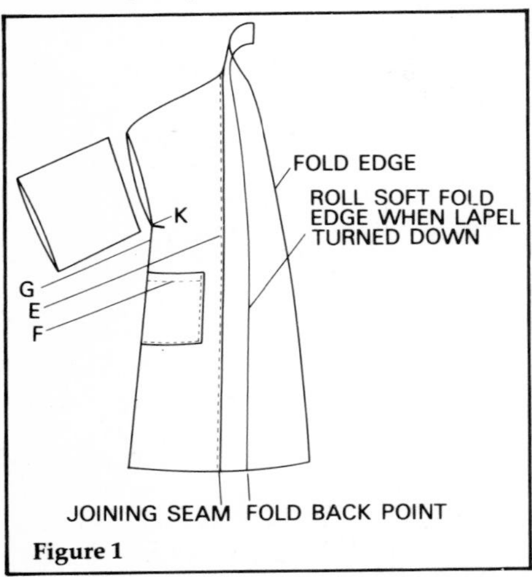

Figure 1

FRONTS TO BACK (1) Right sides together, position and tack shoulder seams to match pattern. **(2)** Machine stitch side seams from hem through pocket edges to just below the sleeve, finishing at point G. Seam allowance: ½ in (1.2 cms). **(3)** Neaten side seams by zigzag. **ASSEMBLE COLLAR (1)** Join centre back seam of collar. **(2)** Gently steam press the seam open. **(3)** Fold the collar lengthwise with right sides together and stitch across the hems: seam allowance ⅜ in (1 cm). **(4)** Trim seam allowance at corner and turn through. **(5)** Hand tack the raw edges of the collar and facings together from neck to hem: seam allowance ¼ in (6 mm). **(6)** If making up in velvet, steam and brush the completed collar. **ATTACH COLLAR (1)** With right sides together, place the centre of the completed collar piece to the centre back neck on the body piece. **(2)** Carefully tack the collar into position matching up points A-C-D and hem point H. **(3)** Starting from the centre back stitch collar to one front finishing at point H: seam allowance ⅜ in (1 cm). **(4)** Repeat for second front. **(5)** Steam press the seam allowance in towards the body. **(6)** To keep the seam allowance in position after pressing, prick stitch by hand from shoulder to point H at hem. (Note point E, Figure 1). **HEM (1)** Neaten the hem allowance by overlock or zigzag. **(2)** Turn up the hem and hand fasten with a light stitch. **(3)** Steam press and brush if making up in velvet: brush only the extreme edge. Do not press on the hem join. **ASSEMBLE SLEEVES: (1)** Stitch both sleeve seams: seam allowance ½ in (1.2 cm). **(2)** Press seams open. **(3)** Turn up cuff allowance as for hem. **INSERT SLEEVES (1)** Take 1 sleeve and, starting with the sleeve seam at the **exact** point K at the underarm seam, tack into the armhole matching up the centre of the sleeve and the shoulder seam, seam allowance: ½ in (1.2 cm). **(2)** Stitch sleeve to armhole finishing firmly at the exact point K. Do not stitch over the side seam allowance. **(3)** Repeat for second sleeve. **(4)** Complete the side seams by stitching from E to the exact point K. **(5)** Finish with a firm machine or hand tack. **(6)** Press the sleeve seam allowances in towards the body. **ASSEMBLE LINING (1)** Cut out lining and stitch together as for body: seam allowance ½ in (1.2 cm). **(2)** Neaten raw edges by zigzag. **(3)** Press lightly on wrong side. **ATTACH LINING (1)** Press under ⅜ in (1 cm) seam allowance all round lining except for hem. **(2)** With wrong sides together, position centre back of collar/facing. **(2)** Tack down each side to hem. **(3)** Slip stitch just over original stitching. **(4)** Turn the hem of the lining up twice, first turning ¼ in, second turning 1 in (2.5 cm). **(5)** Stitch by hand or machine. **(6)** Fasten lining at hem to side seams only, leaving the back lining loose. **SLEEVE LINING (1)** At the under arm point catch together (on the wrong side) the lining and body seam allowances. This holding stitch must be quite loose – about ½ in (1.2 cm) in length. This will prevent the body lining from falling down into the sleeves. **(2)** Complete by catching the sleeve lining to the cuff turn, slip stitching about ½ in (1.2 cm) below neatened edge of the cuff turn. **TO FINISH (1)** After completion of all sewing, place coat on a dress stand or hang from a shaped and padded coathanger. Apply steam – the steam from a domestic kettle will do – and gently brush and set the coat in the pile direction using a pad of velvet as a brush.

CUTTING LAYOUT
(not drawn to scale)
Fabric width:
36 in (90 cm)
Surplus fabric can be used for patch pockets.

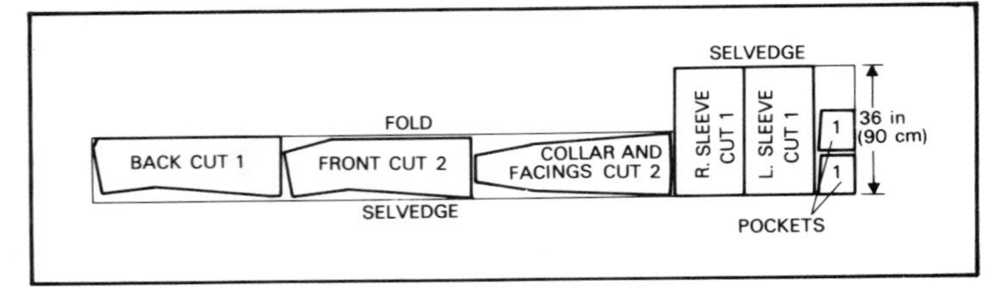

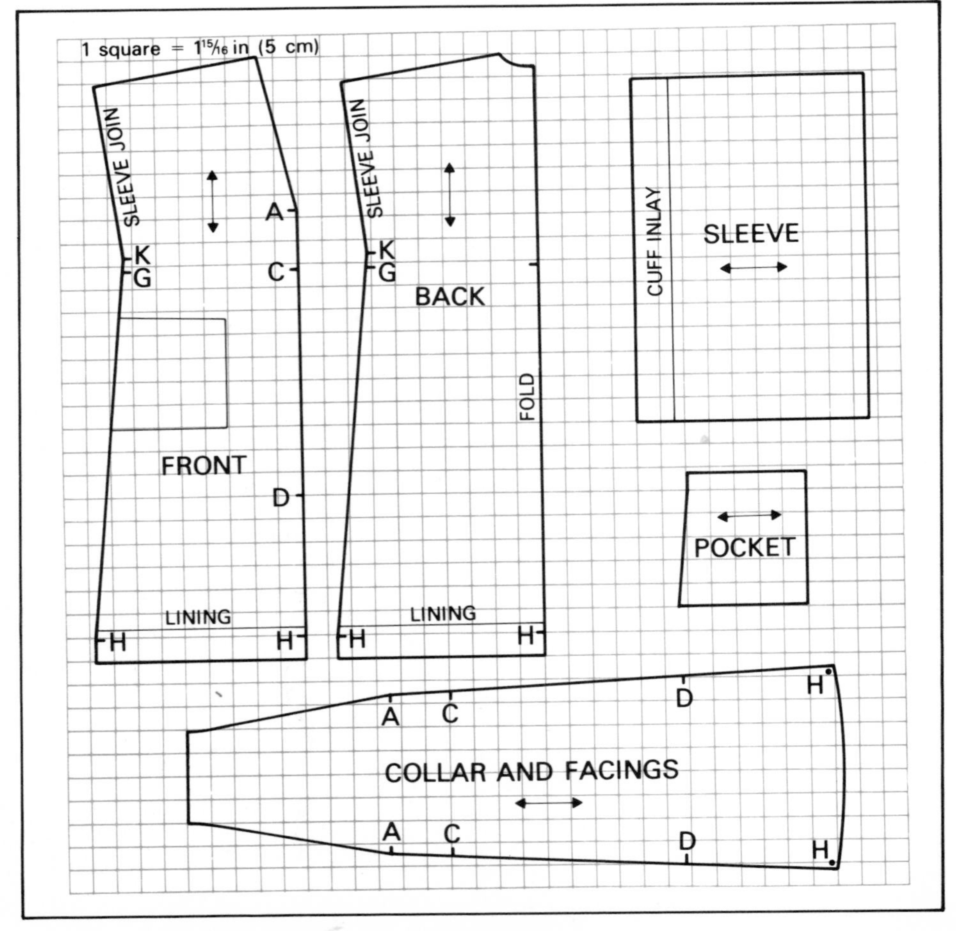

Fabric and Surface Design

Evening coat in a Bianchini fabric. *Gazette du Bon ton,* 1921.

Introduction

Cut your coat to suit your cloth is an adage that all designers take very much to heart. For while a good design made up in an indifferent fabric is likely to sink without a trace, good fabric can *lift* an entire design collection, and carry it all the way to the top. The better the cloth, the better the coat. Here the relative merits of natural and man-made fibre fabrics are not at issue. Performance can be quantified, but it is more elusive qualities that fashion designers look for in fabrics – exciting colours, interesting textures, imaginative patterns and prints – in short, wearable art. The master designer Paul Poiret put it this way – *'Art consists of creating a new form of language which permits one to say in a new way what men have felt from time immemorial. A designer has as many languages as he has fabrics with which to sing of the beauty of women.'*

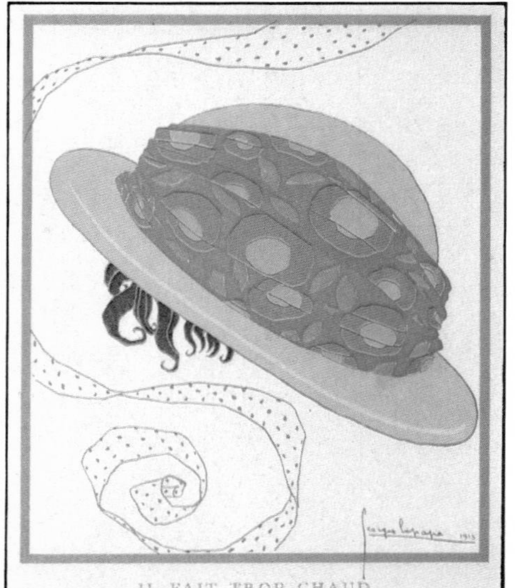

Hat by Paul Poiret, trimmed with Martine fabric
Gazette du Bon Ton, August 1913

In Japan, where antique kimonos decorated by renowned artists are prized as national treasures, the tradition of fabric as wearable art is centuries old, and the oriental influence played an important part in the European decorative arts movements of the 1890s, inspiring, in England, the first Liberty prints. Folk influences were also important, the most dramatic being Diaghilev's *Ballets Russes* which in 1909 introduced Europe to a sensational new spectrum of colours and patterns. In 1911, after a visit to the Wiener Werkstatte and other art schools in Germany, Poiret opened his Martine School of Decorative Arts in Paris, producing naturalistic prints in striking shades that he used to trim his couture designs and accessories. Soon afterwards he commissioned the artist Raoul Dufy to create the first collection of modern *designer fabrics* – art fabrics that add a new dimension to fashion. Dufy's work for Poiret and for the firm of Bianchini Ferier revolutionized the textile industry, and surfaces have been as important a part of fashion design as shapes ever since.

So what is a 'good' fabric, and how can you make the most of it? You can take this advice from Erté, who has designed both fabrics and fashions – 'It is extremely difficult to think up interesting shapes for dresses made in printed fabrics. The printed fabric strikes one's eye in such a way that any detail in the design itself is completely obscured, so it is best to create very simple clothes.' You can study the work of those designers who have a special way with fabrics – Schiaparelli and Zandra Rhodes with their instinct for dramatic novelty prints, Ungaro who can mix prints like a salad, Saint Laurent with his fine eye for colour, Issey Miyake who uses bias cutting to bend the straight lines of woven fabrics into flowing and elegant curves. But best of all, read on to see what these experts have to say about three different facets of surface design.

Fabric designs by Raoul Dufy for Bianchini Ferier Paris 1928.

Susan Collier and Sarah Campbell

Sisters Susan Collier and Sarah Campbell are England's premier fabric designers. Susan Collier was Design and Colour Consultant to Liberty of London Prints for their scarf, furnishing and dress fabric collections, and has designed prints for Yves Saint Laurent, Daniel Hechter, Cacharel, Bill Gibb and Jean Muir, and Sarah Campbell created much-acclaimed collections of designer fabrics for Soieries Nouveautés. In addition to dress fabrics, the Collier Campbell Studio designs exclusive co-ordinated collections of wallpapers, ceramics, furnishing fabrics and sheetings for top international stores including Au Printemps in France.

Susan: 'As children we were both rather 'artistic'. We liked making patterns – couldn't help making patterns, in fact. Our parents always had lovely visual things like illustrations, brightly coloured china and beautiful cloth.

But when I came into the industry back in 1961 or so, all the art seemed to have gone out of fabric design. Prints were either Swedish or Pucci, and there was no ***life*** in them at all. There was nothing around that even faintly resembled the work of painter-designers like Sonia Delaunay. The very idea of painting was far, far away from fabric design. It was as though, because fabric designs had to go through the mechanical process of being repeated – put onto a roller and rolled onto the cloth – they had to be organized in the simplest possible way. ***Too*** organized, because everything that mattered got killed. So I felt there was an opening – a chance for someone to design nice, warm, lively, lovely fabrics, because no one else was doing it.

At the time, I had no idea that most of the textile industry was based on the assumption that nobody wanted nice things. Since then, we have spent a great deal of time and effort proving that, although that may be an industrial view, it isn't the view of ***people***. We don't distinguish between ourselves and our customers. We're you, and you're us. And we know that the lovely visual things we grew up with always made us feel good. In dark moments, and in undark ones, something beautiful to look at – pleasure sliding across the eye – always makes one feel better. You may not focus and realize it's because a print is nice. You just feel a little bit better, and that's what matters.

One of the most important things we've introduced in our work is the idea of *not* making the repeat appear. *Sarah: William Morris, for example, coped with the repeat by making it clear. He organized nature into a rigid pattern, and turned it into a positive design feature.* We cope with the repeat by cheating it – by disguising it – so it seems ever-moving, and never feels claustrophobic or predictable. *It looks like it's going to go on and on, and you're never going to find the same mark twice. For us, that's one of the great pleasures of design.*

Another pleasure is that the hand is always 'left in' our work. If you look at the fabrics in the mixed print suit you can clearly see that they've been designed by a human being – unlike so many fabric and china patterns, that don't look like there's ever been a hand or paintbrush at work. It looks nonchalant, but it's not easy. The better you can draw, the better you can do it. It takes years and years, and why shouldn't it – it's an art.

Colours are the greatest pleasure – all the colours of the rainbow. They're delicious. When my son Louis was one, he liked the colours in one of our furnishing fabrics so much that he licked his way all along the sofa. We've always tended to use the same kinds of colours, although some colours do come up and dominate for a period. Then we need a rest, and go in another direction, and another set of colours tends to dominate. It doesn't really have anything to do with the 'fashion colours', although we accept that that's the way marketing is done. *We never think about the international fashion colour conspiracy. We just use the colours we love, and do what we like.*

We don't just knock out ideas and finish with them. Our designs tend to be a progression, a natural evolution. We do one thing, then our interest is taken up by an aspect of that thing which we go on to develop, and so on. And the same ideas keep coming back – the same obsessions if you like. They aren't dealt with in one blow, by any means. I could pick a diamond lattice or a basket-weave motif and trace it back through all my designs,

and for Susan a rose is the motif that comes up again and again. The random placing of colour is another thing that recurs frequently. I always go back to it, to see how many times I can do it and trick everyone into thinking that the pattern doesn't repeat.

As a studio we are deeply involved with fabric design, we take it very seriously. We feel committed to our designs over a very long period, committed to looking after them from the thought through the drawing, the colouring and the cloth. We always design with the cloth in mind. *And it's not only the cloth, it's the process as well, and the effect it has on the cloth and the design. A design that works for one fabric won't necessarily work for another. The design used here for the skirt in the mixed print suit looked perfect on silk and very good on cotton, but it didn't work at all on wool. You have to allow for these differences in a very disciplined way.*

As far as fibres go, our very favourites are silk and cotton, and after them the very fine wools. There's no doubt that fabrics that have life in them – natural fabrics – give you an awful lot back. I'm very keen on the fact that you can wash them yourself, and if you treat them well, they can last forever. There's something comfortable about them, like a friend. All the things you have should become friends, and if they're things that can't become friends, then you shouldn't have them.

We're always interested to see how our fabrics are used. Sometimes there are wonderful surprises, but I've also been more depressed seeing them made up than I can say. *The worst thing is an outfit that makes you think they've bought three metres too little – and not just because of the royalties! A design shouldn't look pinched or mean – you have to give the fabric enough scope to really work for you.*

The shape is terrifically important, and the Chinese jacket and wrap dress you see here are copies of two of my favourite old friends that I've had for years. I love cloth, colour and pattern, and I love being in them – and that can best be done with very simple shapes. I don't mean ugly, frumpy shapes or huge, floppy, cloak-like shapes – just simple shapes that simply let you ***be***. These clothes are designed for living, they're flexible. You can dress them up and dress them down. My daughter Sophie had a wonderful phrase – brown body-bandages – to describe those dreadful inflexible suits. I don't like corsets, tailored clothes or anything else that's restrictive. Half the world may feel comforted if they're kept in order, but that's not ***our*** half. My favourite clothes are those that keep me in touch with myself. All you can be in a tailored garment is neat, pretty neat at best. But you can't lounge, and relax, and live. I like clothes that you can put a belt round, a blouse over, and trousers under. I like the lightness of layers, they make you feel free. And I like not having to think about whether I'm hot or cold, or 'morning' or 'evening', I like to be able to dress and undress as I go. If the print and shape look good, I feel all right. And these old friends have always done that job beautifully.

The dress is a good, simple shape that can adapt itself well to summer and winter fabrics. The black print suit is a strong and simple treatment of one of our winter

fabrics, and it could be simpler still if the skirt were in plain black. We tend to design fabrics in groups – several complementary fabrics linked through colour and pattern – that can be put together in exciting combinations. We've chosen the four fabrics in the mixed print suit to show you what we mean. But how to do it for yourself?

I once made the mistake of saying that if you choose the fabrics you like, the fabrics are bound to go together through your own taste. Then one day we were going down the Kings Road and saw someone who had obviously taken that advice – and proved it to be untrue. So now I say that when it comes to combinations, you must take great care, and you must have a guiding eye. You might also bear in mind the fact that people tend to like sparser prints near the face. Not everybody does, but there is a strong tendency to keep the neck area clear, and put the strong patterning elsewhere. In that sense, the check fabric would not have been right for the blouse, and the blouse fabric would not have been right for the jacket.

Also, although the prints in the suit are quite different, you can see that they're being held together by the red binding – it's got the colour and the strength. It's rather like an outline, except that it's really an in-line, because it pulls everything together. There has to be a strength, even if it's tiny, because it allows the other colours freedom to fly.

Or think of it this way – look at Old Master paintings of flowers, which most people find immensely pleasing. The flowers are enormously abundant, on a darkish ground with the occasional butterfly and snail, all held in with a beautiful basket or vase, or a tablecloth. It's a perfect expression of pattern on pattern, with a clearish outline round the whole. That's the kind of relationship you should apply to your choice of prints. *You need rest and abundance. If you start with an effusion, you should put it with something that gives you a little clarity and rest. That's what happens in the paintings, and in the combination we've chosen here.*

So much has to do with the care you give things, and today no one quite knows what caring for is all about. *We've lost the knack.* We've lost the need, and above all we've lost the help. The household servants, who used to care for everything so well, plumping and polishing and darning and dyeing and starching and sewing. Even Mr and Mrs Pooter in *Diary of A Nobody*, who had extremely modest means, had a housemaid. Today Mrs Pooter's housemaid has a life of her own, so machines – and surfaces – have got to do the helping. We all have to pay, and we have two basic choices – between tacky, undemanding things that are cheap but don't last, and nicer, more expensive things that go on and on working for you, but which you have to work for too. Good fabrics are good companions and faithful servants, investments that appreciate with time.

When I see a woman wearing prints we did for 1973, as I frequently do when I go to the theatre, I'm delighted. Because they've served her very well, and they still look great. Regardless of what's happened in fashion since 1973, she's got it out because it's her best friend, and it hasn't let her down. And I like it particularly because, after all this time, they're still her special occasion clothes. She knows she will look good, because looking good comes from feeling good. *And that, really, is the most important part of our work – giving you something to enjoy!'*

Chinese Jacket Instructions

DETAILS Light check jacket shown with *cloisonné* necklace from Liberty, shoes by Charles Jourdan, cowl neck blouse and soft skirt. Dark print jacket and matching side slit skirt shown with red *bustier* top, mules by Charles Jourdan and earrings from the Neal Street Shop, London. Red feather hair comb from Henri Bendel, New York.

MATERIALS **Fabric required for jacket,** 3 yd (2.75 m) of 36 in (90 cm) wide fabric or 2 yd 27 in (2.60 m) of 45 in (114/115 cm) wide fabric and **for lining,** 3 yd of 36 in (90 cm) wide fabric. **Recommended fabrics for jacket:** pure silk crepe, polyester crepe de Chine, fine wool crepe and wool challis. **For lining:** lining fabric should be as fine as possible, and must not be heavier than the outer jacket fabric. Fine silk is ideal. **Made here:** light check jacket in two Collier Campbell prints on polyester crepe de Chine, lined with silk. Dark jacket in a Collier Campbell print on pure wool challis, lined with silk.

MEASUREMENTS One size, to fit **bust** 33–38 in (84–96.5 cm). **Back length of finished jacket:** 38 in (96.5 cm).

PREPARATION **(1)** Cut out jacket pattern pieces from a single layer of fabric following cutting layout. **ASSEMBLE SLEEVES (1)** With right sides together, join sleeve seams with a 3/8 in (1 cm) seam. Press seams open. **ATTACH SLEEVES TO BODY (1)** Take one sleeve piece and, with right sides together, place shoulder point A on sleeve to shoulder point on body piece. Seam back sleeve to back point B and front sleeve to front point B. **(2)** Repeat for second sleeve. **JOIN SIDES (1)** Sew up side seams from hem to sleeve join, fastening firmly at points B. **CUT LINING (1)** Take the jacket paper pattern pieces and modify by cutting off the ⅜ in (1 cm) seam allowance on the hem and cuffs. **(2)** Now make a ⅜ (1 cm) seam allowance up from the cut edges. **(3)** Place the modifed pattern pieces on a single layer of lining fabric and cut following cutting layout. Make up as for jacket. **ATTACH LINING TO BODY (1)** Lay the lining hem to the body hem with right sides together, and match up the side seams. **(2)** Stitch lining and body together; seam allowance ⅜ in (1 cm). **(3)** Press the seam allowance upwards. The seam that joins lining and body should be ⅜ in (1 cm) up from the finished hemline. **(4)** Lay the lining and body parts flat and push the sleeve linings into the sleeves. Make sure that the lining and body seams match up. **STITCH CUFFS (1)** With the hand

between the body part and the lining, pull one sleeve through the front opening and stitch round the cuff on the wrong side, exactly as for hem. **(2)** Pull the completed sleeve back into position. **(3)** Repeat for second cuff. **NEATEN BODY (1)** Lay the jacket flat and bring the raw edges of the body fronts and linings together. **(2)** Pin or tack the edges together from the centre back neck to the hem on both sides, matching up the notches at points C. **(3)** Make up the 2 rouleau ties and tack into position, one on the left front edge, and one on the right front edge, approximately 18 in (46 cm) down from centre back neck. **ATTACH COLLAR (1)** Keeping right sides together, place a single layer of the collar piece, point X, to the centre point X of the back neck. **(2)** Tack or pin the collar to point C, then to the hemline on both sides. **(3)** Stitch from X to hem on both sides with a ⅜ in (1 cm) seam. **COMPLETE COLLAR AND FACING (1)** Starting at point X on back neck, fold the collar piece to the inside to form the facing. **(2)** Turn under the ⅜ in (1 cm) seam allowance sandwiching the raw edges, and slip stitch from X to hem on both fronts. **(3)** Turn up the seam allowance at hem of collar piece, slip stitch and press lightly. **TO FINISH (1)** Make up sash with a seam allowance of ⅜ in (1 cm).

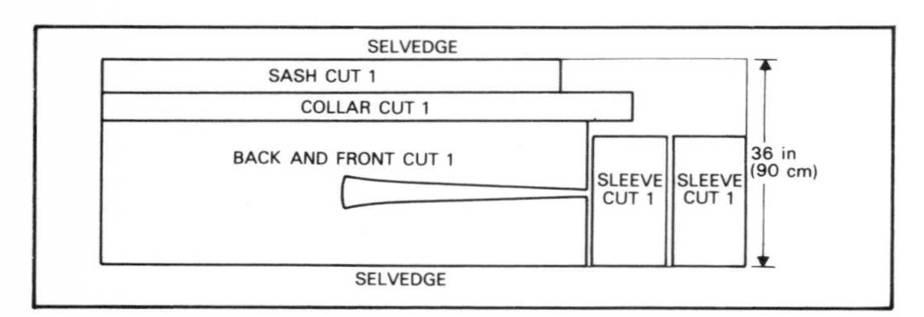

CUTTING LAYOUT FOR JACKET AND LINING
(not drawn to scale)
Fabric width: 36 in (90 cm)

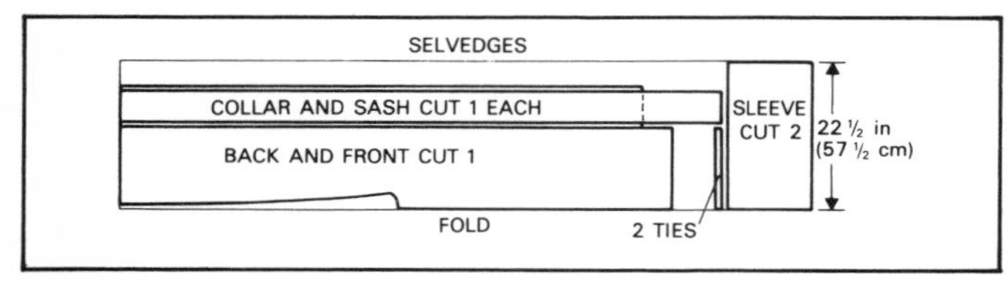

CUTTING LAYOUT FOR JACKET AND LINING
(not drawn to scale)
Fabric width: 45 in (115 cm)

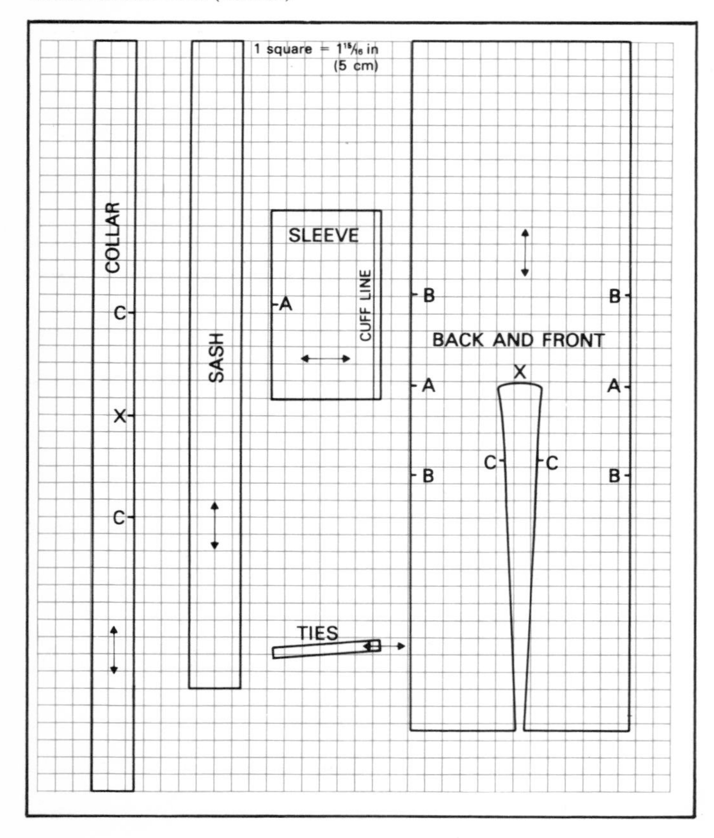

Side Slit Skirt Instructions

MATERIALS **Fabric required:** 1 yd 10 in (1.17 m) of 54 in (138 cm) wide fabric **or** 2 yd 19 in (2.32 m) of 36 in (90 cm) wide fabric. **Recommended fabrics:** suitable for most dress-weight fabrics including cotton lawn and poplin, polyester, viscose, silk, wool challis and light wool-cotton blends such as Viyella. Avoid stiff or bulky fabrics that will not gather up under the elastic waistband. **Trimming:** 1 piece of 2 in (5 cm) wide elastic 22–25 in (56–63.5 cm) in length for waistband. Determine final length by stretching the elastic around your waist until you hit on a comfortable fit. Allow for the fact that joining the ends will reduce the length by 1 in (2.5 cm).

MEASUREMENTS One size, to fit **waist** 24–26 in (61–67 cm), **hips** 34–36 in (86–91 cm). **Length** from top of waistband to finished hem, 38 in (96.5 cm). Length can be adjusted to suit personal requirements; remember to alter amount of fabric required if necessary.

PATTERN VARIATIONS The skirt can be let out to fit **waist** 27–29 in (68.5–74 cm), **hips** 37–40 in (94–101.5 cm) by extending the cutting line 1½ in (3.8 cm) out from the top of the waistline on both pattern pieces, as in Figure 1. From this point, the new cutting line should extend in a straight line until it reaches the original hem line. Side seam allowances should be tapered, as for the smaller size; see Note below. Increase the length of elastic required for the waistband to suit personal requirements.

PREPARATION **(1)** Cut out pattern pieces following cutting layout. **(2)** Neaten side seam allowances by zigzag, overlocking or by turning in the edges ⅛ in (6 mm) and machining flat. **NEATEN WAISTLINE (1)** Turn the top edge of the waistline in to the right side ⅜ in (1 cm) and machine stitch edge on back and front parts. **JOIN SIDE SEAMS (1)** See Note above and join side seams, leaving open the side slit in the left seam and a space for the pocket bag, if you are including one, in the right seam. **MAKE WAISTBAND (1)** Join ends of elastic to make a waistband, and neaten the ends by machining them down flat. **ATTACH WAISTBAND TO SKIRT (1)** Turn the skirt and the waistband to their right sides. **(2)** Keeping the waistband flat, slip it over the top edge of the skirt. The bottom of the elastic waistband should come 1 in (2.5 cm) below the finished top edge of the skirt. **(3)** Pin the skirt and elastic waistband together, making sure that the fullness is evenly distributed. **(4)** When the pieces have been pinned together, place machine foot in position on the elastic, ½ in (1.2 cm) above its bottom edge. **(5)** Now stretch elastic gently and stitch waistband and skirt together, keeping seam ½ in (1.2 cm) from bottom edge of waistband. **TO FINISH (1)** Turn back side slit edges and press. **(2)** Starting from hem, stitch up one side of slit, across the side seam ¼ in (6 mm) above the top of the slit, and down the other side to hem. **(3)** Turn up hem, first turning ¼ in (6 mm), second turning ½ in (1.2 cm) and press. **(4)** Working from right side of fabric, stitch hem ⅜ in (1 cm) up from turned-up hem edge. **(5)** Make up pocket bag, and insert in right side seam. **(6)** Press. **(7)** Make up optional waist sash, and press.

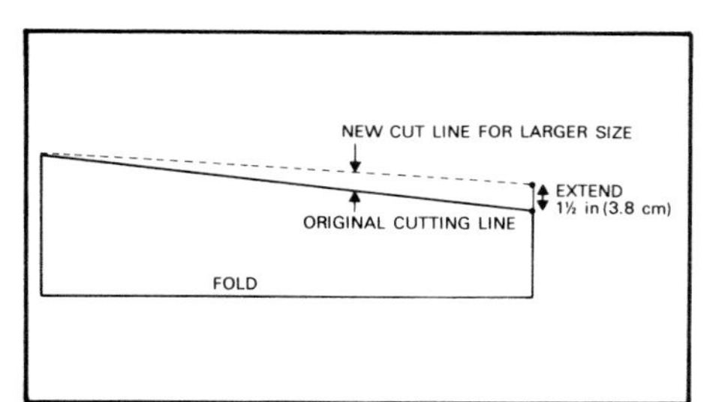

Figure 1

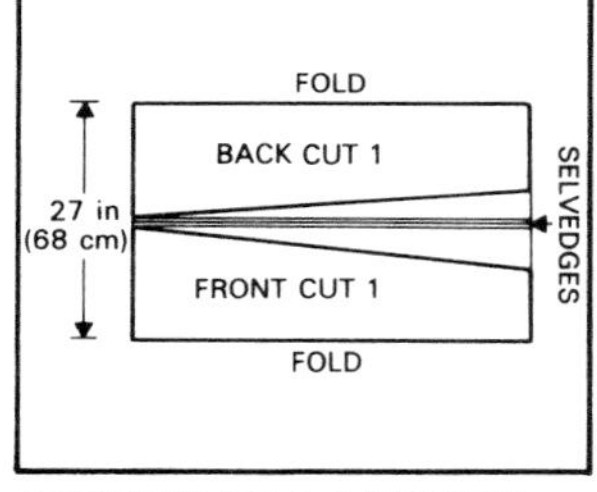

CUTTING LAYOUT
(not drawn to scale)
Fabric width: 54 in (138 cm)

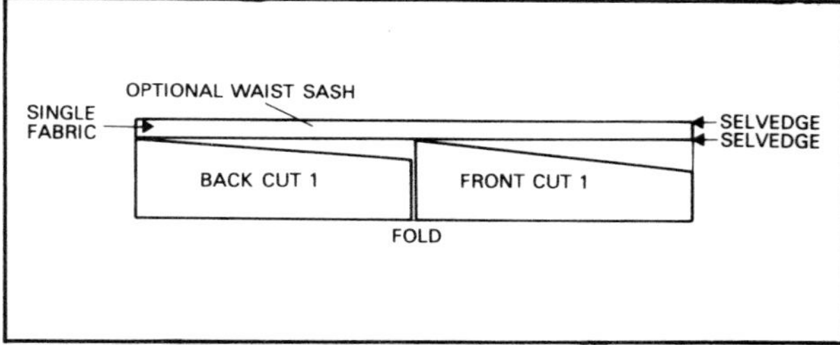

CUTTING LAYOUT
(not drawn to scale)
Fabric width: 36 in (90 cm)

STYLING You can transform the look of the skirt by varying the height of the side slit. Lower the slit – all the way to the hem if you like – for a hobble skirt effect, or raise it very high indeed and wear the skirt over fancy tights or slim trousers. You can accentuate the side slit by edging it with decorative machine stitching, braid, hand-embroidered motifs or hand-painting. Add patch pockets, or insert a pocket bag in the right side seam – see pocket bag instructions for Soft Skirt in *Transcultural Clothes* section (p. 7).

NOTE The side seam allowances are tapered, not straight. The side seam allowance is 1½ in (3.8 cm) at hem rising to 1 in (2.5 cm) at waistline.

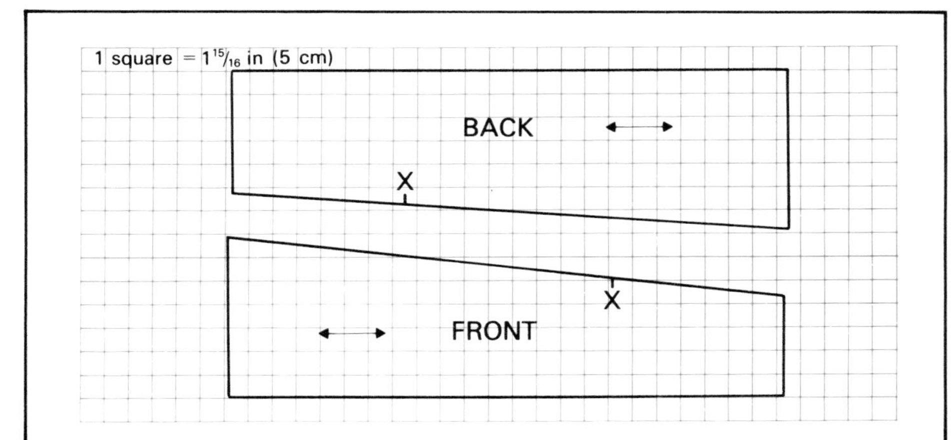

Wrap Dress Instructions

DETAILS Shown with trousers by French Connection and satin mules by Bellesco from Bally. Photographed at the Savoy Hotel, London. Also shown with tights by Aristoc and shoes by Charles Jourdan. Photographed at Dukes Hotel, London.

MATERIALS **Fabric required:** 2 yd 30 in (2.59 m) of 58 in (150 cm) wide fabric. **Recommended fabrics:** suitable for a wide range of dress-weight fabrics including silk, cotton, polyester, viscose and fine wool. Avoid thick or bulky fabrics. **Trimming:** 1 snap fastener and 2 buttons. **Made here:** in a Collier Campbell print on polyester crepe de Chine.

MEASUREMENTS One size, to fit **bust** 33–38 in (84–96.5 cm), **hips** 34–38 in (86.5–96.5 cm). **Back length** of finished dress: 47 in (119 cm). Length can be adjusted to suit personal requirements; alter amount of fabric required if necessary.

PREPARATION **(1)** Cut out pattern pieces following cutting layout. **ASSEMBLE AND ATTACH PATCH POCKETS (1)** Take 1 pocket piece and neaten edges by zigzag. **(2)** Press the pocket facing over to the wrong side and topstitch into position with 2 rows of machine stitching: first row 2 in (5 cm) down from top edge, second row ⅛ in (3 mm) above that. **(3)** Press the sides and bottom edges ⅜ in (1 cm) to the wrong side. **(4)** Pin or tack pocket into position on the dress front and topstitch around ⅛ in (3 mm) from edge. **(5)** Finish with firm stitching at corner pocket openings. **(6)** Repeat for second pocket. **ATTACH FACINGS (1)** With right sides together, lay 1 facing piece to 1 dress front, matching the notches A-B-C. **(2)** Stitch A-B-C, seam allowance ⅜ in (1 cm). **(3)** Trim seam allowance if necessary, then turn and press, locking in the raw edges. **(4)** Neaten the edge of facing by zigzag or by turning in a small seam and topstitching the edge. **(5)** Repeat for second front and facing. **JOIN BACK AND FRONTS (1)** With right sides outside, pin or tack shoulder and side seams together. **(2)** Stitch with a ⅛ in (3 mm) seam allowance. **(3)** Turn through and stitch again, seam allowance ¼ in (6 mm), to lock in the raw edges. **(4)** Press. **ASSEMBLE COLLAR (1)** Lay the 2 collar pieces together with right sides inside. **(2)** Stitch sides and top edge together, seam allowance ¼ in (6 mm) **(3)** Turn through and press. **ATTACH COLLAR (1)** With right sides together, tack or pin the collar into position on the neck edges between points A, sandwiching the collar between the facings from A to shoulder. The facing shoulder edge should be turned ⅜ in (1 cm) in to the wrong side. **(2)** Stitch the collar from A to A, seam allowance ⅜ in (1 cm). **(3)** Turn facings through and press lightly. **(4)** Neaten the back neck by turning in a small seam and finishing by hand. **(5)** Handstitch the facing shoulder edge to the shoulder seam. **ASSEMBLE AND ATTACH SLEEVES (1)** With right sides outside, join sleeve seams, turn through and lock in the raw edges, exactly as for side and shoulder seams. **(2)** Zigzag the cuff edge, press over the cuff facings and topstitch exactly as for pockets. **(3)** Neaten the armholes and sleeve heads by zigzag. **(4)** Tack or pin the sleeves into the armholes, matching up the notches. **(5)** Stitch with a ⅜ in seam allowance and press lightly. **ASSEMBLE AND ATTACH TIES (1)** Make up the 2 ties with a small seam, turn through and press. **(2)** Attach 1 tie firmly to the front edge of the under wrap and the second tie to the side seam at waist position. **MAKE UP TIE BELT (1)** Join the belt pieces to make a long tie belt. **(2)** Fold with right side inside, then stitch across both ends and down the length with a small seam, leaving a small opening for turning through. **(3)** Turn through to right side, close the opening by hand stitching, and press. **TO FINISH (1)** Turn up the hem, first turning ½ in (1.2 cm), second turning 1½ in (3.8 cm). **(2)** Turn under the facing ends and finish by hand or machine. **(3)** Attach button to point D on left front shoulder, and attach a smaller button to the same threads on the wrong side, to act as a stay. **(4)** Stitch a cotton button loop to the right lapel edge, as shown on the pattern. **(5)** To close the neck, sew a press fastener on to the fronts, just below points A. **(6)** Press the dress ensuring that the collar edges and front edges are level. Then, with an easy stitch, topstitch the collar edge to the join at A approximately 1/8 in (3 mm) from edge. **(7)** Topstitch both fronts from A to hemline.

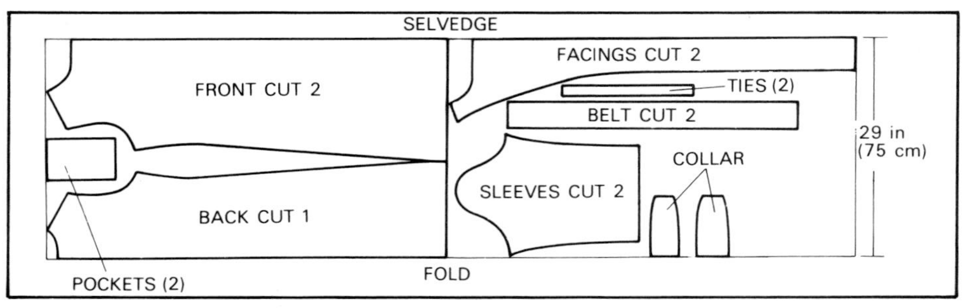

CUTTING LAYOUT (not drawn to scale)
Fabric width: 58 in (150 cm)

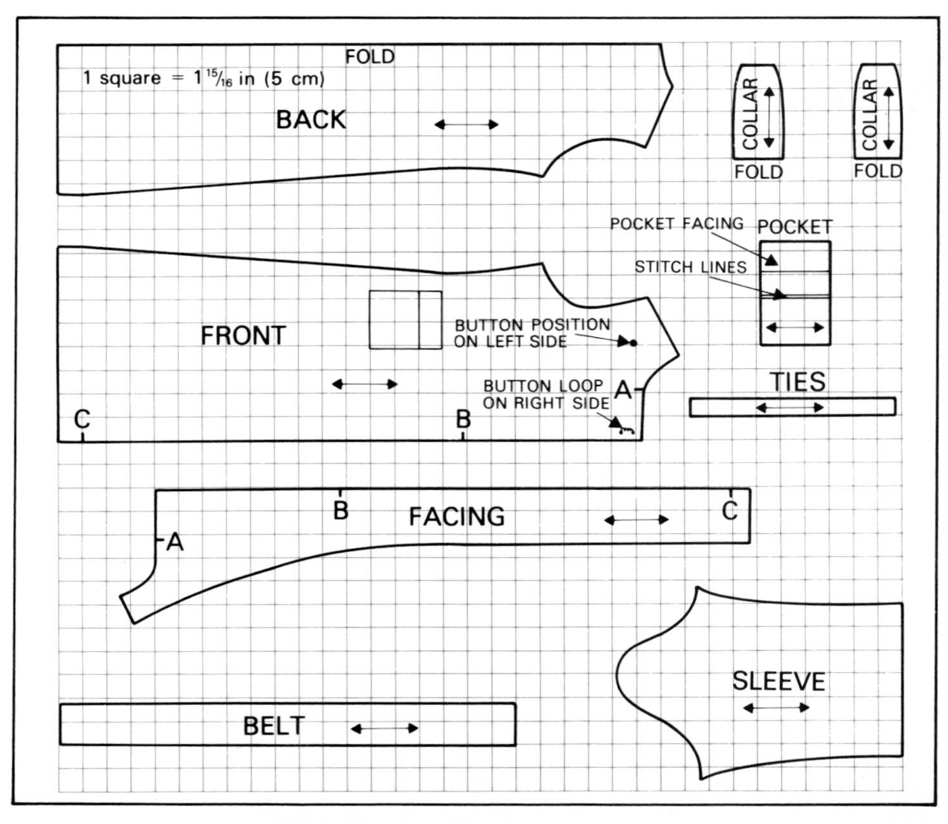

Hazel Chapman

Hazel Chapman is Education Advisor to English Sewing Ltd.

'When the need arises, we all like to feel that we can turn ourselves out in something that's just that little bit special – something that's really unique. You don't have to be a fashion designer or a brilliant seamstress to do it, and you don't have to spend a lot of money either. *Everything is within everyone's means* – all you need is a little imagination.

For example, if I see a gorgeous fabric, I'll have some without thinking about the price. Because I only buy a tiny piece or remnant, and use it for a neckband, a binding, a sash or an insert on the bodice, and go on to make the most of it with contrast topstitching, covered buttons, ribbons, braid – there's really no end to what you can do. It needn't even involve a great deal of sewing. For instance, you can add a ready-made guipure lace collar – or one you've made yourself – to a perfectly ordinary polyester jersey dress and end up with something that really looks very special.

An imaginative use of trimmings is an ideal way of individualizing your clothes, and to me the most exciting possibilities lie in decorations like the thread work on the kimono, shirt and skirt here, all done with free machine embroidery.

Traditional methods of embroidery, lace-making, *appliqué* and so on involve a vast amount of time-consuming handwork. Today, when there are so many demands on our time, the emphasis needs to be on things that can be done quickly. For speed and effectiveness machine embroidery takes a lot of beating, and there's the price factor to bear in mind as well. Thread work is an ideal way of using up all the remnants you always have left on bobbins and reels, an excellent way of getting every bit of value out of each reel of thread.

For many years, machine embroidery was rather frowned on in embroidery circles. There was a general feeling that sewing machines could only produce row after row of perfect geometric stitches with no originality in them at all. That's simply not true, because technically you can do almost anything on a sewing machine that you can do by hand, even canvas work. *It's up to you to provide the originality*, and the scope is tremendous – everything from tiny, dainty motifs to big, brash, bold designs. Whatever you do, you always have that essential speed at your fingertips. I find that particularly attractive because, for me, dressmaking isn't all pleasure. I do it because I want something with my individual touch, and I want it *now* – not in six weeks' or six months'!

As far as equipment is concerned, nothing could be simpler. You need a sewing machine, an embroidery frame to put your work in, fabric, some thread – and that's that. Most of you are likely to have a sewing machine of your own or access to one. It doesn't matter whether it's the newest, most sophisticated model or an old one. As long as it has an electric motor so that both your hands are free, there are lots of beautiful things you can do without having to buy any special equipment or attachments. Just follow the instructions for free embroidery work in your sewing machine handbook.

With machine embroidery there always seems to be a fear that the machine is going to run away with you, that you won't be able to get it to do what you want. It's all a matter of practice and patience. The most important things are to remember that *you* are in control, and not to get stiff or rigid. Keep some fabric in a frame to practise on regularly, particularly just before you get down to the real work of embroidering a garment. It's rather like the limbering up exercises an athlete does – it helps your movements to become more fluid and fluent, and builds confidence.

I was first attracted to machine embroidery because of its potential in terms of working with colour and texture. You can build up lovely layers of shades and superbly textured surfaces very quickly. I keep all my yarns and threads in candy jars, all the reds together, the blues, the greens and so on. I find that arranging them in different combinations immediately triggers off lots of design ideas.

The kimono is an example of strong colours chosen for dramatic effect. The design was inspired by fireworks and sprays of light, and it's just lots and lots of curved

lines in bright, provocative shades. I marked a few basic guidelines on to the silk with tailor's chalk, and then filled in with the stitching – a raised satin stitch, using a tufting foot that can be fitted to most modern machines. You cut the stitches with scissors to get the candlewick texture – it's best to stitch a small area, cut, then start stitching again. Embroidered kimonos, particularly silk kimonos like this one, always make me think of luxury, mystery – and months if not years spent plying the needle by hand. But working with a machine, in short bursts – which is the best way to work on garments this size – the embroidery of the kimono took me, all in all, about twenty hours.

The shirt is a reproduction of an antique Transylvanian peasant shirt, the sort that is traditionally decorated with cross-stitch hand-embroidery that takes months to complete. I could have *appliquéed* braid on to it, or done machine stitching to imitate the handwork. But I thought it would be more interesting to use a modern technique – *dissolved fabric work* – on a very traditional shape. The design, which was based on nasturtium leaves, was machine-embroidered onto *acetate* fabric with *cotton* thread. When the embroidery was completed I trimmed off the surplus fabric leaving less than a centimetre edge round the stitch-work, put the embroidered fabric into a pan of acetone, and left it in the open air until the acetate backing fabric melted away. You must always test the fabric before you start the embroidery, to make sure it will dissolve completely. And you *must* always do the dissolving outdoors, because of the fumes. When the backing fabric has disappeared you're left with just the embroidered motifs – they're rather like pieces of lacework. In this case I measured the areas of the shirt and skirt that I wanted to cover, marked them on to the acetate fabric and filled in the areas with motifs that all linked up, like chain mail. So, when the backing fabric had dissolved, all I had to do was tack the panels and strips into position. The embroidery on the shirt and skirt took up just twelve standard reels of thread. You could never buy a top quality braid or lace of that length at the price that works out to per reel, let alone an exclusive one. It costs your time of course, but that's part of the fun. And the trimming you end up with is absolutely unique.

When people come along to embroidery exhibitions they so often say – 'Oh, I wish I could do something like that!' The point is that you *can* – all you have to do is sit down and start. You can make every mistake in the book, because everyone does in the beginning. And in the end, you don't have to please anyone but yourself.'

DETAILS Kimono shown with shoes by Bellesco from Bally and mauve satin gloves from Harvey Nichols. Photographed at the Hilton Hotel, London. **Made here** in natural cream Ninghai silk from MacCulloch & Wallis, London, lined in lilac silk. For kimono pattern and instructions, see *Transcultural Clothes* section (p 9).

Transylvanian Peasant Shirt Instructions

DETAILS Shown with soft skirt, shoes by Bally and earrings from Detail. Photographed at the Savoy Hotel, London.

MATERIALS **Fabric required:** 3⅛ yd (2.90 m) of 36 in (90 cm) wide fabric. **Recommended fabrics**: suitable for soft handle cotton, fine wool and wool crepe. Avoid pile fabrics and fabrics with a face finish. **Trimming:** 2 pieces of ½ in (1.2 cm) wide tape 32½ in (84 cm) long, 2 pieces of ¼ in (6 mm) wide elastic 12 in (31 cm) long. **Made here:** in cream wool crepe, embroidered with 12 standard reels of Sylko thread.

MEASUREMENTS One size, to fit **bust** 34–36 (86–91 cm). **Back length** of finished garment: 28½ in (71 cm). Length can be adjusted to suit personal requirements.

PATTERN VARIATIONS Side slits can be added to ease the fit over the hips.

PREPARATION **(1)** Cut out pattern pieces following cutting layout. **(2)** Neaten all seam allowances by zigzag or overlocking. **ATTACH FACING TO NECK OPENING (1)** With right sides together, place the neck facing to the centre front neck opening A–B. **(2)** Stitch down the neck slash A to B then up to A on the other side, using a small stitch. **(3)** Carefully clip V point, turn facing to the inside and press. **(4)** Neaten the edges of the neck facing by zigzag, or hand-finish. **(5)** Fasten neck facing to body with 1 or 2 rows of machine stitching. **ATTACH NECKBAND (1)** Press neckband along its centre line, wrong sides together. **(2)** Attach the collar to neckband of body leaving a ¼ in (6 mm) seam allowance over front neck slash. **(3)** Stitch collar to neck, turn and stitch the ¼ in (6 mm) neckband ends. **(4)** Turn the collar out and neaten inside of neckband by machine or hand. Snip the seam allowance on the body before finishing if necessary. **JOIN SIDE PANELS TO FRONT AND BACK (1)** With right sides together, place the side panel to the front points at C, D at shoulder and E at back. **(2)** Stitch C–D–E: seam allowance ½ in (1.2 cm). Press seams open. **(3)** Repeat for the other side. **ASSEMBLE SLEEVES AND GUSSETS (1)** Take 1 gusset and with right sides together place point 0 on the gusset to point 0 on the sleeve and point N on the gusset to point N on the sleeve. **(2)** Stitch from O to **exact** point N. **(3)** Stitch the sleeve seam from the cuff to the exact point N. **(4)** Fold the gusset to corner and complete the sleeve seam by stitching N–O. **(5)** Gather the sleeve head with a loose stitch between the ½ in (1.2 cm) seam allowance, pull up the gathers so that the back and front sleeve head parts 1–X and X–1 each measure 8½ in (21.5 cm), X being the shoulder point. **(6)** Repeat for second sleeve and gusset. **GATHER CUFFS (1)** Place the ½ in (1.2 cm) wide tape on the inside of the fabric to form a channel to take the elastic. **(2)** Stitch tape to sleeve leaving the ends neatened but open to take the elastic. **(3)** Elasticate cuff. **(4)** Repeat for second cuff. **JOIN SLEEVES TO BODY (1)** With right sides together, place the shoulder and sleeve centre points X together. **(2)** Pin the back and front sleeve head seams into position, matching up 1–O–K on the sleeve to 1–O–K on the body part. **(3)** Stitch back and front sleeve seam from X to the **exact** point K. **(4)** Complete the side seam assembly by stitching from the exact point K to the hem. **(5)** Repeat for second side. **TO FINISH (1)** Press up the hems of the body and sleeves, and finish by hand or machine. **(2)** Press.

CUTTING LAYOUT
(not drawn to scale)
Fabric width:
36 in (90 cm)

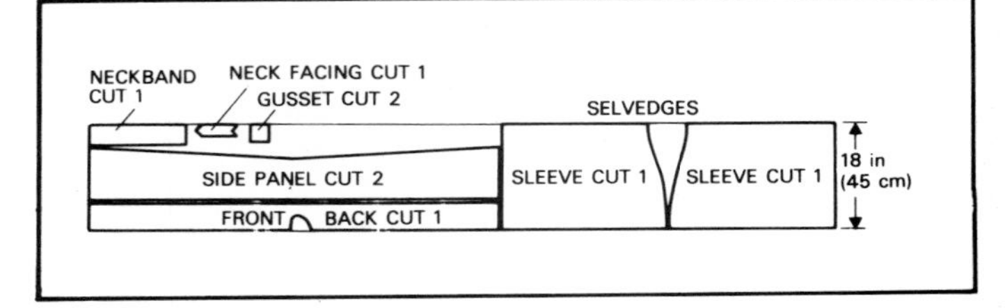

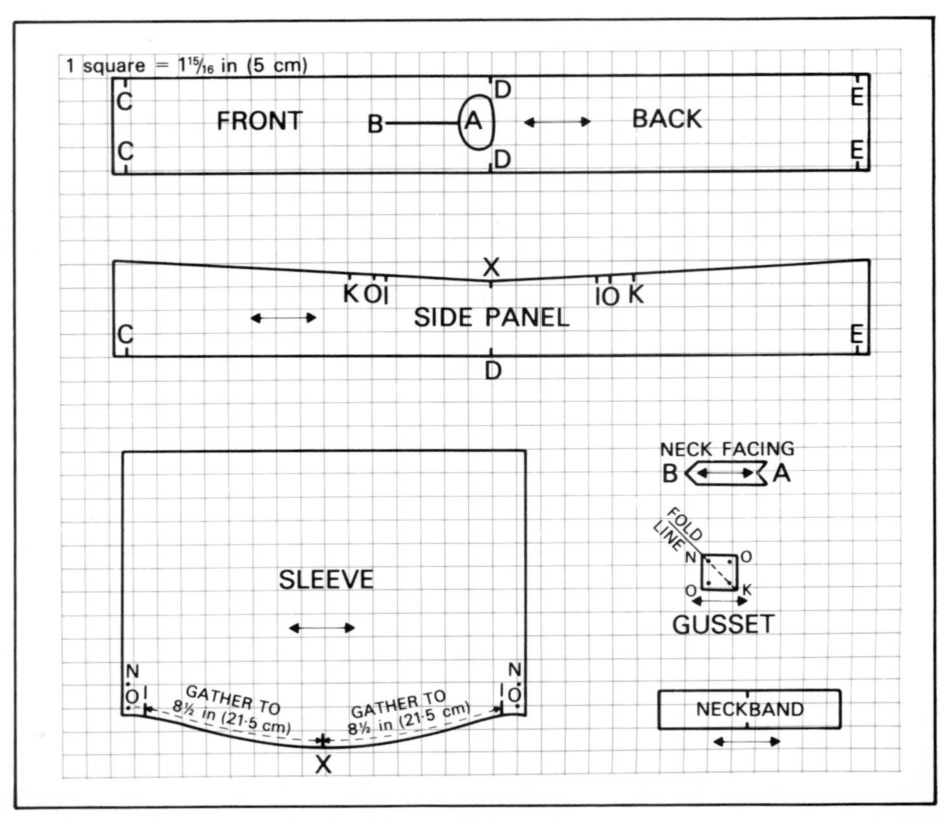

David Reeson

David Reeson studied painting at Loughborough College of Art and Design in Leicestershire, and illustration at the Royal College of Art, London. He teaches in London and is art director of *The Fashion Guide*.

'I like people to be original and creative in the way they wear clothes, and as I see it, the most satisfying way to be different is to wear designs of your own.

Fashion designers concentrate on shapes, but as a graphic designer I'm more interested in what you can do with the surfaces of shapes – how you can use colour, line and space to create designs *on* designs. And for that, I find that hand-painting is an ideal technique.

Often it's the only way to get exactly what you want. For instance, I've always liked shirts with bright colours and strong motifs, but I never seemed to find any to buy except in second-hand sales and on holiday. Finally I realized that the kinds of designs I was looking for were very like my own drawings and paintings. So now I collect plain silk and cotton shirts and paint them myself.

Looking back, it was an obvious solution, but the fact that it took me some time to come up with it is something of a comment on the place art and originality have in everyday life. So many of the fabrics on the market today are 'undesigned' – recoloured re-issues of prints that have been going for years and years. They're mechanical rather than artistic or creative, dull and tired rather than fresh and new. The same goes for all those scarves and accessories scattered with CCs, CDs and LVs. There's so much of it about that it's all too easy to get into the habit of settling for what's available. Too easy to accept rather than innovate, to get into a rut. That's why art – anything creative that you do yourself – is so important. It introduces change, vitality, new ideas and images – all the things that make life interesting. Just as there's an art to living, there must be a place in life for art. For that reason, I don't distinguish between the work I do on fabric and the work I do on canvas or paper. I think of the clothes as an extension of drawing, or as wearable paintings. And I think this kind of approach should be carried over into carpets, ceramics, everything we live with. There's no reason at all why art should just be kept in a frame and stuck on a wall.

I prefer to paint on natural fabrics, particularly silk because it drapes beautifully, takes the colours well, and *holds* them – which is very important. I like to work with white and cream fabrics because they don't interfere with the dyes, so you can keep the colours looking transparent. Even the palest coloured fabric will distort the dye shades, but you can turn the distortion into a positive feature of the design. Or you could start with a coloured fabric, put on white dye, then put coloured dyes on over the white. It's always better to work with a fabric that has a flat weave, because the dyes tend to run along

the grain on a twill weave. Whatever type and colour of fabric you use, be sure to wash and press it before you begin, to remove any finish on the cloth. And always experiment on a bit of the fabric – or on an inner hem if it's a made-up garment – before you start, because no two fabrics or dyes react in the same way.

There are some marvellous liquid dyes in very strong colours that you can use like inks, but the drawback there is that you have to paint a fixative on afterwards, leave it to dry and then wash it out, so it isn't realistic for anything larger than a scarf. All the clothes here were painted with creamy, water-soluble dyes that you fix by ironing. You can use them thick, or thin them down and use them like watercolours. The fact that they can be used either way gives you a chance to produce some very interesting textures. Personally, I find the colours are not as interesting as the liquid dyes, but you can get round that by mixing your colours yourself. I never use ready-made colours straight from the pot.

I started by painting window blinds and scarves, because they were closest to my usual medium. In designing on the flat, you concentrate on colour, pattern, the type and size of the fabric. But as soon as you move on to clothes, even the simplest garments, you have to design in three dimensions. You have to think about the way the fabric is cut, the position of the seams, the general style and shape of the garment and the placement of your designs in relation to the whole. The intensity of your design has to complement the shape – sometimes you overwhelm, sometimes understate. You have to

approach each garment in a different way, as you can see from the clothes I've painted here.

The kimono is a wonderful shape for hand-painting – there are lots of uninterrupted surfaces and the garment moves and flows exceptionally well when it's worn. It was a huge area to cover, and the space had to be filled in an imaginative way. Because of the size, I wanted a very definite design – one that would 'take over' the shape. The size also dictated the variety of pattern. While a scarf in just the small dashes could be quite nice, on something as large as this it's a good idea to have a change of scale. The overall pattern helps to define different spaces, and the concentration of black gives the design some weight at the bottom, which it needs.

The poncho blouse is a basic square, an ideal shape to work with since you don't have to worry about extra darts or seams. The design is an abstract pattern suggested by foliage, a variation of the painting on the kimono, but the treatment is much lighter. Because of the size of the top, there's variety of pattern but not much change of scale. And the leaves are open and airy at the hem, to avoid giving the impression of bulk across the hips.

I've always liked borders because of the way they frame space, and strong border patterns can be just as striking as patterns that cover the entire surface of a garment. The border on the long skirt was inspired by Egyptian friezes of highly stylized feathers, sun discs and flowers. This is the only design of the four in which the pattern repeats, but the primary-bright colours keep it lively and add movement.

The sundress, by Antony Kwock, was the most challenging of the shapes to paint because of the spiral seam that dominates the design. I didn't want the seam to cut through the middle of the motifs, so I worked out my design on the flat on paper first, to determine the placement. The seam is unusual and surprising, so I wanted to incorporate an element of surprise in my design as well, something that would accentuate the turning of the spiral and of the body. From the front you see what look like pink clouds, but as you turn to the back you see they're pink birds.

Of course, there are no hard and fast rules. This is the way I've done it, but there's no reason why your treatment of the same garments should be similar to mine in any way. I suppose that's what I like most about hand-painting – the fact that it gives you the chance to wear things that no one else in the world has got, things you can put *your* name on. So often you draw or paint in isolation, and no one ever sees your work. This way, you can turn your wardrobe into an art show. Better still, you can wear something today, wash it, paint on some new details, press it up and there you are – something new to wear. You can change it every time you wear it – and there can't be a better example of creative dressing than that!'

DETAILS Kimono shown with earrings and bracelet from Detail, evening mules by Charles Jourdan and belt from Liberty. Poncho blouse shown with shoes by Bally. Border-painted skirt shown with shoes by Charles Jourdan. All photographed in the Park Lane Hotel, London. **Made here:** kimono in light cream Antung silk from MacCulloch & Wallis, London. Poncho blouse and spiral sundress made in Tootal's Casablanca spun viscose. Long skirt made in Tootal's Flanesta viscose. **Patterns:** patterns and instructions for all garments are to be found in other sections of the book. For Japanese kimono, poncho blouse and skirt see *Transcultural Clothes* section (p 9, 13, 21); for Antony Kwock's spiral sundress, see *Design and Designers* section (p 81).

Design and Designers

Bruce Oldfield

My designs have always been very simple, ever since I left college. And that's why I think I took off so quickly. When I started it was all very grand evening wear, and I think everybody was waiting for someone to come along and do something simple. The time was right for *sleek chic*.

Modern. That's a word I like to describe my designs. The Americans are very into a modern look – often out of necessity rather than actual desire, I suspect, because American clothing is manufactured, not made. So everything has to be pared down to the lowest common denominator: simpler than simplest. They can't do more than three or four seams, they can't do handwork. American designers have made a virtue of it; they've evolved a style out of necessity. I like clean lines and simplicity, but I don't do them out of necessity – with me, it's desire. One of the reasons I work as I do is that I don't have to make too many concessions. I design as I want to, not as I must.

My designs are simple, understated, chic – and sexy. My things are always fairly *aware* of the body, even if they're not cut tight, tight, tight and slashed up to *here*. When I say sexy, I don't mean exposing everything. I mean *hinting* at what's there rather than flashing it. Allure. Mystery. I don't think of this dress as something for a nineteen-year-old girl. A nineteen-year-old can wear it, of course, but it's intended for someone more sophisticated. Someone who knows how to . . . *behave* . . . in it.

I don't want someone to walk into a room in one of my designs and have people say 'that's a fabulous dress' – I want them to say 'that woman looks fabulous'. I understate the design, and the *person* comes through. It's whether or not the woman looks good that really matters. She mustn't feel compromised by what she's wearing. I like a woman to look good, feel relaxed, be confident – and she will if her clothes show *her* off to her best advantage.

Of course, women don't always know what suits them. That's the lovely thing about having a lot of private clients, as I do. You can say 'No, I don't think that's *you*' and talk them out of it. You also get to know all the little *bêtises* they have about their bodies. They have them about everything – the neck, the collarbones, the upper arms, the legs. It's the same with shoulders – a lot of women who have big shoulders hate them. Personally, I like a good pair of shoulders, I think every designer does. If you have a good pair of shoulders everything you wear tends to look good, because clothes usually hang from the shoulders. Good shoulders are a *great* asset. So I can tell my clients things like that, and I learn from them too. For example, a lot of women won't wear bare dresses once they get to a certain age. Thirty. What they want is a little jacket or a coat that *isn't* a fur coat, something to slip on in the evenings, to cover up. In the past I didn't think about it much, but I do things like that now because I realize they're important.

Most of my designs are worn from five o'clock till whenever, but very few are uncompromisingly 'evening'. They actually have a daywear simplicity about them, they're just made in more luxurious fabrics, and often very rich, strong colours. Red is the colour I like most. Not necessarily a red dress, but something with a reddish tinge – pinks, purples, maroons. I find I come back to them every season. Sometimes I combine two colours on a dress – teal blue with scarlet, or hot pink with tropical blue. I use a few prints, but my heart's not really in them. I like to get hold of a plain cloth and *do* something with it. When I first started I did a lot of hand-smocking, which involved taking a plain fabric and putting surface interest *onto* the fabric as well as *into* the shape of the garment. Wonderful! You can't do the same kind of thing with prints – they're too limiting. And border prints are the most limiting of all.

Another thing that's limiting is using just one fabric. When I first started off, to make life easier for myself and because I had a penchant for it, I worked exclusively in jersey. It suited me and my small business, because at that stage I couldn't have too many ball games happening in one ball park. It's been difficult ever since, because you get labelled. If you ask someone who doesn't look closely at what I do, they'll say 'Oh yes, Bruce Oldfield, soft jersey dresses', which is a bore. I still use a lot of jersey, but I use a lot of other fabrics as well, as you see with this design. You can make it in jersey or woven cloth – in silk crepe de Chine, in georgette, but it doesn't have to be silk. You can make it in rayon crepe, you can even make it in very fine wool.

I like this design because you can do anything in it, go anywhere in it, from lunchtime to bedtime. The shape is slim, but not too tight – it's an easy dress to wear. The neckline is very flattering, very soft. It's simple, it's elegant – it's what *you* want it to be.

It also shows what I mean about doing something with a plain piece of cloth. The concept of the dress is very simple, one piece each for the front and back. It's the simplest possible way of achieving that shape. If you look at the pattern, you'll see it's almost a T-shape, but the shoulder is extended up, and you gauge it back down to form the pleats on the shoulder. It all works *together*. You could put in an armhole seam, but it simply wouldn't look as good. It wouldn't be a beautifully simple design – just a cheaper one. The dress doesn't take very much fabric as it is. By chopping in the seam you lose a lot of style, and just save a few pence. You can always tell if something's been skimped. That's not the kind of concession I want to make.

As I said, this is a *very* easy dress to wear, and you can do a lot with it. If you want to dress it up, I'd suggest a broad belt or sash, very *important* earrings, flashy combs in the hair, quite strong on the make-up – good red or pink lips. Then a nice pair of patterned stockings – fishnet or the spotty ones – and the *highest* heels. If you want to dress it down, try little belts, maybe sash the waist with a bright scarf. Add coloured tights, court pumps, a little mannish hat – *not* a cocktail one – and a bright shoulder bag, worn diagonally across the chest.

And now the only thing I have to tell you is what I tell my model girls, just before they go onto the catwalk at the collections: *You're supposed to feel good in these designs – that's the whole idea behind them. So go out and show them how good you feel. Go out and knock 'em dead!*

Split Sheath Instructions

DETAILS Shown with hat by David Shilling, shoes by Manolo Blahnik, earrings by Artwear from Browns and bracelet from Emeline.

MATERIALS **Fabric required for dress:** 2 yd (1.85m) of 56 in (142 cm) wide fabric; **for sash and rouleau trim,** 1 ⅛ yd (1 m) of 34 in (86 cm) wide fabric. **Recommended fabrics for dress:** jersey, silk crepe de Chine, crepe georgette, fine soft wools; **for sash and rouleau:** silk satin or same fabric as dress. **Made here:** in Racine's jersey trimmed with pure silk satin.

MEASUREMENTS One size, to fit **bust** 32–36 in (81–91 cm), **hips** 34–38 in (86–96 cm). **Height** of side slit: 17½ in (44.5 cm).

STYLING It is not necessary to neaten the edges when using a non-fray jersey fabric. For other fabrics, neaten edges by zigzag or turning in. Adjust height of side slit to suit personal requirements.

NOTE: All seam allowances are ⅜ in (1 cm) unless otherwise stated.

PREPARATION (1) Cut out pattern pieces following cutting layout. Note that rouleau must be cut on the bias. **ASSEMBLE DRESS (1)** With right sides together, stitch and press open the centre back seam. **(2)** Stitch the side seams, leaving open the side slit on the left side only. **(3)** Stitch shoulder seams. **(4)** Press open. **PLEAT NECKLINE (1)** Pleat the neck circumference to measure 34 in (86 cm) as indicated on pattern. The 5 pleats at right back should face the **centre back** seam. The 6th pleat at the shoulder should face towards the **front,** concealing the shoulder seam. The remaining 5 pleats on the right side should face towards the **centre front. (2)** Repeat for left back and front. **(3)** By hand or machine, stitch a stay stitch around pleated neck circumference to hold pleats in position and ensure that the neck edge will measure exactly 34 in (86 cm). **ASSEMBLE ROULEAU (1)** Join the ends of the rouleau strip to form a circle. **(2)** With right sides together, place the join to one shoulder. **(3)** Pin or tack the rouleau round the neckline taking care not to stretch it. **(4)** Stitch rouleau to neck edge, just covering the stay stitch. **(5)** Fold to inside and turn up ¼ in (6 mm), then hand catch to machine line, locking in the raw edges. **TO FINISH (1)** Turn up hems, sleeves and side slit edges ⅜ in (1 cm), and finish by zigzag, hand or easy tension machine stitch, taking care not to stretch the edges. **(2)** Press smoothly on wrong side with a cool iron. **(3)** Join silk satin pieces to make sash 3 × 65 in (8 × 165 cm) or as desired.

CUTTING LAYOUT
(not drawn to scale)
Fabric width: 56 in (142 cm)

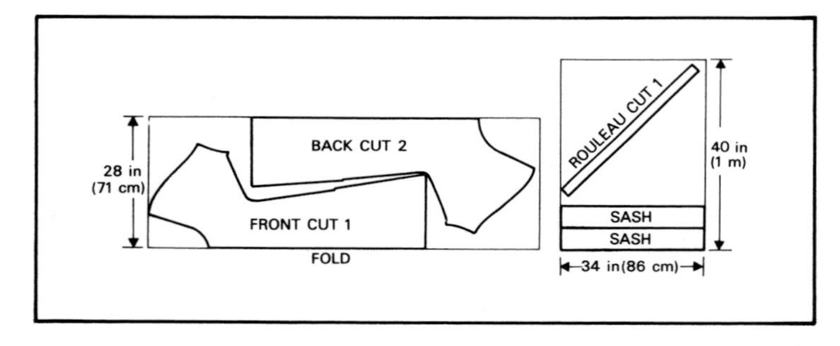

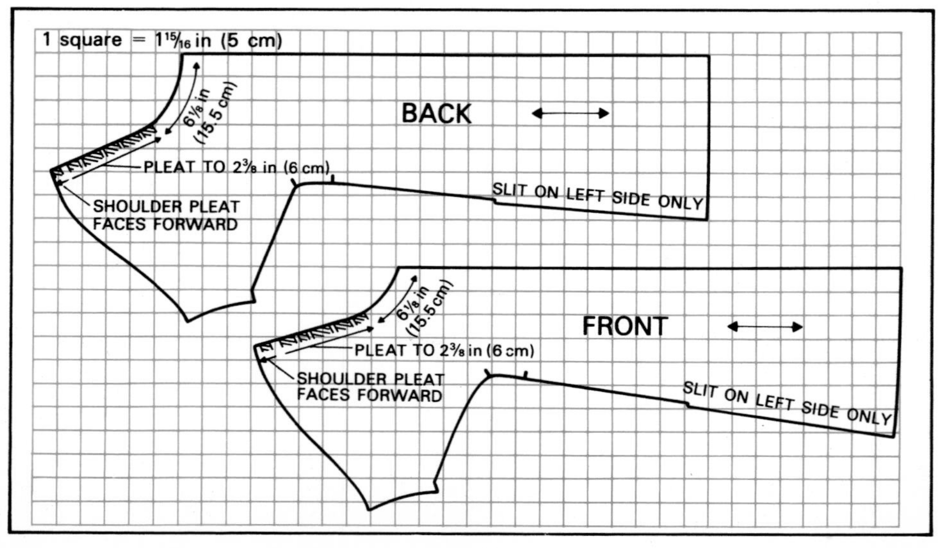

Gordon Luke Clarke

Do you know what makes a woman beautiful? *Character – and style.* Style doesn't come from the old formula of 'breeding', and character isn't something you learn by the age of eighteen. It comes in time, and it comes from within.

I like a woman who looks like she's her *own* woman. A woman who is together, and has a style of her own. It's not something you can buy – it's something you must *feel*. It can just be small things – the way you turn up a collar or belt a jacket, the way you carry yourself.

Attitude is the strongest and best accessory, because the way you look depends on the way you feel. I've been to lots of fashion parties where everyone was dressed to the nines or desperately over-dressed, and very often the person who came off best was dressed very simply, even in jeans and a T-shirt. It wasn't what she wore – it was *how* she wore it – that mattered. Her attitude and her style.

A woman used to buy *couture* because it gave her an instant status. It showed her husband or father could afford to buy expensive clothes; it implied she had taste because she was dressed by a smart designer. Now women's confidence comes from other sources than a label. It comes, as it should, from themselves and the lives they lead.

Gordon Luke Clarke and his daughter Moana

An outfit that has been well designed should be something that you can wear anytime, anywhere. That's what designing for the way women live today is all about. I never design a little cocktail dress or a suit to wear to the races – my clothes can take you to a ball or to the beach. Clothes must function *for* the woman, adapt to her needs – not the other way round. They should never dictate to her, or limit her in any way.

Clothes must make a woman feel good in them. Every woman has a different shape, walks in a different way, strikes different poses. A woman knows how she looks best naturally – even if she isn't always aware of it. Clothes musn't force a woman into one shape or one pose. They must give her the freedom to be herself, she must be able to give them the shape that *she* wants.

You don't need hundreds of clothes to look good. I've just done a collection based on only four designs – a pair of trousers, two jackets and a skirt. I carried them into different fabrics, patterns and colours, and varied the combinations until they added up to a complete repertoire of looks. You can do exactly the same thing. The secret of creative dressing is learning how to put things together.

Accessories are very important. They're the means by which a woman can make her clothes project any look or mood she likes. A woman has many facets, and clothes should always be designed with that in mind.

What are these different facets of femininity? You could call it magic – a special kind of alchemy – that comes from the interplay of what I see as the four elements of Woman:

> *Fire* is the temptress, sexy and passionate.
>
> *Water* is the cool sophisticate, the spirit of elegance.
>
> *Earth* is warm, reassuring and dignified, an aspect of the maternal.
>
> *Air* is giddy, witty and effervescent – the soul of flippant chic.

Every woman has these elements – they make up her moods, the way she feels, the way she looks. She should never be ashamed or afraid of them, particularly today when her life is her own.

I've chosen my *Domino* designs to show you exactly what I mean. I see the first *Domino* variation with the matching shirt jacket and trousers as Fire, the second with shirt jacket and skirt as Water, and the third with jacket, triangle and pale trousers as Earth. They're all cut from the same pattern, all made up in the same colours, and they're completely different – as all women are.

The same simple accessories have been used throughout, but they add to each look in a different way. Fire wears the white fabric flowers in her hair, Water scatters them over a picture hat, Earth wears them as a corsage. The next accessory is a scarf in soft black gauze, six and a half feet long and 15 inches wide. Fire twists it into a provocative bandeau bra, Water ties it into an elegant *bustier* top. Then there's the triangle that Earth wears as a graceful pointed *bustier* – it can also be worn over the head, round the shoulders, around the waist tied in front or at the side. And what would Air wear? I see her running down a beach – maybe dressed entirely in triangles of different sizes, tied in clever ways.

My collections start with the silhouettes. I've always remembered what Hardy Amies told me – 'It's what you can afford to take off a garment that makes it stylish, not what you *have* to put on it.' I start by designing a beautiful, simple shape, rely on colour for dramatic effect, and add the details at the very end.

Colour is wonderful, it's a whole vocabulary of mood and meaning. You can do so much with colour alone. My favourites are the jewel colours, clear and rich, and I like black because there are so many dramatic ways of wearing it. When I first started in design, there were lots of rules about mixing colours – blue and green should never be seen, and so on – that were completely wrong. I love mixing colours and I love mixing patterns too. Look at ethnic societies – they wear everything with everything and still look amazing. Ethnic combinations would have to be toned down or refined for a collection, but they show that you can wear prints with prints, colours with colours. In fact, you can do anything – if you do it with pizazz.

Most of all, I love being a designer – for me, design is everything. But all I can do is set the stage – *you* have to do the performing. If you have style and character, I know you'll be fantastic.

Domino Instructions

DETAILS First shirt jacket shown over matching triangle worn as *bustier* top and contrast trousers, with suede evening mules by Charles Jourdan. Photographed at the Royal Horseguards Hotel, London. Second shirt jacket shown with matching trousers, shoes by Bellesco from Bally, diamond and pearl necklace by Asprey and fabric flowers from Beale & Inman Ladies Department. Photographed at the Hilton Hotel, London. Also shown with matching circular skirt, shoes by Bally, hat by Beale & Inman Ladies Department and champagne by Moët et Chandon. Photographed at the Savoy Hotel, London.

MATERIALS Fabric required for shirt jacket only: 2 yd (1.85 m) of 54 in (138 cm) wide fabric *or* 2½ yd (2.30 m) of 36 in (90 cm) wide fabric. **Fabric required for shirt jacket and trousers together:** as shown on cutting layout: 5½ yd (5.05 m) of 36 in (90 cm) wide fabric. **Fabric required for shirt jacket and skirt together:** 6 yd (5.50 m) of 54 in (138 cm) wide fabric. **Recommended fabrics:** suitable for a wide range of dress-weight fabrics including cotton, silk, viscose, polyester and fine wools. **Made here:** First shirt jacket, skirt and triangle made in printed cotton poplin by Beeren Fabrics of Holland. Second shirt jacket and matching trousers in printed spun viscose by Walraf Textildruck. **Trimming:** 3 buttons.

Shirt Jacket

MEASUREMENTS One size, to fit **bust** 34–38 in (86–96 cm). **Back length** of finished jacket: 25 in (64 cm).

STYLING Can be worn open as a jacket, buttoned up as a shirt or tied in the front as a calypso-style bolero.

PREPARATION (1) Cut out pattern pieces following cutting layout. **(2)** Note seam allowances on pattern. **(3)** Neaten all seam allowances by zigzag, overlocking or by turning in the edges ⅛ in (3 mm) and machining flat. **PREPARE FRONTS (1)** Press up neatened hem allowance ½ in (1.2 cm) to the wrong side. **(2)** Press over the ¼ in (6 mm) allowance on front edges to wrong side. **(3)** Press back the 1½ in (3.8 cm) front facings and topstitch from neck to hem. **(4)** Repeat for second front. **BACK (1)** Press up hem as for Fronts. **(2)** Pin the centre back inverted pleat into position and press in the pleat to hem. **(3)** Secure the top of pleat with a topside square tack as shown in Figure 1, approximately 3 in (7.5 cm) down from neck edge. **ASSEMBLE BODY (1)** With right sides together, stitch fronts to back. Press side seams open. **ASSEMBLE PATCH POCKETS (1)** Take 1 of the neatened patch pocket pieces and press

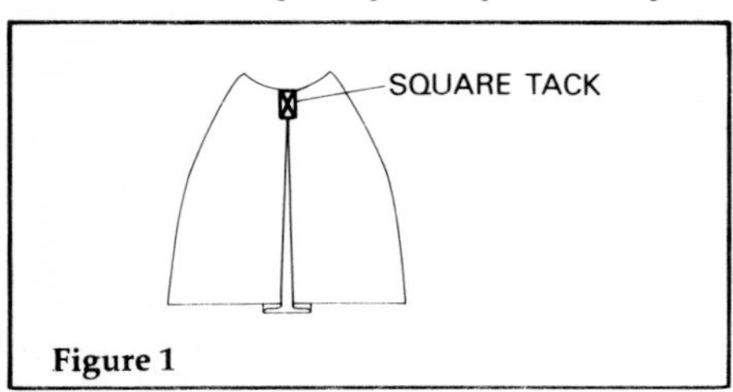

Figure 1

in the centre inverted pleat. **(2)** Tack or pin the pleat into position and turn down the top facing to the wrong side. Press. **(3)** From the right side, stitch the facing into position with 1 row of stitching 1¼ in (3 cm) from top edge. **(4)** Secure top edge of pleat by square machine tack as for Back pleat. **(5)** Press over the side and hem allowances to wrong sides. **(6)** Repeat for second pocket. **ASSEMBLE BREAST POCKET (1)** Take breast pocket, turn in the top facing 1 in (2.5 cm) to the wrong side and press. **(2)** Secure with 1 row of machine stitching ⅞ in (2 cm) down from top edge. **(3)** Press the side and hem allowances ¼ in (6 mm) to wrong side. **ATTACH PATCH POCKETS (1)** Lay the 2 patch pockets into position on the fronts. **(2)** Stitch the sides of the pockets with 2 rows of machine stitching, first row on the edge, second row ¼ in (6 mm) inside first row. **(3)** Topstitch the bottom hem of the shirt jacket, taking in the bottom edge of the pockets, first row ⅛ in (3 mm), second row ¼ in (6 mm) inside first row. **ATTACH BREAST POCKET (1)** Lay the breast pocket into position on the left front and secure with 2 rows of machine stitching, first row on edge, second row ¼ in (6 mm) inside first row. **ATTACH SLEEVES TO BODY (1)** Make up each neatened sleeve piece: seam allowance ½ in (1.2 cm). Press seams open. **(2)** Turn up sleeve hems as for body hem. **(3)** Check that the sleeves are correct for armholes then stitch sleeves to body with a ½ in (1.2 cm) seam allowance. Press seams open. **ASSEMBLE COLLAR (1)** Fold top collar piece along its length right sides together and stitch ends with a ¼ in (6 mm) seam allowance. **(2)** Trim down the seam allowance, turn through to the right side and press. **ATTACH NECKBANDS (1)** Sandwich the assembled collar piece between the 2 neckbands with right sides together, making sure that the top collar matches the notches on the neckband at points 0. **(2)** Machine stitch through the 3 layers past point 0 and round the neckband ends at point E. **(3)** After stitching, trim off half of the seam allowance on the rounded ends. **(4)** Press completed collar and band. **ATTACH COLLAR TO NECK (1)** Start at centre back neck. With right sides together pin the centre of the single layer of collar band to the centre back neck. **(2)** Continue to pin or tack the collar band to the front edges of the shirt neck. **(3)** Stitch the single layer of neckband to neck edge and finish ends carefully. **(4)** Turn collar so that the second band lies inside the neck as a facing. **(5)** Turn under the ¼ in (6 mm) seam allowance and finish by hand, locking in all raw edges. **TO FINISH (1)** Make buttonholes and attach buttons as indicated on pattern. Press. **(2)** If desired, make waist sash from surplus fabric.

CUTTING LAYOUT
(not drawn to scale)
for shirt jacket and trousers.
Fabric width: 36 in (90 cm)

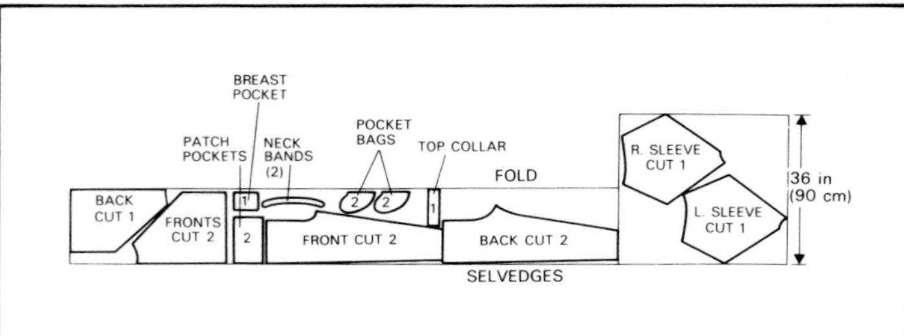
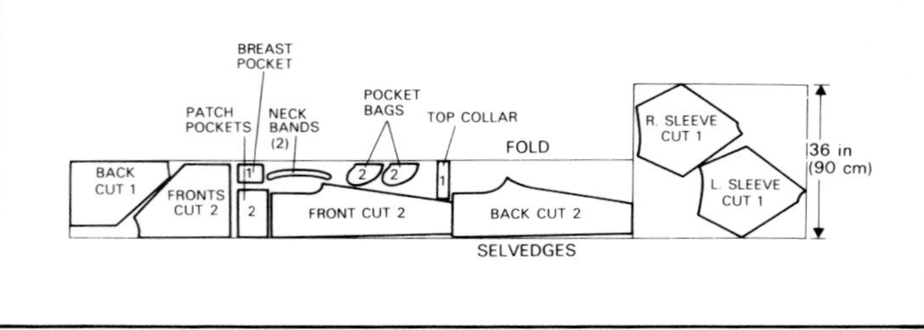

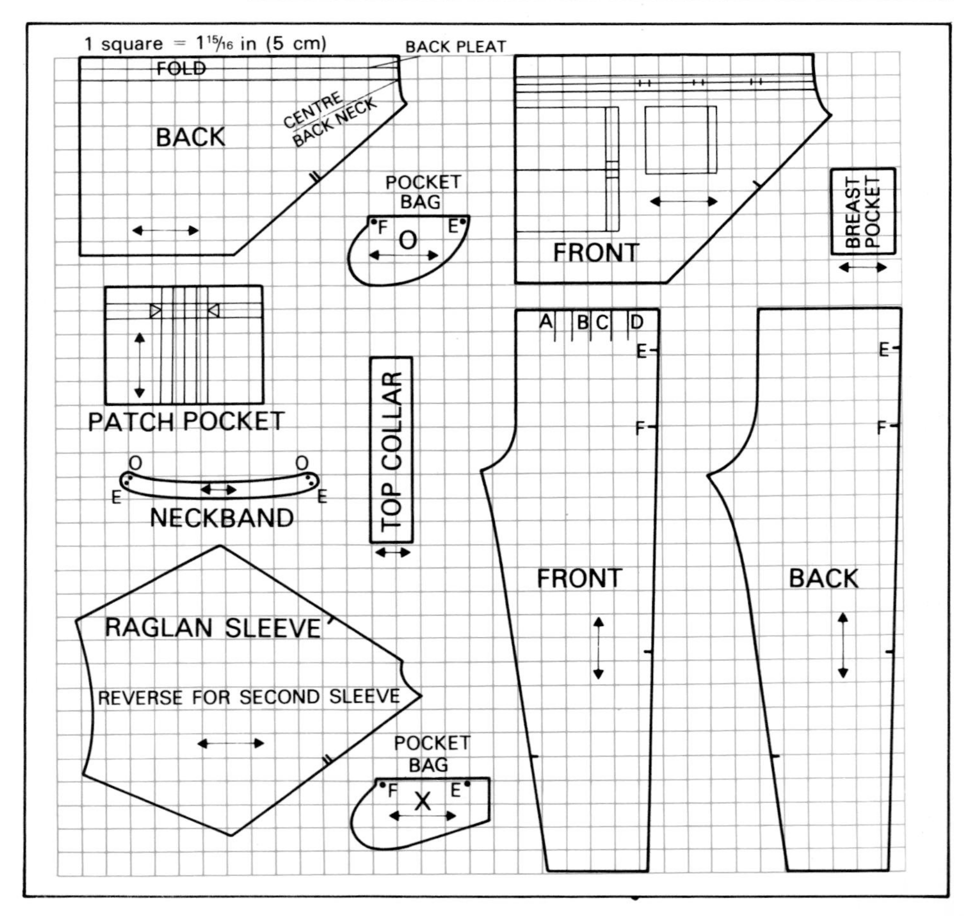

Pleated Trousers Instructions

MATERIALS Fabric required for trousers only: 1½ yd (1.37 m) of 54 in (138 cm) wide fabric **or** 2 yd 22 in (2.40 m) of 36 in (90 cm) wide fabric. **Trimming:** ¾ yd (70 cm) of ¾ in (2 cm) wide elastic for waist.

MEASUREMENTS One size, to fit **waist** 24–28 in (61–71 cm), **hips** 34–36 in (86.5–91.5 cm). **Finished length:** 44½ in (113 cm) from top of waistband.

PREPARATION (1) Cut out pattern pieces. **(2)** Neaten all edges by zigzag, overlocking or by turning in the edges ⅛ in (3 mm) and machining flat. **(3)** Note seam allowances on pattern. **ASSEMBLE POCKET BAGS (1)** Take 1 smaller piece marked O and place in position on the larger piece marked X. The right sides should be inside and the hem and notches matching. **(2)** Stitch from point E round the pocket bag to F leaving the pocket mouth open. Seam allowance: ¼ in (6 mm). This method of pocket-making leaves a single layer of fabric to be attached to the waist seam. **(3)** Repeat for second pocket. **MAKE UP LEGS (1)** Take 1 back and 1 front piece and, with right sides together, stitch the side seams ⅝ in (1.5 cm) from edge, leaving open the pocket mouth between the notches. Press seam open. **(2)** Repeat for second leg. **ATTACH POCKETS (1)** Take 1 finished pocket bag and place in position on wrong side of 1 trouser leg, matching up the notches on the pocket mouth and the opening on the side seam. The small piece O should be sandwiched between the pocket piece X and the front trouser piece. **(2)** Stitch the pocket facings to the side seam facings. Finish with a firm machine tack at top and bottom of pocket mouth. **(3)** Repeat for second leg and pocket. **FRONT PLEATS (1)** Take 1 leg piece and pin out the 2 pleats A–B and C–D on the wrong side. **(2)** Stitch pleats down from top of trousers 2¾ in (7 cm). Machine tack seam ends firmly. **(3)** Press pleats on the wrong side, pushing pleat allowance towards the side seam. **(4)** Repeat for second leg piece. **JOIN LEGS (1)** With right sides together, pin the 2 leg pieces together and stitch the crotch seam ⅜ in (1 cm) from edge. For extra strength, stitch this seam twice. **(2)** Press. **COMPLETE WAISTBAND (1)** Fold the waistband allowance over 1⅜ in (3.5 cm) to the inside of the trousers. **(2)** Stitch 1 row as near to the top edge as possible. This will make the top edge of the trousers firm, and prevent the waistband allowance from rolling over. **(3)** From the right side, stitch the waistband into position 1 in (2.5 cm) down from the top edge, locking in the single layer of pocket facing. **(4)** Before completing this row of stitching, thread the elastic through the waistband, and join the elastic to the required circumference. **(5)** Complete the stitching, locking in the elastic. **TO FINISH (1)** Turn up hems ⅝ in (1.5 cm), and finish with 1 or 2 rows of stitching. **(2)** Press.

Circular Skirt Instructions

MATERIALS Fabric required for skirt only: 4 yd (3.70 m) of 54 in (138 cm) wide fabric. **Fabric required for skirt, shirt jacket and triangle together,** as shown on cutting layout, 6 yd (5.50 m) of 54 in (138 cm) wide fabric. **Trimming:** ¾ yd (70 cm) of 1 in (2.5 cm) wide elastic for waistband.

MEASUREMENTS One size, to fit **waist** 24–28 in (67–71 cm). **Skirt length:** 34 in (86 cm) from bottom of waistband. Length can be adjusted to suit personal requirements.

PREPARATION (1) Cut out pattern pieces following cutting layout. **(2)** Neaten edges by zigzag or other method. **ASSEMBLE POCKET BAGS (1)** With wrong sides together, stitch the 2 pocket bags with a ¼ in (6 mm) seam. **(2)** Turn through and stitch round the edges from the wrong side, locking in the raw edges. **(3)** Neaten pocket openings by zigzag. **(4)** With right sides together, stitch the side seams of the skirt with a seam allowance of ½ in (1.2 cm), leaving open the pocket mouths. Press open. **ATTACH POCKETS (1)** Take 1 finished pocket bag and sew the pocket mouth facings to the side seam facings in the usual way, firmly tacking top and bottom of pocket opening. **(2)** Repeat for second pocket bag. **PREPARE WAISTBAND (1)** Join the waistband to a circumference of 36 in (90 cm). **(2)** With right sides together, pin the waistband join to the top of the side seam and stitch waistband to skirt top with a seam allowance of ¼ in (6 mm). **(3)** Fold the waistband to the inside of the skirt, turn under the seam allowance and fasten the inner band to the machine line by hand from the inside, or by machine topstitching through from the right side, leaving an opening in the band for insertion of elastic. **(4)** Thread the waistband elastic through the band, and machine stitch very firmly to the correct waist circumference. **(5)** Complete the waistband, locking in the raw edges. **TO FINISH (1)** Finish hemline by zigzagging, then by turning up the hem ½ in (1.2 cm) and finishing with 2 rows of machine stitching. The hem can also be finished by overlocking. **(2)** Press.

Triangle Instructions

STYLING Make up the triangle as large as the fabric will allow. It can be worn over the shoulders, tied in front as a shawl, draped over one hip and tied at the waist, or worn as a *bustier* top tied at the back, with one point hanging down in front over, or tucked into, a waistband.

PREPARATION (1) Cut out, following cutting layout. **(2)** Turn the 3 edges in twice, and finish with 1 or 2 rows of machine stitching.

CUTTING LAYOUT
(not drawn to scale)
Fabric width: 54 in (138 cm)

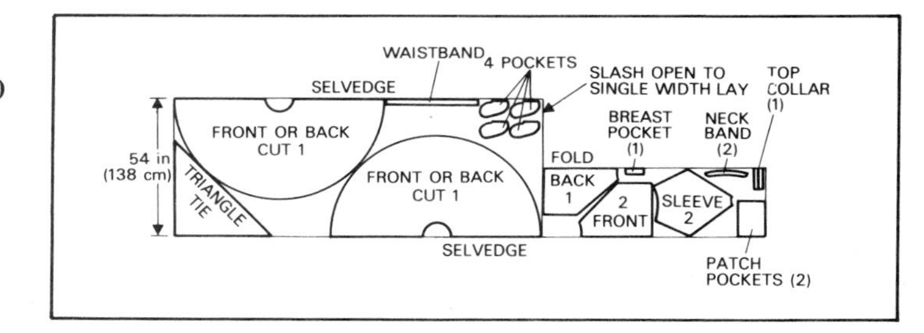

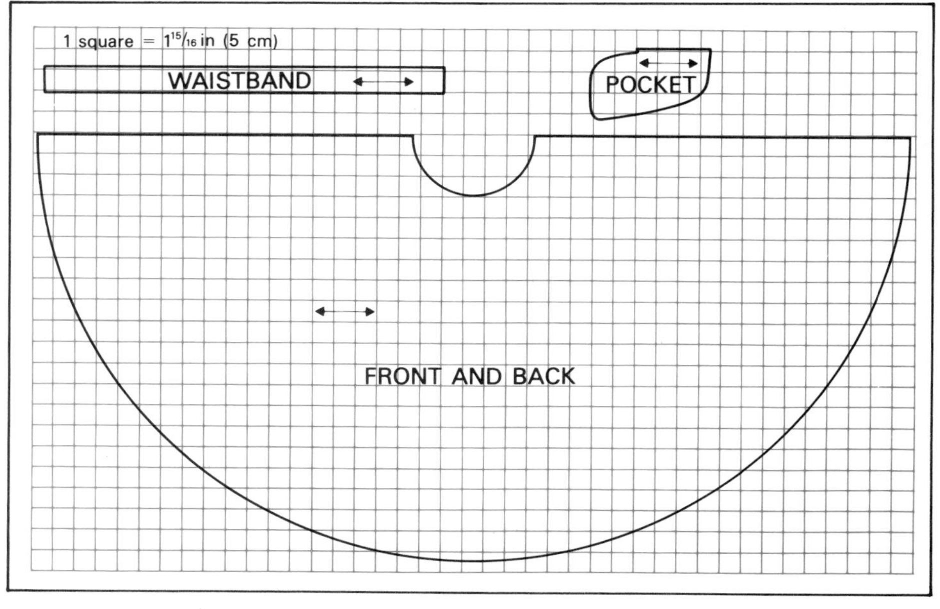

Les Lansdown

For me, there's a world of difference between day and night. You spend the day working, hopefully at something you enjoy – more often at something you don't. That's why the nights are so important. They give you a chance to relax, to enjoy life and to feel special. You can leave the limits of day behind, and be what you would like to be. I consider myself very lucky, because I enjoy my work. So it's important to me that my evening clothes should be able to make people who are not so fortunate feel happy and look good. I like clothes that are clean-lined, comfortable and wearable. An almost-classic look, with a special touch of magic. All my designs have a fluid quality, the lines flow and suggest movement – because movement is what makes a woman attractive. And I think a woman should be attractive to men. But you should never overdo it, never flaunt yourself. That's raw sex appeal, not style. It's important to retain the *mystery*, to hint rather than be explicit. Otherwise there's nothing left with which to tempt–and temptation is what it's all about.

Tarty clothes are nasty and common and cheapen the woman who wears them, but ruffles and lace can be just as misleading. I like women to be feminine but not to play on being feminine in the obvious way, with pink bows and sulky petulance. It diminishes them in the same way that tarty clothes do. I like a woman to sparkle, to be romantic, but most of all I like a woman to be herself. Because a dress can only be an accessory to her personality.

Femininity has many faces. Sometimes I think of women as cheetahs – slinky, silently disturbing, subtly mysterious, and I was thinking of beautiful cat-women when I designed my sarong wrap. The bare shoulder is very sexy, but never anything less than elegant. From the front it looks serenely classic, but when you turn it around you see there's much more – or less – than first met the eye. It leads the man one step further, reveals a bit more of the mystery and magic, arouses the interest. The tiny rouleau strap that goes under the scarf is quite invisible, setting the man the provocative question of what holds the top in place. Just between us, the answer is 'tension'. When you put the top on, pull the ties down at the back before crossing them over and bringing them to the front. The tension of the downward motion eases the top into place and keeps it there comfortably.

The colour scheme was suggested by daffodils against a grey sky, and many of my colours and design themes are inspired by nature. I use hand-printing, hand-painting and beading on many of my evening designs, but I always scatter the motifs over the surface. It's much more stylish than an over-all placement; it also suggests movement and depth, and adds extra interest.

I also think of women as being softly lustrous and graceful, like a Lalique vase, which inspired my second design. The dress has long clean lines, enriched with gold piping. The collar makes the neck look longer, the straight lines flatter the figure into looking taller, and the magyar sleeve relieves the straight lines with a gentle curve.

Some designs look best in certain colours, but the Lalique dress would look superb in just about any colour you could choose. Personally, my first three choices would be garnet, turquoise and lemon yellow. I don't like hard colours, and my tastes run in two quite different directions. On the one hand, I like colours that are strong but warm – like fuchsia, cornflower blue and turquoise. But I also like subtle shades, like sable or peach. And you can't beat black for pure style. Black is usually thought of as a winter colour or an evening colour, but for day wear in high summer, there isn't a smarter colour under the sun.

Whatever the style, a timeless quality is always a part of good design. Good designs don't have to rely on trends for their appeal. If you fish something out of your wardrobe that you wore a few seasons ago and think it looks dreadful, it probably looked just as dreadful *then*. I always want my designs to pass the test of time. But more than anything, I want them to give pleasure.

pin or tack the piping all round the edge of the bib from neck point A to neck point B. **(7)** Stitch the piping to the bib with an exact ¼ in (6 mm) seam, and press lightly. **(8)** Make up the 3 2 in (5 cm) rouleau loops and tack them into position on the bib front as shown on pattern. **ATTACH BIB TO FRONT PANEL (1)** Lay the right side of the bib to the **wrong** side of the front panel piece, matching up the neck slash. **(2)** Stitch round the neck slash edge: seam allowance ⅛ in (3 mm). **(3)** Nip corner of seam allowance at bottom of neck slash if necessary, then turn the bib front through so that it lies flat to the outside of the front panel. **(4)** Lightly tack the bib to the front panel, taking care not to mark the fabric. **(5)** Sew the bib to the panel by hand, from the **wrong** side, taking care to only catch the seam allowance so the stitches don't show on the right side. **JOIN SIDE PANELS TO FRONT PANEL (1)** with right sides outside, pin or tack side panel pieces to front panel, matching up the notches. **(2)** Stitch the 2 panel seams from the shoulder to the hem line with a ⅛ in (3 mm) seam. **(3)** Turn through to the wrong side and,

Lalique Dress Instructions

DETAILS Shown with evening sandals by Charles Jourdan, earrings and bangles by Butler & Wilson, hair comb by Tailpieces, gold belt to order by Rony of London. Cuffs hand-painted by David Reeson. Photographed at the Park Lane Hotel, London.

MATERIALS Fabric required for dress: 3 ⅞ yd (3.55 m) of 45 in (115 cm) wide fabric. **Recommended fabrics:** silk crepe de Chine, polyester, polyester crepe de Chine. **Made here:** in double weight pure silk crepe de Chine. **Trimming** ¼ yd (25 cm) of organdie or organza for mounting the collar and bib front.

MEASUREMENTS One size, to fit **bust** 35–38 in (89–96.5 cm), **hips** 37–40 in (94–101.5 cm). **Back length** from shoulder to finished hem: 61 in (153 cm). Length can be adjusted to suit personal requirements, alter amount of fabric required if necessary.

STYLING The panel seams can be piped with contrasting fabric, as shown in photograph. A fabric sash can be substituted for the wide leather belt.

NOTE All seam allowances are ⅜ in (1 cm), unless otherwise stated.

PREPARATION Cut out pattern pieces following cutting layout. **ASSEMBLE FRONT BIB AND ROULEAU BUTTON LOOPS (1)** With right side up, lay the bib piece on to the mounting fabric. **(2)** Tack the two fabrics together just inside the raw edge, within the seam allowance. **(3)** Trim the mounting fabric to the exact size of the bib piece. **(4)** Fold the bias piping strip in half down its length with right side outside and press lightly. **(5)** Lay the pre-pared bias piping strip on to the right side of the bib front. **(6)** Keeping all raw edges together,

CUTTING LAYOUT
(not drawn to scale)
Fabric width: 45 in (115 cm)

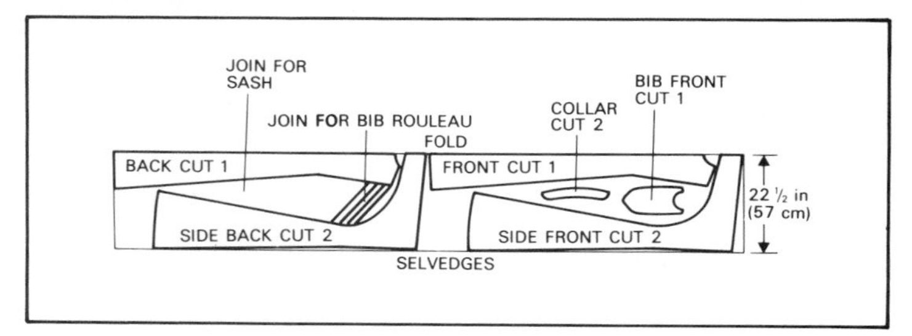

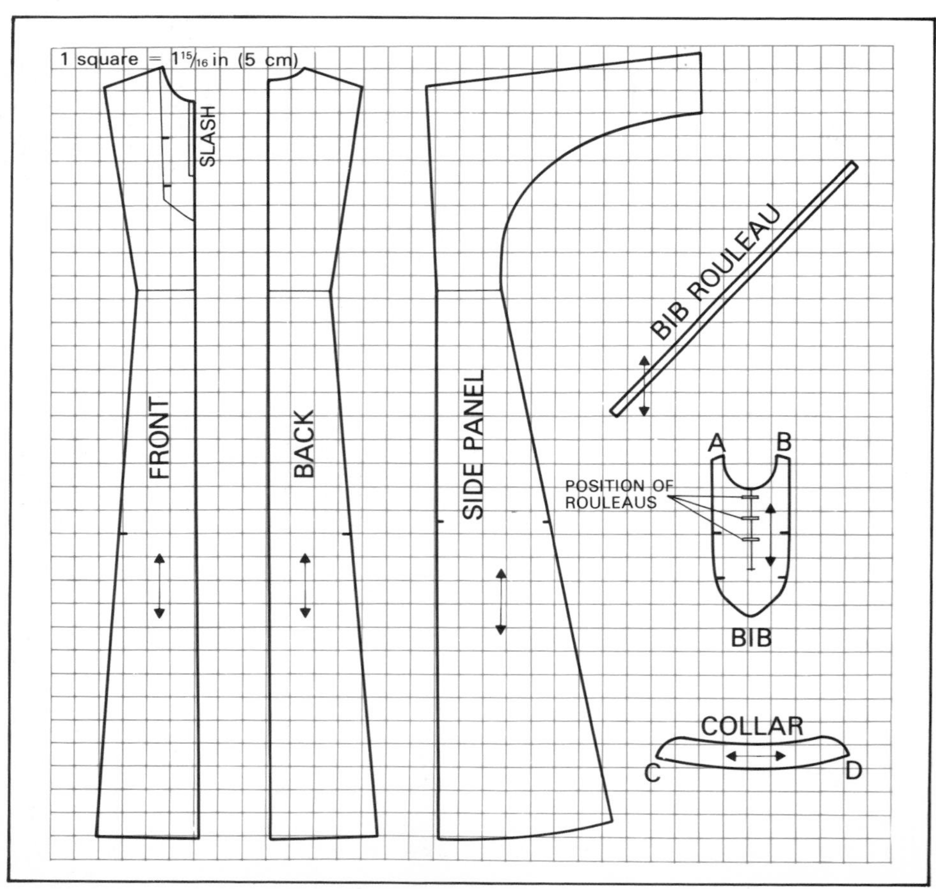

with the stitched edges together, press the edge flat then machine again with a ¼ in (6 mm) seam, locking in the raw edges. **(4)** Repeat exactly for the back panel pieces. **JOIN BACK TO FRONT (1)** With right sides outside, lay the back piece to the front piece matching up the notches. **(2)** Pin or tack into position for sewing. **(3)** Stitch the shoulder seams from neck point to cuff and the side seams from cuff to hem with a ⅛ in (3 mm) seam. **(4)** Turn through and complete as for front and back panel seams. **ASSEMBLE COLLAR (1)** Take 1 collar piece and attach it to the mounting fabric, as for bib front. **(2)** Lay the 2 collar pieces together with right sides together and sew a ¼ in (6 mm) seam around the outer edge from D to C. **(3)** Turn the collar through to the right side and press the stitched edge lightly. **ATTACH COLLAR (1)** With right sides together, tack a single layer of the collar to the neck edge, matching up the notches. **(2)** Stitch collar to neck edge with a ¼ in (6 mm) seam. **(3)** Turn under the inside collar seam allowance ¼ in (6 mm) and slip stitch to the back of the machine stitch line. Press lightly. **TO FINISH (1)** Turn up the cuff and hem edges with 2 ⅛ in (3 mm) turns. Machine stitch with 1 or 2 rows of stitching. **(2)** Sew on 3 buttons to fit rouleau button loops.

Sarong Wrap Top Instructions

DETAILS Shown with Balinese trousers, jewellery from Butler & Wilson, hair comb from Tailpieces and evening sandals by Charles Jourdan. Photographed at the Park Lane Hotel, London.

MATERIALS Fabric required for unlined top: 3⅝ yd (3.35 m) of 45 in (115 cm) wide fabric. **Fabric required for lining:** if a lining of the same fabric is required, add 5 in (13 cm) to the above measurement. **Trimming:** 1 ready-made tassel approximately 8 in (20 cm) long. **Recommended fabrics:** fine soft fabrics that drape well, such as silk georgette, voile, silk crepe de Chine, light wool challis and fine jersey. **Silk georgette and voile must be lined.** Avoid loosely woven fabrics and crisp or hard fabrics with no drape quality. **Made here** in silk georgette; Balinese trousers in Dormeuil silk crepe de Chine.

MEASUREMENTS One size, to fit **bust** 32–36 in (81–91.5 cm), **waist** 24–28 in (61–71 cm).

PATTERN VARIATIONS For lining, cut 2 front pieces. After cutting, place the 2 pieces together and proceed as if for 1.

STYLING This sarong wrap has a slender rouleau strap concealed by the long scarf that drapes from the shoulder with its longest point nearest to the centre of the back. To wear, slip your arm through the rouleau strap and wrap the side ties around your waist twice before tying at front with a knot or bow.

NOTE All seam allowances are ½ in (1.2 cm) unless otherwise stated.

PREPARATION (1) Cut out pattern pieces following cutting layout. **(2)** Take front piece and draw up in 1 in (2.5 cm) or insert two small tucks between points X–X to shape bust. **BIAS STRIPS** Prepare bias strips, by folding each strip lengthwise with right sides outside. **ATTACH BIAS STRIPS TO BODY (1)** Take one edge of the doubled strip and, with ⅛ in (3 mm) seam, stitch to the seam line along the left armhole on front from H to point 2, right sides together. **(2)** After machining, trim the seam allowance on the front piece to ⅛ in (3 mm). **(3)** Turn over the fold edge of the strip to the wrong side of front and fasten down to the original stitch line with a hand stitch. **(4)** Repeat for the right side H–X–1 and the hem 3–4, making sure that the three piped edges are of even width. **PREPARE SHOULDER (1)** Gather or pleat H–H on front to measure 3 in (7.5 cm) and secure with easy machine stitch along edge of seam allowance. **PREPARE SIDE TIES (1)** Fold ties with right side inside, machine across the wide base then up the tie to the narrow opening. **(2)** Clip corner seam allowance, turn through and press. **ATTACH TIES TO BODY (1)** Turn in the seam allowance at the open end of tie and insert edge 1–3 of body piece up to the seam allowance. **(2)** Topstitch to secure and lock in all raw edges. **(3)** Repeat for second tie and edge 2–4. **ATTACH SCARF TO SHOULDER (1)** Fold scarf piece with right side inside, machine and turn through as for side ties. **(2)** Pleat the open end of the scarf to the size of the prepared shoulder piece H–H. **(3)** Make up a rouleau strap 13 in (33 cm) long. **(4)** Close the shoulder seam, locking in one end of the rouleau strap 1½ in (3.8 cm) from the neck edge. If the fabric is very fine, the shoulder can be closed by a French seam (sew the two fabrics together on the right side, then turn and sew on the wrong side, locking in the raw edges). If the fabric is not very fine, close with an inside seam and bind the raw edges. **(5)** Now stitch the loose end of the rouleau strap firmly to the body piece at point R. **TO FINISH** Attach tassel to the longest point of the scarf.

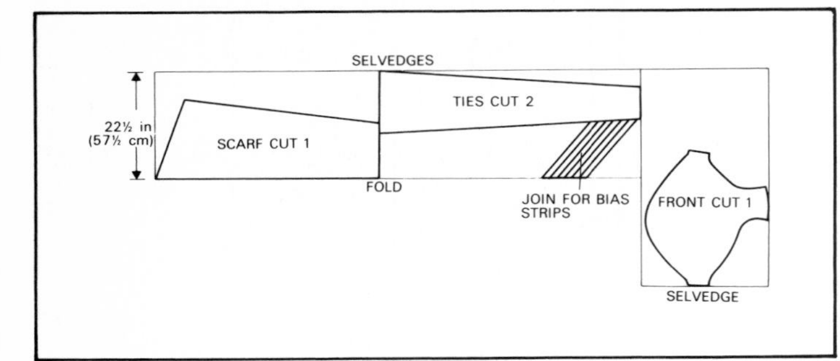

CUTTING LAYOUT
for unlined top
(not drawn to scale)
Fabric width: 45 in (115 cm).

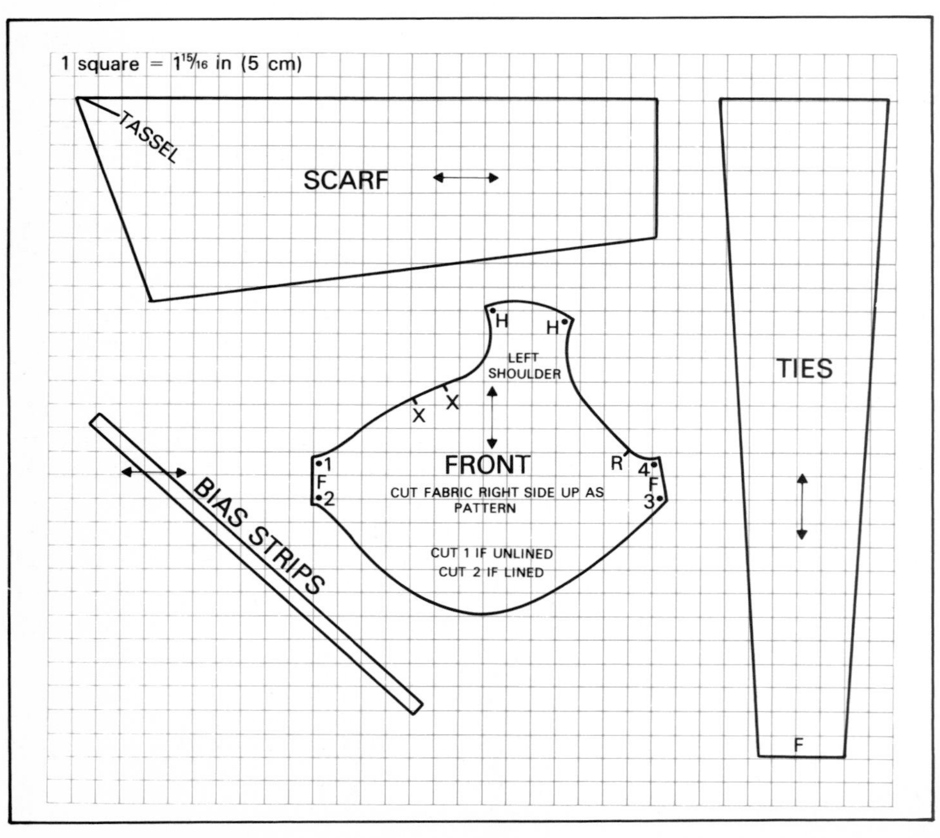

Victor Herbert

We created the first evening track suit, and I think we can claim quite fairly to have made active sportswear into high fashion for day and evening. But our designs are very different from any preconceived ideas you may have when I say 'active sportswear'. I don't know if you've noticed gymnasts performing their exercises in their whites – they all look sleek, rather like birds, I think. That's the idea of our track suits, and all our designs.It's not the track suit that you'd see on big overweights jogging round the park, but that bit – that essence of a track suit – that only looks superb on the most refined athlete.

My shapes are very simple, but the pattern pieces themselves are borderline masterpieces – there's a lot of ingenuity in the cut of our clothes. Then we carry the designs through in exotic fabrics – fabrics you wouldn't have expected. For example, we've made a swimsuit in a stretch fabric, covered it with metallized sequins, and turned it into something that isn't just for swimming in. It's for lounging, for dancing in, for wearing down the high street with a side-split skirt. We've used the same fabric for a body stocking – or body suit, as I prefer to call it. The suit has very tight sleeves, it *hugs* the body; above all it's very sensuous and sexy. It has a very deep V-neck plunging at the front, and it's cut to accentuate the length of the leg. The clothes may not look like sportswear when I'm finished, but the sportswear angle *is* there. And because of it, everything I design is very wearable – they're things that you've probably worn many times before, but in a different form.

I don't design to the English norm. And – this will give a lot of people a lot of hope – I could *not* be a degree student in this country. I would not get through the course, because my approach is atypical – I'd get stuck like a piece of bone in the machinery.

For example, I don't know how – academically – to cut a pattern from a block. For a normal pattern cutter, a pattern cut from a block has to look just the way he's been told it should look, if he's done it correctly. If it's different, it's wrong. When I cut a pattern, I make it up and look at it, and decide whether or not it's what I *want*. I'm not concerned with whether it's academically 'correct'. I have very high standards, standards that don't allow me to make something poorly. So if my pattern isn't perfect first time, if there are any bumps or wrinkles or if the fabric is pulling, then I take the whole thing apart and, through a process of trial and error and elimination, carry on until I get it right. As a result, my patterns aren't stale re-runs – they're full of interest. Take a coat – standard pattern cutters use one basic coat pattern, and just vary the details. When I make a coat, I don't go back to the pattern for my last coat and use it as a foundation – I start again. If you want a different coat, you *can't* use the same pattern!

Because of my persistence, I've developed an analytical approach to problem-solving. And since I *can* solve problems, people think of me as a technician, a good pattern cutter, rather than as a creative designer. But as I've explained, academically I don't know how to cut a pattern, and I don't want to. It wouldn't do me any good to know, it would just get in the way. I didn't set out to defy all the rules of cutting and making a garment. I set out to design what I've thought up in my head – to *make* what I *thought*. I'm fortunate in not having any preconceptions about how something 'should' be, so I can concentrate on what I want it to be. That's what true creativity is.

I've always been fairly earthy in what I do – I like good basic shapes, and I've never been a sticker-on of things. If one could generalize, the Italians seem to be absolutely fantastic at producing beautifully designed, high quality knitted and woven fabrics, which they put into very wearable simple garments. They're not very imaginative with their clothes, but they're very snappy. The French have become over-inventive. They've got a lot of support, so they can afford to have fun, make the shoulders as big as they like, be as daring as they want – it's much more sensational. The British tend to be the poor relations. With the possible exception of people like Jaeger, we can't match the quality of Italian fabrics, and we've never trained our people to cut as well as the Italians and French. So we've become a nation of stickers-on of things – masters at cluttering and chucking things on the top.

Think of design in the Sixties and early Seventies – they'd stick sequins, prints and colour everywhere, throw as much as they could onto one garment. The shape and quality of the cloth didn't matter as long as there were nine million bits on it. There was nothing really chic about the clothes – they were just very flamboyant and over-indulgent. It's filtered down into the art schools, so that all the students today learn to put on as many buttons and pockets as they can, as many flaps on every pocket, and as many ugly collars on everything as possible. All my designs have pockets, but you can't see them. To me, a pocket is for putting your hand in, or keeping your money in. I don't make a feature out of it because I don't think it matters. But it does to most fashion people in this country – it's where the buttons and pockets are that's important to them, not the shape.

So these are my thoughts when I'm designing, and I try to avoid all the pitfalls as I see them, in terms of continuing to go along with what has previously been considered to be good taste, good judgment and good tailoring. I've chucked it out the window. Instead, I've adopted a simpler approach – you have a number of pieces of cloth, then you put them together somehow, and let the cloth do the work.

The type of fabric, and its quality, override all the other decisions that I make. I think of fabrics in a kind of slang language. When I get hold of something that's *'sloppy'*, like fine *panné* velvet, I immediately think – beware! It's uncontrollable. It's like quicksilver, it's going to move around a lot. I know I'd better not make anything that requires accurate straight stitching, or anything that's figure hugging. On the other hand, there are fabrics I think of as *'felty'*, like a felted carpet or the fabric that scouts' hats were made from. In that case I think in terms of simple tailored or structured clothes, for which felty fabric is ideal. But I don't think of tailored garments when I get hold of a fine botany wool worsted fabric. I'd never make a fine woollen worsted suit that had all the traditional tailoring details, because it seems pointless. I don't like the idea of all that *gubbins* – a bit of this, a bit of that – put in to make a soft cloth look flat and firm. Why not just go for a firmer fabric?

Where quality is concerned, we only buy the very best. I'm not that keen on taking fabrics made in this country because the quality and design standards simply aren't good enough. In fact there are very few English producers who will make anything less than 1,000 metres of fabric to order, and then they quibble about it and complain bitterly that it's not 10,000 metres. On top of that they make you wait for ages, because they're only interested in big production. I have very little faith in people who shut all of the doors. I don't mind when they shut some of the doors, or most of the doors – but when they shut *all* of the doors, then they've got problems. To see what I mean, you have only to go round Leeds, Manchester, Bradford, Yorkshire and Lancashire, the homes of the wool and cotton industries. Design-wise they're completely depleted – there's nothing left. They're all victims of 'management decisions' and a commitment to quantity instead of quality. There is nowhere for all the students that do textile printing in our colleges to sell and nowhere for them to work in the British textile industry. British textile producers prefer to go abroad to buy their designs, rather than trying to create a design centre here.

I decide on the fabrics six to nine months before I start on the styling. I design by swinging the pencil – I draw the clothes and a person in them, somebody that I know, and I have a bit of fun. There are very few new things in fashion, and if they are new – genuinely new – then they intrigue me. I remember buying fluorescent pink and lime green socks in the Fifties, the Teddy Boy days. Now to me, that was a great innovation – *electric* socks. The Ted suit, the trousers and jackets, weren't new – but the socks were. All-in-one-piece stretch body suits are new, even revolutionary. A whole garment in any colour and any thickness – it's fantastic!

I'm going to show you a bodysuit and a T-shirt, because I know these two things will be around in twenty years' time. The bodysuit is cut high at the front to make the outside of the leg look very long, and the top shapes and accentuates the bust. It's tight waisted, and because it's stretch fabric it pushes in all the extra bits of flabbiness, so you find you've almost got a support stocking on. It's rather like a modern-day corset, but it's very comfortable.

If you have a bit of flair and don't mind making something up in stretch fabrics, these two designs will be very stimulating, and will give you a chance to get involved in some styling. You might have thought a bodysuit is a bodysuit and a T-shirt is a T-shirt and that's that, you can't style them any further. But I say – Oh yes, you can! Here's the basic pattern, see what I've done with it, and have a go yourself.

T-Shirt Dress and Top Instructions

Sewing with Stretch Fabrics

For fabrics with a slight stretch, use a small stitch and make sure that the stitch tensions, top and bottom, are equal. When working with very stretchy or elastomeric fabrics, use a close zigzag stitch and do not neaten seams.

DETAILS T-shirt dress shown with jewellery from Detail, shoes by Bellesco from Bally and tights by Aristoc. Photographed at the Savoy Hotel. T-shirt top shown with skirt by Victor Herbert and jewellery from Detail.

MATERIALS fabric required for dress: 1 yd 15 in (130 cm) of 59 in (150 cm) wide fabric for size 1 dress, 1 yd 24 in (1.53 m) of 59 in (150 cm) wide fabric for size 2 dress. The neck facings and sash have been included in the cutting layout and fabric requirement, but can be made from a contrast fabric as shown in photograph. **Recommended fabrics:** fine silk or silk-wool jersey, cotton or synthetic weft knit single jersey. Be sure to ask specifically for weft knit single jersey. **Made here:** in silk-wool jersey. **Also:** 22 in (56 cm) of thin stay tape or the selvedge of a fine lining fabric.

MEASUREMENTS Size 1 to fit **bust** 31–33 in (79–84 cm), hips 32–34½ in (81.5–86.5 cm). Size 2 to fit **bust** 33–35 in (84–89 cm), **hips** 34–36 in (86.5–91.5 cm). **Back length:** 42½ in (108 cm). Length can be adjusted to suit per-

sonal requirements; increase the amount of fabric required if necessary.

PATTERN VARIATIONS for the T-shirt top, shorten the dress pattern to the desired length, omitting the pocket bags. The sleeves can be shortened to just above the elbow without destroying the balance of the design. To calculate the amount of fabric required for the T-shirt top reduce the amount of fabric given for the dress by the amount that the dress pattern is being shortened.

STYLING If a contrast shade of fabric is used for the neck facing, a decorative piped effect can be obtained by rolling the inner facing to show just above the neck edge before securing.

NOTE All seam allowances are $\frac{3}{16}$ in (5 mm) unless otherwise stated. Press seams open or neaten together with zigzag.

PREPARATION (1) Cut out pattern pieces following cutting layout. Note that the V neckline and facings are cut on a **bias** grain. **ASSEMBLE FRONT SEAMS AND POCKET BAGS (1)** Open out the front piece and lay flat, with right side up. **(2)** Take 1 single pocket bag piece and stitch to the front piece at X–X. Neaten the seam edges together. **(3)** Take 1 matching front panel piece and 1 single pocket bag piece and repeat exactly as for front. **(4)** Repeat for other side. **ASSEMBLE PANEL SEAMS AND POCKET BAGS (1)** With right sides together, pin or tack the front panel seam into sewing position matching up the pocket bag pieces and the bust notches A/A–B/B. The

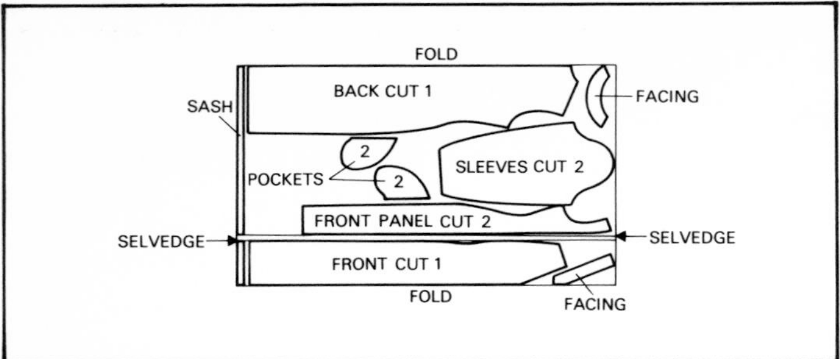

CUTTING LAYOUT

For size 1
(not drawn to scale)
Fabric width: 59 in (150 cm)

For Size 2
Adjust sleeve positioning, then as for size 1

Size 1

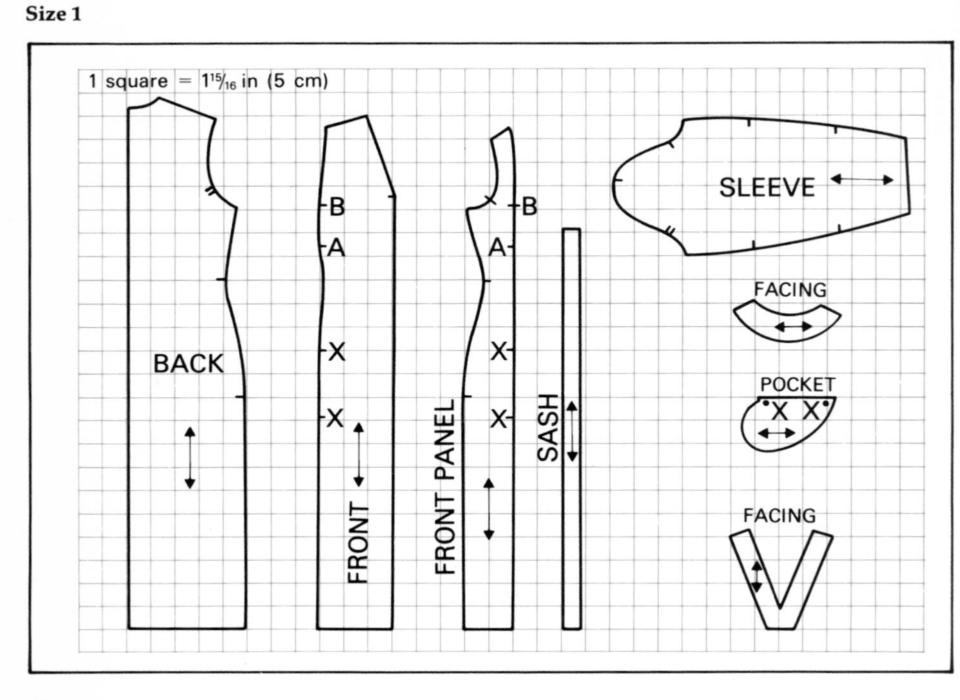

Size 2

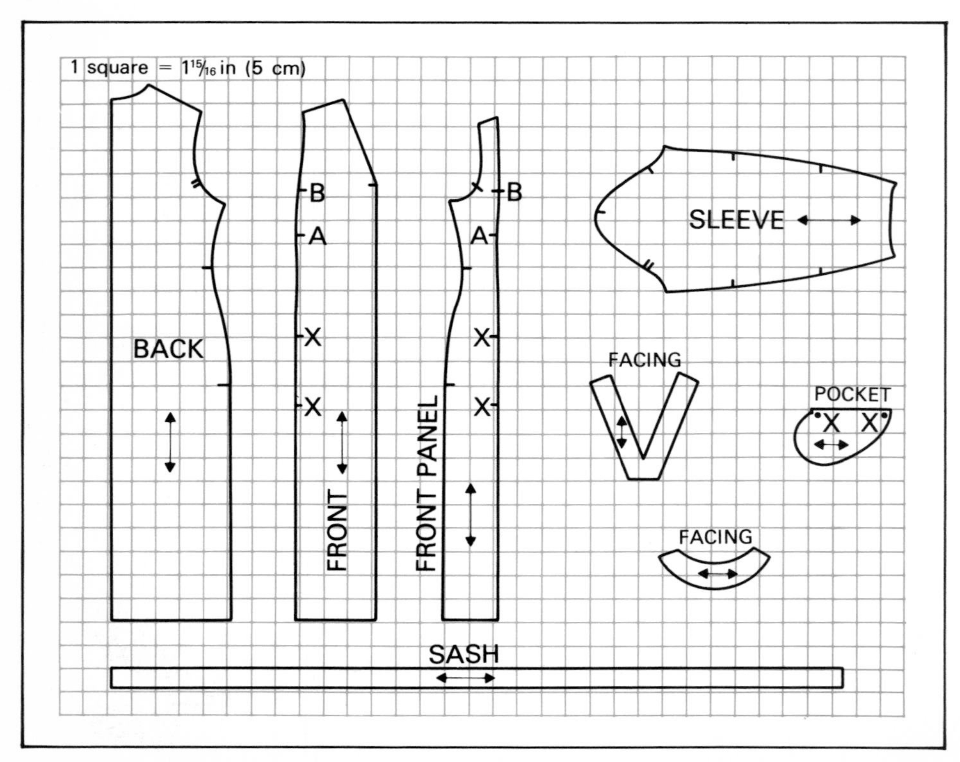

side panel A/B should be stretched on to the front A/B at this point. **(2)** Machine stitch the front panel seam from the shoulder to top of pocket at X, then round the outer edge of the pocket bag and on down to the hem. **(3)** Neaten seams together from shoulder to hem, including the outer edge of the pocket bag. **(4)** Machine tack firmly at top and bottom of pocket opening. **(5)** Repeat exactly for second panel seam and pocket bag. **(6)** Press lightly. **JOIN SHOULDER SEAMS** Lay the completed front piece to the back piece, stitch the shoulder seams, and neaten together. **ASSEMBLE AND ATTACH NECK FACINGS (1)** Tack stay tape around the neck edge on the wrong side of the neck opening. **(2)** Secure the stay tape with a row of machine stitching all round the neck just inside the edge. **(3)** Take the front and back neck facing pieces and, with right sides together, machine the shoulder joins. **(4)** Press seams open. **(5)** Neaten the outer edges of the facing piece. **(6)** With right sides together, lay the completed facing to the neck opening and machine stitch into position around neck. **(7)** Nip the seam allowance at bottom of V and turn facings through to wrong side of neck opening. **(8)** Tack or pin the facing into position, and press lightly to ensure that the seam lies evenly around the neck edge. **(9)** Topstitch the facings into position with 1 or 2 rows of stitching. **JOIN SIDE SEAMS** Join the side seams, neaten edges together and press. **PREPARE AND ATTACH SLEEVES (1)** Take 1 sleeve and, with right sides together, stitch neatened edges of sleeve seam together. **(2)** Turn up sleeve cuff hem 3/16 in (5 mm) to wrong side, then zigzag into position from the right side, making sure that the zigzag stitch goes slightly over the edge to form a decorative shell-edge finish. **(3)** Pin the completed sleeve into the armhole matching up the underarm seams and the back, front and shoulder notches. **(4)** Tack into sewing position, easing in the small amount of sleeve head fullness. **(5)** Machine the sleeve to armhole. Neaten the edges together and press lightly. **(6)** Repeat for second sleeve. **TO FINISH (1)** Turn up hem exactly as for sleeve cuffs. **(2)** Make up sash.

Bodysuit Instructions

DETAILS Shown with earrings by Tailpieces and mules by Charles Jourdan. Photographed at the Park Lane Hotel, London.

MATERIALS **Fabric required:** for size 1, 7/8 yd (0.80 m) of 60 in (160 cm) wide fabric; for size 2, 1 1/16 yd (1 m) of 60 in (160 cm) wide fabric. **Recommended fabrics:** suitable **only for elastomeric stretch fabrics** such as Helanca; be sure to ask specifically for an elastomeric fabric, not just a 'stretch' one. **Also:** a fine ballpoint needle (size 11) and a suitable stretch thread such as Coats Drima; 2 snap fasteners.

CUTTING LAYOUT
For Size 1
(not drawn to scale)
Fabric width: 60 in (160 cm)
For Size 2 As for Size 1, but adjust positioning slightly to fit

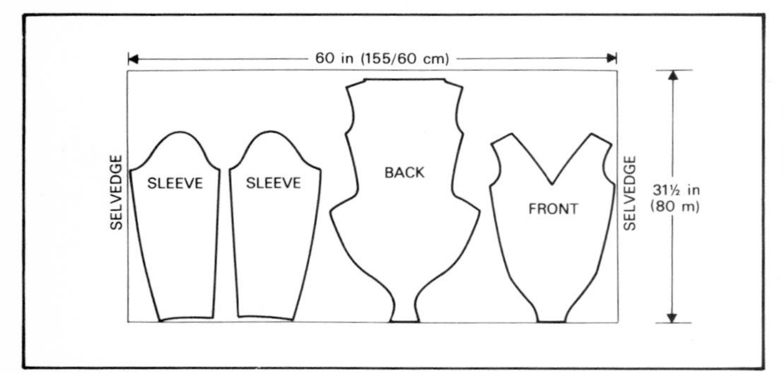

MEASUREMENTS Size 1 to fit **bust** 32–34 in (81.5–87.5 cm), **hips** 34–36 in (87.5–91.5 cm); Size 2 to fit **bust** 36–38 (91.5–96.5 cm), **hips** 38–40 in (96.5–101.5 cm)

NOTE All seam allowances 3/16 in (5 mm) except for gusset, which has a 3/8 in (1 cm) turn up. Stitch all seams with a close zigzag stitch. Do **not** neaten seams.

PREPARATION **(1)** Cut out pattern pieces following cutting layout. **(2)** Turn the front V neck and back neck edges in to the wrong side and stitch down flat. **JOIN SHOULDER SEAMS (1)** Lay the front and back body pieces together with right sides inside and stitch shoulder seams together. **ATTACH SLEEVES (1)** With right sides together and all notches matching, pin the sleeves on to the armhole seams, stretching the body underarms to fit the sleeves. **(2)** Sew in the sleeves. **JOIN SIDE AND SLEEVE SEAMS (1)** Pin the side seams together with right sides inside. Match up A-B-C-D on the front body to A-B-C-D on back body, stretching B-C on the front to fit B-C on the back. **(2)** Stitch side and sleeve seams. **(3)** Turn up cuff allowances to the wrong side and zigzag flat. **(2)** Turn up the allowance at the bottom of the back and front crotch ends and zigzag flat. **(3)** Overlap the crotch ends 3/4 in (2 cm) and secure with snap fasteners.

Size 1

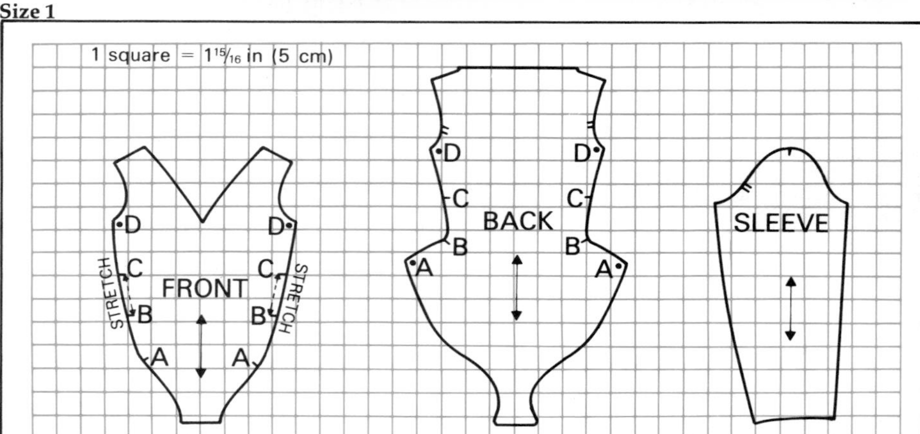

Size 2

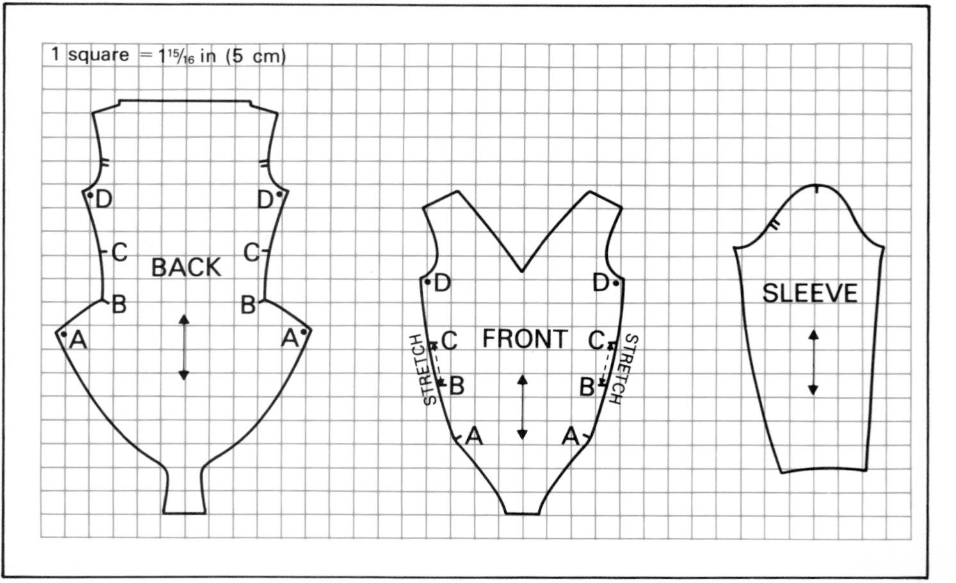

Antony Kwock

I design for a way of life, and for me everything starts with the body. The way you carry yourself, the way you walk, being in good condition from your hair right down to your toes, are all more important than the clothes you wear. After all, there's no such thing as a dress that will make a fat person look thin – it can only make them look less bulky. I don't mean that you should be fanatical about health and cranky about food, or go jogging every day. But if you want to look good, you should do what's good for you. Have a ten minute walk instead of going by bus or car, walk up the stairs instead of taking a lift, look after your hair and skin. Freedom – and style – only come with discipline.

Clothes should dress the body – but not *hide* it. You shouldn't allow clothes to give you a personality. Fantasy dressing is fine for an evening or two – but it's theatre, not life. I like clothes to be simple and workable, to have a smart, clean, classic look – but not a look that never changes. Everyone feels like a change from time to time, so I like to design basics that you can use to create any look you like. Things that can reflect different trends, different styles, different moods, without ever being extreme in themselves.

This spiral sundress, for example, is just like a chameleon. You can make it in cotton for the beach, you can make it in fine wool and wear it with a belt and jacket, you can make it in crepe or satin for evening, draped and caught to one side with a brooch as it is here. You can make it change character completely by varying your hair style, accessories and make-up, and it looks equally good in plain fabrics and in prints, preferably abstract prints.

The distinctive feature of the design is the spiral seam that turns the flat piece of fabric into a tube. You can do more with a tube – put more of yourself into it – because it's not shaped. Being cut on the bias, it moulds to your body when you wear it, draping itself around the natural curves. When you cut something on the bias, the seams are usually tricky to sew. This is like taking a straight line and softening it, giving it dimension by cutting it on the bias and at the same time eliminating the technical difficulty of doing the seam. The lines are carried right around the body. The better the drape of the fabric, the slinkier the dress will be, and it looks better on curves than on absolutely straight figures.

This one dress will go a long way because you can do such a lot with it, and I think versatility is very important. Men can get through a whole year just by switching around the things in their wardrobe; women tend to buy things that can just be worn in one way, and then have to go out and get still more things. The more a woman adapts to the way men dress – in spirit –the closer she'll be to discovering one of the secrets of looking good. When it comes to style, less *is* more!

Spiral Sundress Instructions

DETAILS Shown with jewellery from Butler & Wilson and shoes by Charles Jourdan.

MATERIALS **Fabric required:** 2⅜ yd (2.20 m) of 36 in (90 cm) or 45 in (115 cm) wide fabric. **Recommended fabrics:** fabrics with good handle and drape qualities such as soft silk taffeta and crepe, fine soft satin, lightweight wool crepe and challis, soft cotton and polyester crepe de Chine. Avoid fabrics that are uneven in weave or loose in texture, slubbed fabrics, large printed motifs and thin or stiff fabrics that do not drape or fall well. **Made here:** in pure silk Canton crepe.

MEASUREMENTS One size, to fit **bust** 30–34 in (76–84 cm), **hips** 32–36 in (81–91 cm). **Back length:** finished dress hangs 12 in (30 cm) from ground on a 5′ 9″ (175 cm) figure in 3 in (7.5 cm) high heels.

PATTERN VARIATIONS To lengthen for evening wear, extend the bias hem line B–D to the desired length. To shorten, raise line B–D as required or cut as pattern and shorten at fitting. If altering length, remember to adjust the amount of fabric required.

STYLING You can turn the sundress into a short summer shift, a cocktail dress or an evening dress by varying the fabrics and raising or lowering the hem. Plain colours and all-over prints like the one shown here are ideal. Checks and stripes can also be used, but must be chosen with great care. Checks must be firm, square and evenly spaced both up and across the fabric. Stripes must be regular, firm, and run perfectly straight without distortion. When making up in checked or striped fabric, tack the seams very carefully by hand, checking that all lines match up **perfectly** before machining. To finish, the neck slash and back drape areas A–Y and Y–C can be bound with a suitable binding or braid, and the back neck can be clasped instead of tied. The very experienced reader could investigate the possibilities of seam design by making up the dress with the spiral seam on the **outside,** finishing the seam edges with one of the many fancy edge stitches available on modern machines. Use surplus fabric for stole, sash, kerchief or soft bags.

NOTE All seam allowances are ½ in (1.2 cm) unless otherwise stated.

BEFORE STARTING **(1)** The beauty of this pattern lies in its simplicity, but you must get it right. Before cutting, check the pattern and practise the spiral method of production as follows. **(2)** Lay down the pattern so that A–A1–B are nearest to you in left to right order. **(3)** Pick up corner point A and place it on point C. The neck slash position will now appear to be on a fold. **(4)** Now bring over C1 and place it on A1. Hold these points together and turn the pattern over. **(5)** Complete the spiral by joining B–D on the hem line. **PREPARATION** **(1)** Cut off any tight selvedges from fabric. **(2)** Lay pattern piece in position and cut fabric with **right** side up. **ASSEMBLY** **(1)** Slash neck opening Y–X and neaten by turning in twice and machining flat with a small stitch. A small pleat or dart can be sewn at point of neck opening X–Z to reinforce the end of slash. **(2)** In the same way, neaten the edges A to top of neck slash Y and from Y to a point 1 in (2.5 cm) from C. **(3)** Pick up point A and pin it to point C. **(4)** Now bring over C1 and place it on to A1. **(5)** Tack the resulting seam position A/C – A1/C1. **(6)** Turn the dress over and complete spiral seam to hem, finishing at points B–D. **(7)** Tack seam position for machining. When sewing the spiral seam use a small, easy stitch, and ensure that the seam is not stretched during machining. **(8)** Neaten the inside seams by zigzag and the hem by hand or machine. Some man-made fabrics are virtually non-fray and will not require neatening. **(9)** Reinforce the back join position with a firm stitch. **(10)** Press.

1 square = 1¹⁵⁄₁₆ in (5 cm)

D

C_1

B

C

A_1

SLASH

X

Y

A

CUTTING LAYOUT
(not drawn to scale)
Fabric width: 36 in (90 cm)

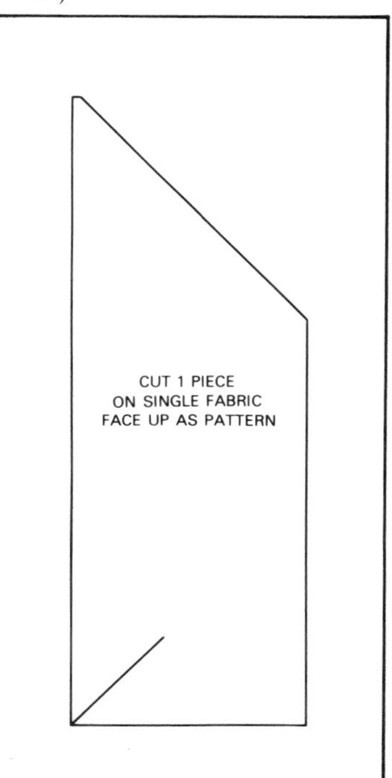

Sara Fermi

I never studied fashion as such – my interest began when I was reading Fine Arts and Art History at Radcliffe College in America. I joined the Harvard Drama Club hoping to act, and ended up helping with the costumes instead. Later on, when I transferred to Columbia University, I worked on costumes for the Columbia Players. But it wasn't until after graduation, when I went to work with Ray Diffen, the theatrical costume designer, in New York that I feel I really began to learn about design.

For an actor, costumes are a very important prop, but few actors have Ray's exceptional sense of what you can *do* with costumes. He had an infallible instinct for creating the perfect effect, for choosing *exactly* the right colours and fabrics. The costumes weren't just 'in period' – they were more real than real, like an Andrew Wyeth painting.

Then I married and it wasn't until the family moved to England and my children were nearly grown up that I found myself drawn back to design. There wasn't any opportunity for theatrical work, and I didn't want to do anything tied to fashion, so the only thing left was – nightgowns! My first design was so simple it didn't even have set-in sleeves. It was all rectangular pieces and gussets, which is the way cut clothing as we know it began centuries ago. As it turned out, nightgowns were a very good thing to start with, because of all the clothes we wear today, nightgowns have changed the least. They're much the same as the long shirts worn as undergarments during the Middle Ages, when people just took the top layers off at night, got into bed in their shirts, and put the top layers back on again next morning.

You could say I learned the history of clothing shape by shape, starting at the very beginning. As I worked, I saw how fashion had gone downhill when ready-to-wear took over. Before then, styles changed very slowly. They evolved naturally, and the shapes were refined by centuries of use. They had dignity and grace, and always enhanced the appearance of the wearer. Even up to the late eighteenth century clothing was still developing beautiful shapes, still improving. But the Industrial Revolution devalued human nobility and destroyed the natural system that produced fine craftsmen and lovely clothes. Modern high fashion shapes are awkward, and they just don't seem designed to make the wearer look *better*.

I never intended to restrict myself to nightgowns, but in the beginning I didn't know what direction my work would take. Then I decided that all the really good shapes had already been invented – it was just a case of bringing them back to life. And so the nightgowns just led naturally into what I call 'historical clothing' – *designs with a tradition*.

For inspiration, I go to paintings every time. I like almost all of Holbein – he's so precise and exact. Jan Van Eyck is another favourite, and you find wonderful detail in paintings of the Flemish school. Then there are the superb shapes of the Italian Renaissance – the high round necklines and slender sleeves of Pietro della Francesca's High Renaissance madonnas, and the low necklines and graceful bell sleeves that you see in the Late Renaissance paintings of Andrea del Sarto. Museums are another excellent source, particularly the Victoria and Albert Museum in London, and the Museum of Costume in Bath.

But my designs aren't mere copies of museum pieces. You have to capture the spirit of traditional designs without turning them into stilted period costumes, and I find my background in the theatre has been very helpful in this respect. In costume design the clothes have to be dramatic and flamboyant, and the shape is the thing that matters most. Now I find that detail and construction are just as important. I want the shape, look *and* feel to be right, so the clothes are like something you've discovered in an old sixteenth- or seventeenth-century chest.

Bringing historical clothing back to life means adapting it to modern tastes, techniques and materials. If the changes you make represent a natural development of the original you can raise a Renaissance neckline, slim down a traditional peasant smock or lengthen a Tudor shirt without destroying the basic shape and style. The fabric you choose is very important, and I insist on natural fibres in everything I make. Cotton, silk, linen and wool have a lifetime of their own. They start off new and harsh, get softer with wear, and develop a personality when they get old. Synthetics certainly don't. But cotton, for instance, just gets better and better until it wears out. You could say it ages gracefully. I find that if I stick to natural fibres I can translate an old design into a modern fabric very successfully. For example, the original cotton fabric used for Victorian nightdresses is too heavy and hot for today's centrally heated homes. But a Victorian nightdress *must* have that starched, crisp Victorian look – that's what makes it so pretty and feminine. Polyester looks cold and feels dead, nylon is out of the question. Apart from everything else, it's not suitable for nightwear because it's not absorbent. Voile is very dainty, but it tends to pull at the seams, and nightgowns *must* have a secure construction. The only fabric that will do is pure cotton lawn. It looks right for the period, and it feels right for today.

The trimmings you use should be as historically accurate as possible; if they're not, that's often where a design of this kind starts to go wrong. For instance, if you put lace onto a Tudor design, it would immediately look fussy and theatrical. Lace didn't come into general use until after Tudor times, and the dignified lines of Tudor clothes only look right with the strong, rich Tudor style of trimming, a combination of embroidered edgings and '*blackwork*' – decorative hand-embroidered stitches in black or dark thread that look like lacework in reverse against the cream background fabric.

As long as the trimming is accurate in principle, you don't have to be afraid of using modern techniques or variations. When trimming my Tudor designs I use machine stitching and furnishing braid, which I prefer to dress braid because the colours are subtler and the quality is better. It's not cheap, but it's well worth spending a bit more on, particularly since you don't need to use a lot of it. It's also worth learning to do simple hand embroidery – feather, chain and buttonhole stitches – to use as a decorative feature instead of machine stitching. It lifts the design into a different dimension, and adds a wonderful richness to it. Personally, I prefer to stick to the Tudor convention of using dark thread and edgings, but you can use other colours very successfully. The cream background tones down the intensity of the shade, so you can use quite bright colours. Pomegranate pink looks very pretty, and apricot has a lovely rich glow. I use modern trimming on my nightdresses as well. Victorian nightdresses were trimmed with *broderie Anglaise*, but it's very difficult to get the gathers even if you gather by hand as they did. So I use pre-gathered *broderie Anglaise* – it's *much* easier, and looks just as good as the original.

The tie front opening on the *Shepherdess* design makes it a perfect nursing nightie, but finding the right ribbon can be a problem. For a start, you just can't find good subtle colours ready made, so I have the ribbons for my collections dyed up specially. For making at home, I recommend choosing a fabric with cream in it and threading the neckline with cream ribbon – it's much better than using a coloured ribbon that doesn't quite match, or a black ribbon 'because it goes with everything'. Three-eighths of an inch is the ideal width, but if you can't find it use half-inch ribbon – the quarter-inch width looks too skimpy. I prefer working with plain fabrics because you can see the lines and shape of the design better, and you can use trimmings with more freedom. When I do use patterned fabrics I always go for the smaller, oldfashioned prints, and Liberty prints, like the ones used here, are my favourites.

Buttons are the fastenings I like best, used with buttonholes or worked thread loops, and I spend ages at the button counters. I get pearl buttons whenever possible, and also use wood and horn, but I stay right away from shiny buttons, brassy buttons and plastic. I also use hooks and bars, or hooks and worked thread loops. I don't like to use zippers because they're a fiddle to put in, they tend to gape when put under pressure, and if they break you've got to take the garment to pieces to put a new one in. With buttons or hooks, only one piece has to be replaced. Whatever fastening you use, the important thing is to get it *on* the strain point, not just where it looks like it ought to be. And if you see something you really like – fabric, lace, braid, buttons – my advice is to buy a lot of it, because they'll probably stop making it!

I take a lot of time researching and working out the details, but paintings and old clothes can't tell you everything you need to know – you also have to work by instinct. It's difficult to describe and all I can say is that it's a *feeling*, like those strange déjà-vu feelings about having been to certain places, or having seen certain things before when you haven't, that I think everyone has at some time. When I was making my first nightdresses I felt almost transported in time. It was a very eerie feeling, but I always try to recapture it. Because when I do, I know that – somehow – the design I'm working on is 'right'.

Of the designs shown here, *Henry* is taken from the Tudor period and *Althea* from the High Renaissance, *Sheperdess* is based on a Regency silhouette and *Amy* is a classic Victorian nightdress. Although they're very different, they have that special dimension that I think is so important. Historical clothes have a timeless quality, but at the same time they give you a strong sense of continuity with the past. They make you think of big rooms with high ceilings, and give you the same feeling that you get in a lovely old home where families have lived for generation after generation, polishing and caring and making it perfect. I think that's why so many people are buying and collecting old clothes now. They appreciate the quality of the work that went into them, they know that the people who made them cared. I'm not a fan of the youth cult and I don't believe that something is only good if it's new. If something is good to start with, it just gets better with time – *like we should*.

Henry VIII Instructions

DETAILS Shown with wool challis cloak by Madge O'Connor. Photographed at the Royal Horseguards Hotel, London.

MATERIALS **Fabric required for dress:** 4⅝ yd (4.25 m) of 36 in (90 cm) wide fabric. **For optional tie belt:** add extra ¼ yd (23 cm) to fabric requirements. **Trimming:** 1⅛ yd (1 m) of 1¼ in (3.25 cm) wide braid, preferably furnishing braid. **Also**: 24 in (61 cm) of cord for neck ties, and 2 or 4 small shank pearl buttons. **Recommended fabrics:** natural wild silk or lightweight cotton poplin. **Made here**: in unbleached Antung wild silk from Pongee Ltd, London.

MEASUREMENTS One size, to fit **bust** 36–40 in (91–102 cm). **Back length** from shoulder to finished hem: 56 in (142 cm). Length can be adjusted to suit personal requirements; increase amount of fabric required if necessary.

PATTERN VARIATIONS **Shirt:** this design can be adapted to make a shirt or tunic by omitting the skirt gussets and cutting the front and back body pieces 30 in (76 cm) long or as required, with or without stitched side vents. It requires 3⅛ yd (2.90 m) of 36 in (90 cm) wide fabric.

NOTE All seam allowances are ⅜ in (1 cm) unless otherwise stated.

PREPARATION **(1)** Cut out pattern pieces following cutting layout. **ASSEMBLE RUFFLES (1)** With right sides together, fold all ruffle strips in half lengthwise and stitch the ends. Turn out and press. **(2)** Using a fine zigzag, edge stitch around the closed edge being sure to stitch **over** the edge, as in Figure 1. Use coloured thread to match main colour of braid. **(3)** Run 2 rows of gathering thread ¼ in (6 mm) from the raw edge and draw up ruffles to 1 in (2.5 cm) less than neck and cuff facings. **(4)** Stitch to facings leaving a step of ½ in (1.2 cm) at each end, as in Figure 2. **(5)** Press with raw edges on **wrong** side of fabric. **PREPARE CENTRE FRONT OPENING (1)** Finish outer edges of front facing by turning under ⅜ in (1 cm). Zigzag or machine flat. **(2)** Right sides together, attach to front opening by stitching ¼ in (6 mm) from slash in front. **(3)** Cut slash in facing to match slash in front, turn to inside and press. **ASSEMBLE SHOULDERS (1)** Sandwich back shoulder between 2 shoulder straps, stitch, and press straps forward to enclose raw edges. **(2)** Press ⅜ in (1 cm) turning in top strap. **(3)** Stitch underneath strap to inside of front shoulder. Topstitch top strap over this seam to cover raw edges. Use a decorative stitch if desired, topstitching back shoulder seam to match. **ASSEMBLE NECK (1)** Gather neck of garment, drawing up to 1 in (2.5 cm) less than the neck facings. **(2)** Stitch neck facing to inside of neck opening with raw edges on the outside, as in Figure 3. **(3)** Stitch the neck braid to the inside of the facing across the **ends** only. Turn

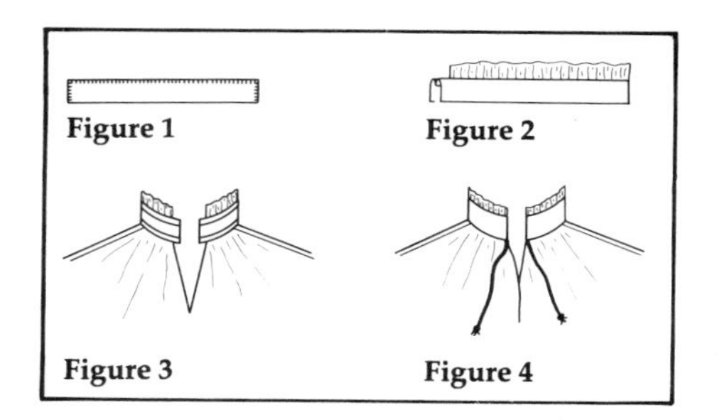

CUTTING LAYOUT

(not drawn to scale)
Fabric width: 36 in (90 cm)

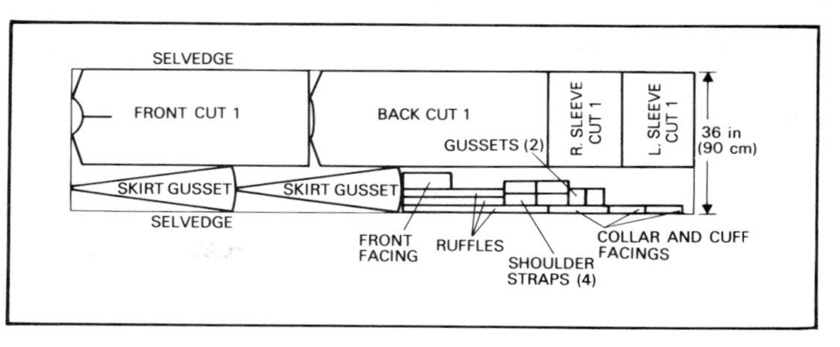

out so that the ends are finished and the braid is on the outside, still detached along its edges. **(4)** Inset the ends of cords under the braid at either side of centre front opening. Topstitch along the top and bottom edges of the braid covering all raw edges of the facing seam allowances and securing ties, as in Figure 4. Match colour of top thread to braid and bobbin thread to facing. **ATTACH SLEEVES (1)** Position sleeves D-D, E-E. **(2)** Gather top of sleeves and stitch to armholes, stopping 2 in (5 cm) above underarm seam. **ATTACH GUSSETS (1)** Zigzag gussets. **(2)** Stitch 2 sides of gusset in right angle between dress body and sleeve, then complete armhole seam. **(3)** Repeat for second arm. **ASSEMBLE BODY (1)** Stitch skirt gussets to each side of front body, starting at the hem. Zigzag seams. **(2)** Stitch front to back, up side seams through to underarm seams, leaving a 3 in (8 cm) cuff opening at end of sleeve. **FINISH CUFF (1)** Clip seam allowance at this point, and finish opening by turning in raw edges and neatening by hand. **(2)** Gather end of sleeve to 1 in (2.5 cm) less than cuff facing length. Stitch facing to inside and attach braid exactly as on neckband. **TO FINISH (1)** close cuff with 1 or 2 small shank buttons and loops. **(2)** Hem dress by hand, ⅜ in (1 cm) first turning, 1 in (2.5 cm) second turning. Press.

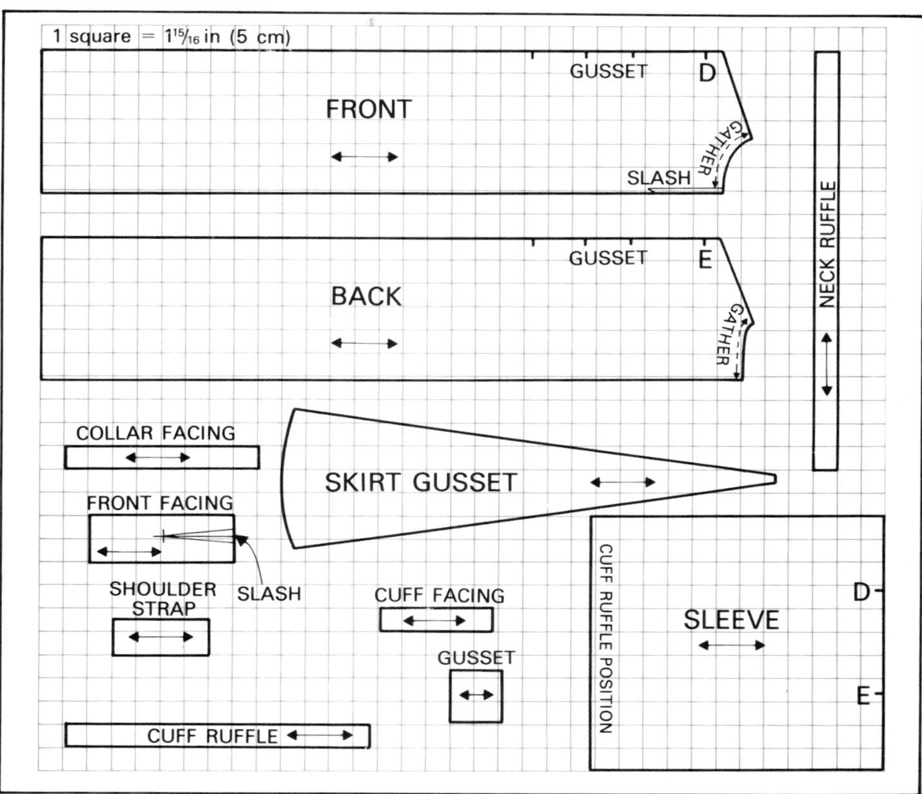

Althea Instructions

MATERIALS **Fabric required for body:** 4 yd (3.70 m) of 36 in (90 cm) wide fabric. **Fabric required for sleeves and yoke:** 2¾ yd (2.55 m) of 36 in (90 cm) wide fabric. **Trimming:** 2 small pearl buttons for cuffs. **Recommended fabrics** for body, cotton lawn or poplin; for sleeves, cotton lawn. **Made here:** body in Liberty print cotton, sleeves in plain cotton lawn.

MEASUREMENTS One size, to fit **bust** 33–38 in (84–97 cm), **back length** from shoulder to finished hem 57½ in (146 cm). Length can be adjusted to suit personal requirements; increase amount of fabric required if necessary.

STYLING Additional appliqué strips can be added to sleeves, as shown in photograph.

PREPARATION **(1)** Cut out pattern pieces following cutting layout. **PREPARE APPLIQUÉ STRIPS (1)** Piece together the bias strips to form one length for the yoke 44 in (112 cm) long. **(2)** Piece together strips for sleeves, two 43 in (110 cm) long, and two 35½ in (90 cm) long. **ASSEMBLE SLEEVES (1)** Press ¼ in (6 mm) turning on one side of the longer appliqué strips and on both sides of the shorter strips. **(2)** Lay one longer strip on to the lower edge of the upper sleeve, right sides up, and with unpressed edge of strip level with raw edge of the sleeve, stitch ¼ in (6 mm) from raw edge.

CUTTING LAYOUT

(not drawn to scale)
Fabric width: 36 in (90 cm)

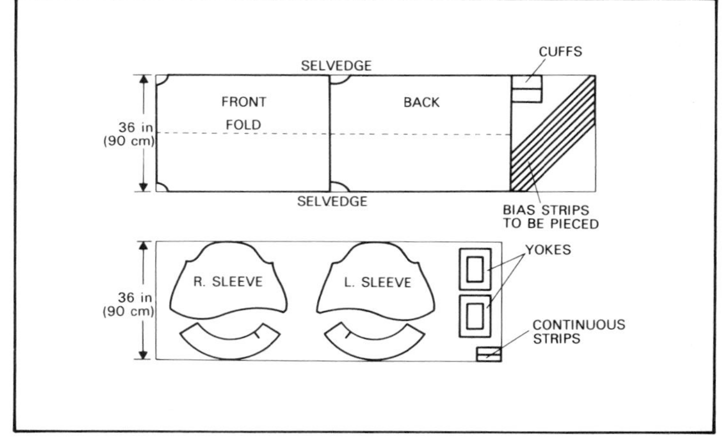

Topstitch the turned edge on to the sleeve. (3) Appliqué shorter strip on the lower section of sleeve 1½ in (3.8 cm) from upper edge. **(4)** Stitch upper and lower sleeve sections together with ⅜ in (1 cm) seam allowance, and zigzag seam allowance. The spacing between the 2 bands should be just over 1 in (2.5 cm). **(5)** Cut opening slit in sleeve as indicated on pattern and finish with continuous strip. **ASSEMBLE YOKE (1)** Press ¼ in (6 mm) turning on both sides of appliqué strip for yoke. **(2)** Stitch strip to outer yoke exactly in the yoke centre and mitre each corner by folding and stitching as you come to it. **(3)** With right sides together, stitch yoke and yoke facing at the neck opening. Turn out and press. **ASSEMBLE BODY (1)** Gather front and back body pieces along top edge. **(2)** Right sides together, stitch to front and back of top yoke. **(3)** Press up turning of yoke facing and slip stitch to seam, covering raw edges. **ATTACH SLEEVES (1)** Gather sleeves at top between notches, draw up to fit armhole. **(2)** Fit sleeves to armhole. Be sure that the fuller part of the sleeve is at the **back. (3)** Stitch sleeves to armholes and zigzag seam allowance to neaten. **(4)** Stitch sleeve seams and side seams of dress together, then zigzag seam allowance to neaten. **CUFF (1)** Fold cuff in half lengthwise, right sides together, and stitch ends. Turn to right side and press. **(2)** Gather end of sleeve to fit cuff. Stitch to outside of cuff. Don't forget to turn under one side of continuous strip for cuff overlap. **(3)** Slip stitch on inside to cover seam allowance. **TO FINISH (1)** Make buttonhole on cuffs and attach buttons. **(2)** Hand hem bottom of dress, first turn ¼ in (6 mm), second turning at least 1 in (2.5 cm) depending on length desired. Press.

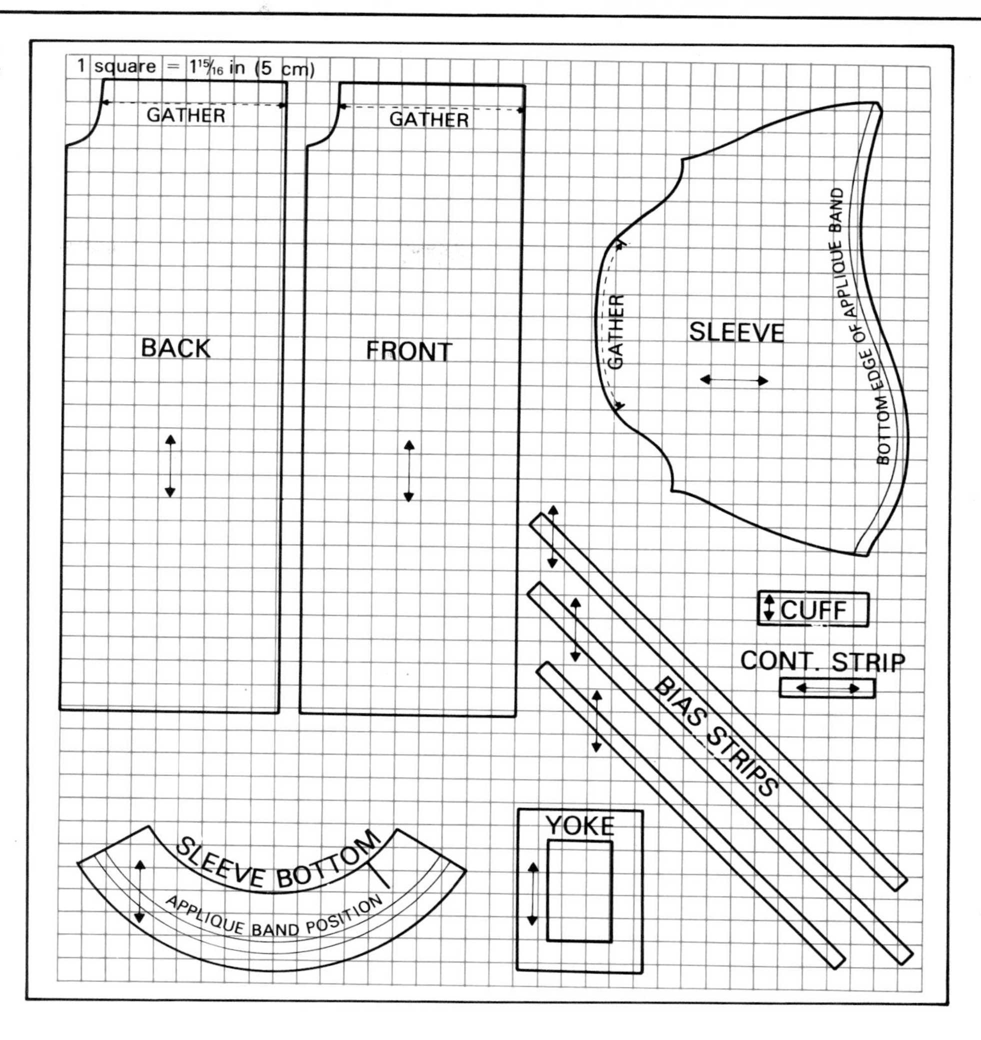

Shepherdess Instructions

DETAILS Shown with sculpted tree from Supotco.

MATERIALS **Fabric required:** 4 yd (3.70 m) of 36 in (90 cm) wide fabric. **Trimming:** 3 yd (2.75 m) of ⅜ in (1 cm) wide ribbon cut into 4 equal lengths. **Recommended fabrics:** fine cotton or cotton lawn. **Made here:** in Liberty Tana Lawn.

MEASUREMENTS One size, to fit **bust** 32–37 in (81–93 cm), **back length:** 56 in (142 cm). Length can be adjusted to suit personal requirements; remember to increase amount of fabric required if necessary.

STYLING The front opening on this design makes it suitable for use as a nursing nightie.

PREPARATION (1) Cut out pattern pieces following cutting layout. **ASSEMBLE FRONT OPENING (1)** Lay the continuous strip round the centre front opening, right sides together, stitch and press to the inside. Finish by

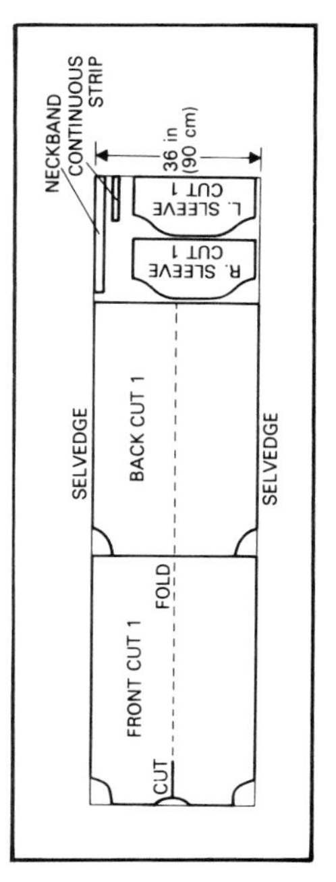

CUTTING LAYOUT
(not drawn to scale)
Fabric width: 36 in (90 cm)

hand. **(2)** Press 2 turnings in on top of front body piece, first turning ¼ in (6 mm), second turning 1¼ in (3.2 cm), and topstitch. **(3)** Stitch another row ½ in (1.2 cm) above this row to form a channel for the ribbon. **(4)** Run ribbon through channel on each side and secure ends at armholes. **PREPARE SLEEVES (1)** Make a buttonhole ½ in (1.2 cm) long in the centre of each sleeve as indicated on pattern. **(2)** Stitch underarm seam of sleeves with a seam allowance of ½ in (1.2 cm), and neaten by zigzag. Press open. **(3)** Press 2 turnings at sleeve ends, first turning ¼ in (6 mm), second turning ¾ in (2 cm). **(4)** Topstitch both sides of hem turning to form a channel for the ribbon. **(5)** Insert ribbon through buttonhole. **(6)** Run gathering threads in top of sleeve heads between notches. **ASSEMBLE BODY (1)** Gather top of back body and pull up to 9 in (23 cm). **(2)** Stitch side seams: seam allowance ½ in (1.2 cm). Neaten by zigzag. Press open. **(3)** Right sides together, stitch sleeves to armholes up to notches with seam allowance ⅜ in (1 cm). Neaten by zigzag. **ASSEMBLE NECKBAND (1)** Press neckband

in half lengthwise, right sides outside. **(2)** Mitre 2 corners by folding 9 in (23 cm) apart, as shown on pattern pieces. **(3)** Stitch, trim and nip corners of mitres to ensure a neat neckband. **(4)** Stitch right side of inner half of neckband between corners to wrong side of the gathered edge of the back body piece. **(5)** Gather up the sleeve heads to 6½ in (16.5 cm) between the notches. **(6)** Stitch the inner edge of the neckband to the wrong side of the sleeve heads: seam allowance ⅜ in (1 cm). The neckband should extend below the ribbon channel by approximately ½ in (1.2 cm). Leave the extension free to turn under to neaten. **(7)** Press under the ⅜ in (1 cm) turning on the outside of the neckband and topstitch down enclosing all raw edges at back and sides of neck. **(8)** Neaten extension ends by hand, including the intersection of band and ribbon channel. **TO FINISH (1)** Turn up hem, first turning ¼ in (6 cm), second turning ¾ in (2 cm). **(2)** Zigzag or finish by hand.

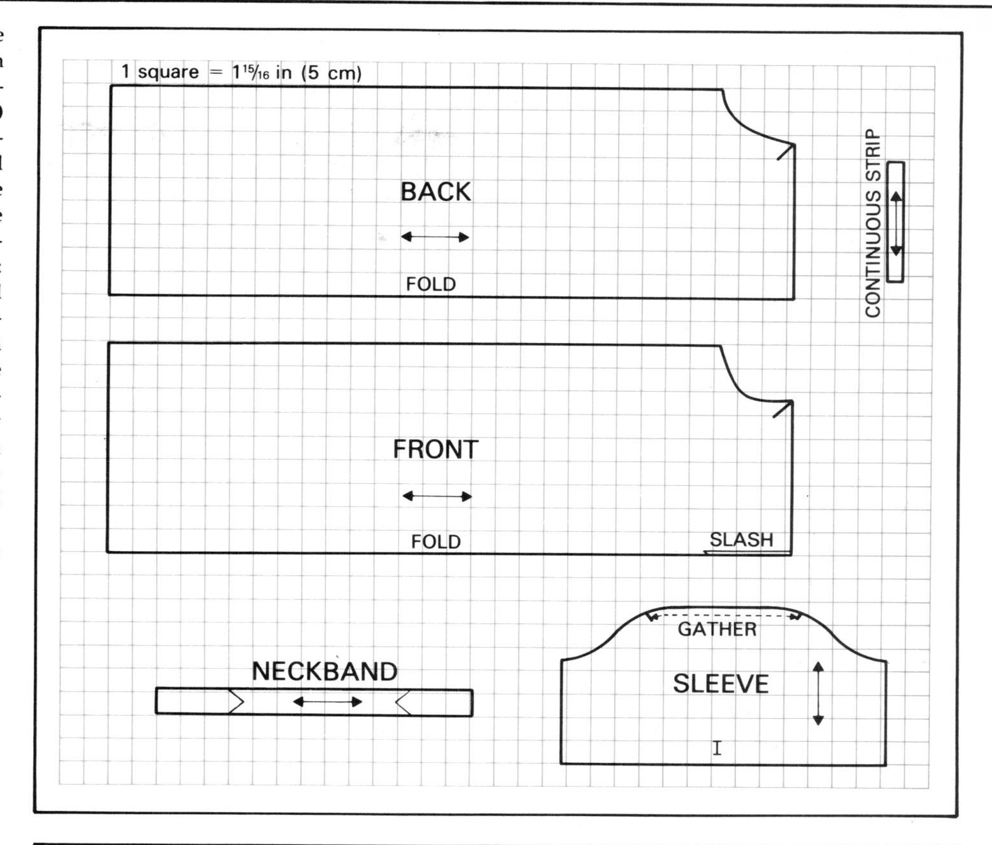

Amy Victorian Nightdress Instructions

MATERIALS Fabric required: 4¼ yd (4 m) of 36 in (90 cm) or 45 in (115 cm) wide fabric. **Trimming:** 2⅛ yd (2 m) of ½–¾ in (13–20 mm) wide pre-ruffled *broderie Anglaise*; 6 small pearl buttons. **Recommended fabrics:** pure cotton lawn. **Made here:** in Tootal's fine cotton lawn, trimmed with English Sewing's pre-ruffled *broderie Anglaise*.

MEASUREMENTS One size, to fit **bust** 34–36 in (86–91 cm); **height** 5′6″ (168 cm); **back length** to finished hem, 55 in (140 cm). Length can be adjusted to suit personal requirements.

STYLING Fine feather stitching can be added to neckband, cuffs, yoke edge and front opening.

NOTE All seam allowances are ⅜ in (1 cm) unless otherwise stated.

PREPARATION (1) Cut out pattern pieces following cutting layout. **(2)** Slash and nip front opening of front bodice. **TUCK FRONT BODICE (1)** Starting 1¼ in (3 cm) in from front bodice armhole seam, stitch 12 tucks on **each** side of centre cut. Each tuck should be ¼ in (6 mm) wide and 4 in (10 cm) long, with ½ in (1.2 cm) between each line of stitching, as shown in Figure 1. **(2)** Press pleats towards armholes. **TRIM FRONT FACINGS (1)** Stitch *broderie Anglaise* to one edge of each of the 2 facing pieces. **(2)** Stitch the other edge of the facing pieces to the neck slash, with right sides

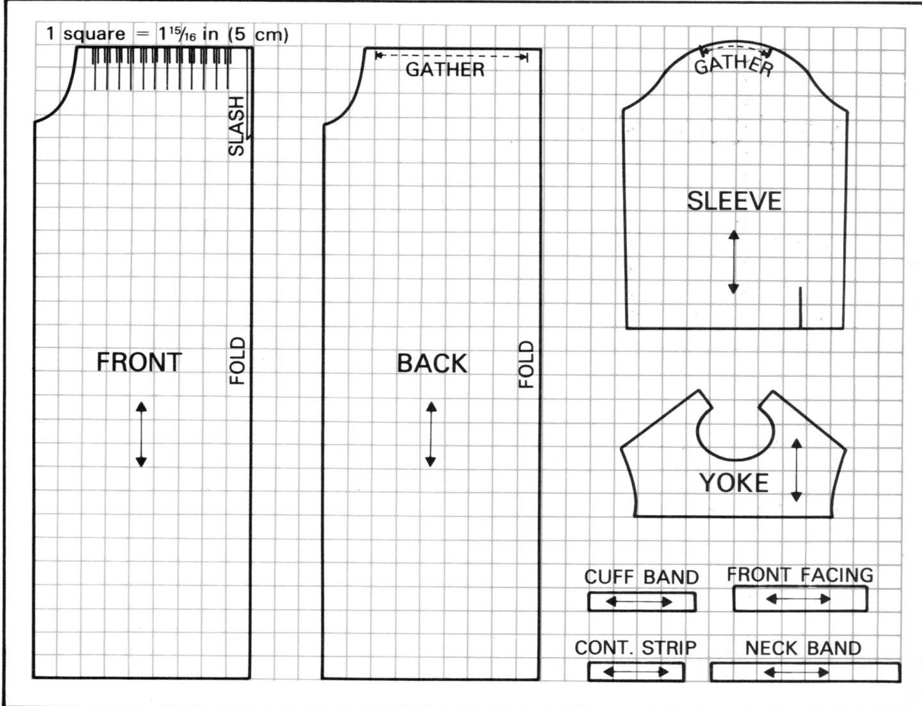

CUTTING LAYOUT
(not drawn to scale)
Fabric width: 36 in (90 cm)

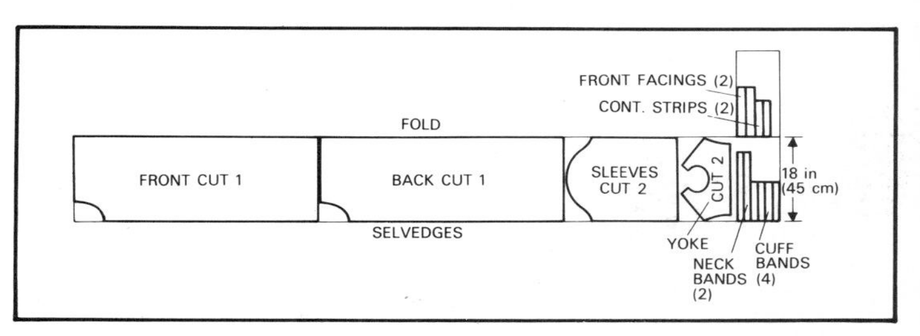

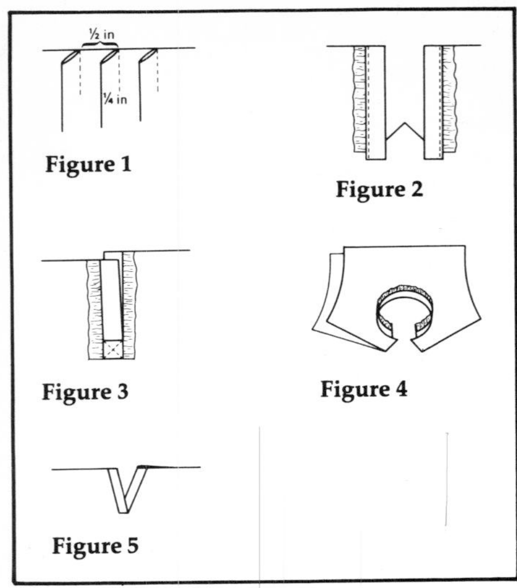

Figure 1
Figure 2
Figure 3
Figure 4
Figure 5

together as shown in Figure 2. **(3)** Press facing flat to the right side of bodice top and topstitch down. **(4)** Lap over the stitched facings so that they lie on top of each other, as in Figure 3. **(5)** Neaten and topstitch at bottom of neck opening. **ASSEMBLE COLLAR AND YOKE (1)** Sandwich *broderie Anglaise* between the two neckband strips, and stitch. **(2)** Finish ends, turn out and press. **(3)** Starting from centre back of yoke, sandwich completed collar between yoke pieces as shown in Figure 4, leaving an equal 'step' at each front end. Turn out and press lightly. **BACK BODY (1)** Gather top edge of back body to fit yoke. **(2)** Right sides together stitch to outer back yoke edge only. **(3)** Slip stitch or machine yoke facing edge to cover seam allowance. **FRONT BODY (1)** Stitch front body on each side of centre opening to the outer front yoke. **(2)** Finish inside as for back yoke. **SLEEVES** Right sides together, **(1)** Stitch sleeves to armholes, gathering and drawing up surplus fabric between notches to ensure a perfect fit. **(2)** Zigzag raw edges. **ASSEMBLE CUFFS (1)** Enclose the sleeve opening vent with continuous strip, as in Figure 5. **(2)** Repeat for second cuff. **(3)** Make up cuffs exactly as for collar band, with *broderie Anglaise* sandwiched into outer edge. **FINAL ASSEMBLY (1)** Stitch nightdress sides and underarm sleeve edges together. **(2)** Zigzag seam edges. **(3)** Gather ends of sleeves to fit cuffs, stitch outer cuff piece to sleeve on outside, then slip stitch on inside to finish, enclosing seam allowance. **TO FINISH (1)** Make buttonholes, 1 at neck in yoke, 3 on front, 1 on each cuff. **(2)** Attach buttons. **(3)** Make 2 turnings in hem, first turning ⅜ in (1 cm), second turning 1 in (2.5 cm). Machine or slip stitch to finish.

Machine Knitting

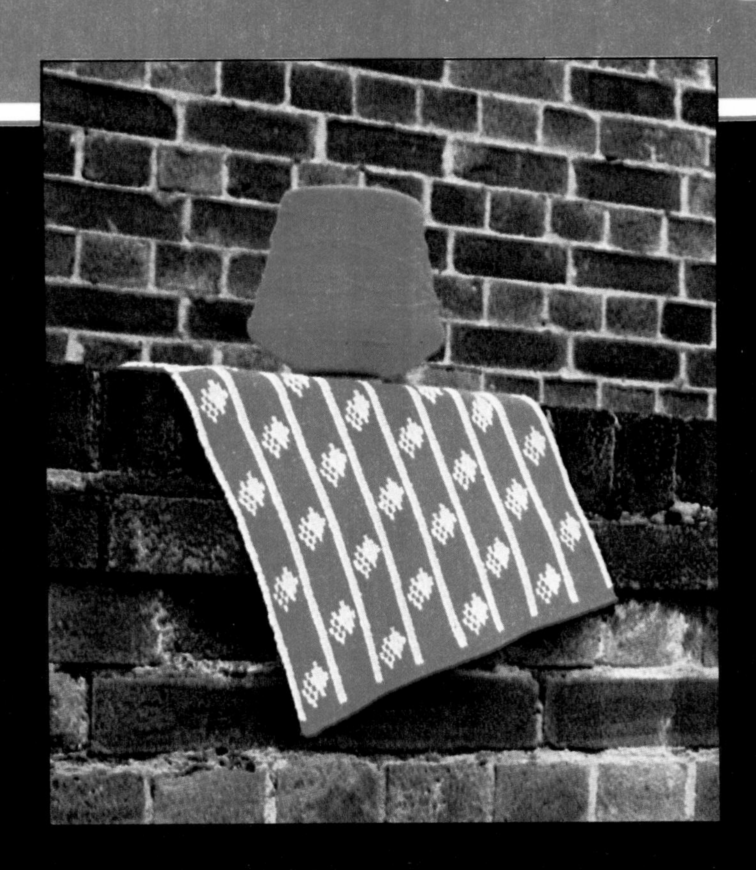

Introduction

Coco Chanel was the first designer to realize the potential in machine knitting – *'You see, my dear,'* she used to say, *'I built my fortune on an old jersey.'*

Chanel adored the comfort and casual elegance of classic men's knitwear and she cast around for a fabric that could help her to bridge the gap between masculine style and feminine chic. Seizing on the supple qualities of machine-knit fabrics and the way in which they lend themselves to handsome geometric patterns, she made them up in simple shapes coloured with her favourite beiges, greys and black. The result was the 'jumper suits' that took the Twenties by storm, shown here.

Today's knitting machines place at your fingertips refinements undreamt of in those early days of high-fashion knitting. In the Twenties few yarns were suitable for machine working and textures were limited to 'smooth', 'rough' and 'ribbed'. Now, there are hundreds of yarns that can be used singly or together to produce everything from filmy lace to shaggy pile, and on an advanced machine like the Knitmaster Electronic SK500 almost anything is possible – you can invert colours, 'mirror' an image or double the length and width of a motif simply by pressing a button. You can even produce your own lavish signature fabric by autographing the special paper that goes straight into the machine instead of punch cards. You can 'play' a knitting machine as artistically as you would a piano – the only difference is that you will be creating clothes instead of chords.

Technically, machine knitting combines many features of traditional fabric weaving and handknitting – you can weave as on a loom with the Weavemaster attachment, and you can duplicate many handknitting stitches in a fraction of the usual time. But these similarities can be misleading, because machine knitting is not just a quicker, simpler way of producing hand crafts: it is a separate creative medium that offers exciting design possibilities not to be found in any other fashion field.

To show you what they are and how to make the most of them, we asked Knitmaster to create a special Creative Dressing Collection of machine knits, and as the collection took shape in their studios we talked to the Knitmaster Design Team, headed by Iris Bishop, about their work.

Designing Machine Knits

Fabric Design

Fabrics play a special role in machine-knit designing, as Maggie Dyke, who created all the fabrics in the collection, explains:

'I find knitting and weaving the most satisfying of the textile arts because *they give you the chance to really create something*. If you have to begin with a piece of cloth that someone else has designed, you're limited from the start. But with machine knitting you can work with the colours and textures yourself, make your own fabrics, and carry them through into your own fashion ideas.

That's particularly important because *fashion design and fabric design have to work together to achieve the best results*. The size of the pattern on the fabric has to be in proportion to the size of the garment, the patterns have got to be placed *on* the garment in an imaginative way, and so on. But fashion designers often impose their ideas on a fabric without stopping to think whether or not they're suitable, so the design and the fabric clash. By working through the process yourself from start to finish, you can make sure that the shape leads on naturally from the fabric, and makes the most of its special features.

Think the whole thing through from start to finish, *look* at everything you do from every point of view. Ask yourself if the fabric is going to work with the garment you want it to, and if not, why not. Is the fabric working out as you intended? Or is it doing something else that could be equally appealing? You'll soon find yourself looking at fabrics in a different way, getting away from rigid ideas that certain fabrics, for example, have certain "right" sides. In tuck stitch you normally use the side facing you, in Fair Isle stitch you concentrate on the side that's away from you. But if you turn them round you'll see that you can use the other side just as successfully – it all depends on the effect you want to achieve.

It's best to begin by deciding on the kind of garment you want to make, and on the colours and textures you want to use. That will usually suggest some kind of stitch pattern, and you can experiment with different cards and colours until you find the combination that's right. Very often a fabric that isn't suitable for one sort of garment is perfect for another, so I've found it very useful to build up a reference file of fabric samples along with details of the stitch, tension and yarns I've used. I note down the details as I go – it's an infuriating thing to do when you're actually knitting, and dying to get on to the next piece, but it saves you spending hours trying to work out how you did it later on.

I've taken a few pages from my portfolios to show you what I mean. I've always found tuck-stitch patterns the most fascinating of the stitches used in machine knitting. You can work with a tremendous range of yarns in different weights and textures – everything from fine sewing cotton to mohair – and the result is a wonderful variety of fabrics. These three tuck-stitch samples give a good idea of the potential.

Two were knitted on the Knitmaster 120 chunky machine, one in cream Jaeger mohair, the other in a black 4-ply yarn and Silverknit lurex. The third, in Silverknit 3-ply Adah cotton and fine silver lurex, was done on the Knitmaster 328 machine, using one of the standard punch cards in conjunction with a needle arrangement. To vary the pattern, I altered the needle arrangement or the striping sequence every three or four inches.

Used imaginatively, attachments have a lot of design

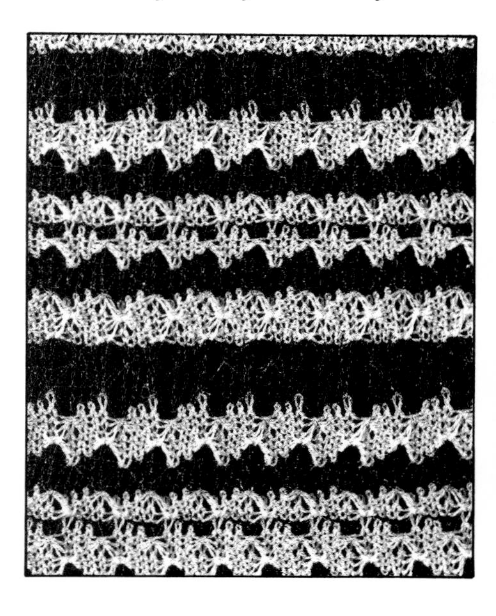

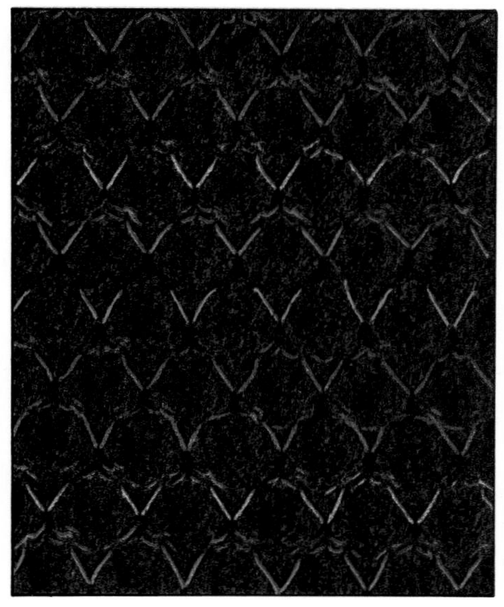

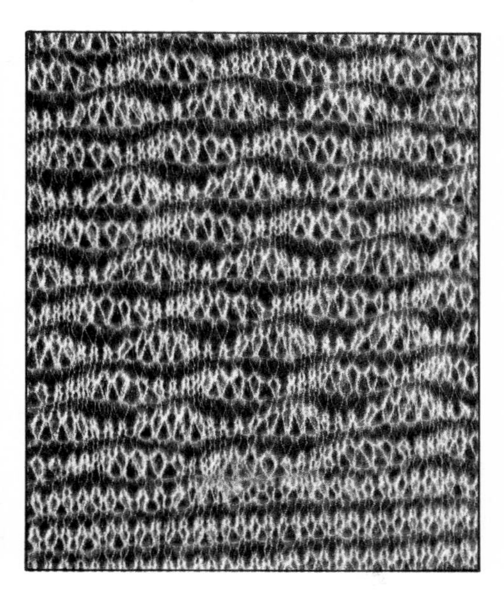

potential, too. For example, ribbing attachments are often used only for rib stitches, but you can use them for pile knitting, tuck rib, pin tuck, double jacquard and circular knitting.

Both pink fabrics were knitted on a Knitmaster 328 with ribber and striper, using the same punch card. The pink-and-white fabric is a drive lace pattern in 2-ply Dakota and Thirties cotton. The pink-and-grey fabric is a double jacquard – it's exceptionally light in weight for a double rib knit because I used light Silverknit 2-ply Dakota yarns. Scalloped edging is made by starting the pattern directly after completing the cast-on procedure, then pulling the points down slightly while pinning the fabric out for pressing.

The sixth sample gives you a different look at a classic Fair Isle fabric. Fair Isle knitting is usually done in plain yarns but I used Pingouin Volutés 4-ply yarns, one of which is injection dyed to the same shade as the second yarn. The result is a much subtler pattern and I've gone on to show how it can be varied.

Machine knitting gives you unique creative opportunities. Handknitting is a sometime thing, something to pick up and put down, and do while you're watching the telly. In machine knitting you work very quickly and intensely, so you really have a chance to develop and discover yourself.

The most important thing – the whole way through the process – is to think about what you're doing, very very carefully. Keep an open mind. And enjoy yourself, because *in fabric design you might not get quite what you wanted first time – but you can never really do anything that's wrong!'*

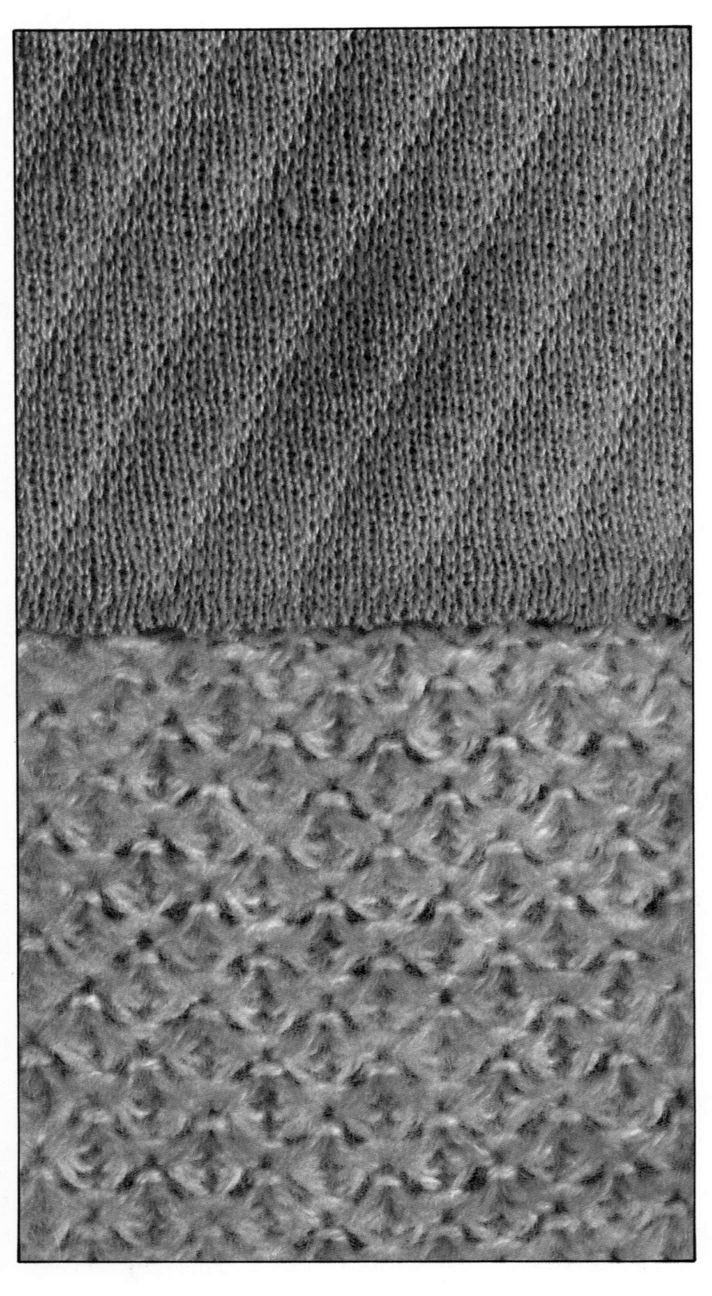

Fashion Design

In the studios, the fabrics that Maggie has created for the collection are passed to designers Susan Guess and Laraine McCarthy, who will interpret them into fashion shapes. As professionals working in tandem, they share out the fabrics and develop them into fashion designs that reflect two very different but complementary styles.

Here again, colour and texture are the starting-points – and for Susan the tactile dimension of knitting is the most appealing. 'I love the texture and softness of knitting,' she says, 'and the way it *moves* and *flows* over the body. It's not static, like cloth. Knitted fabrics do so much that you don't have to use fancy shapes to achieve beautiful results.'

Colour is the cue for Laraine's designs. 'I intended to be a painter,' she explains, 'but I soon found that I wanted to be useful, to create something with a purpose, rather than just indulge myself. Now I find I can do both, because I work with colours and the play between shapes and colours, much as I did when I was painting. Only what I'm doing now is practical as well as decorative.'

Every fabric has its own logic, which must be puzzled out before the designer's sketches can begin to take shape. 'With experience you develop an intuitive sense about fabrics,' says Laraine, 'but I always like to have a large piece of fabric to start with when I'm sketching, so I can see exactly how the fabric hangs and drapes and moves.'

'In itself, no fabric is difficult to work with,' adds Susan. 'Difficulties only arise if the fabric is unsuitable for the design you want to use.'

As we present the Creative Dressing Collection, you can see what the designers did, and why.

Zebra Coat

(Fabric by Maggie Dyke, design by Susan Guess)

Maggie: The stitch pattern I've used is *pile knitting* – with a difference. Standard instructions for pile knitting tell you to use a thick 4-ply yarn as the main yarn, but I prefer to use two 2-ply yarns instead of the one 4-ply because that gives you two main yarn colours, which makes the fabric much more interesting.

Standard instructions also tell you to use an invisible thread as the background yarn but the result is a stiff fabric that's hard to handle because the invisible thread is made of nylon, so it's very rigid. I've used a Thirties cotton, which is like a very fine sewing cotton, instead of the nylon. It produces a much softer fabric and the background yarn becomes a *positive* feature of the design, instead of something you're trying to hide.

The pattern was inspired by drawings of zebras but the black-and-white colour contrast seemed too harsh for the soft texture of the fabric, so we did it in brown and cream instead – the same colours as a Grevys zebra.

Susan: I kept the design as simple as possible so it wouldn't play against the complicated fabric. The lines are straight and clean, the collar comes right up below the chin, and the length of the coat draws out the proportions so you really do look six inches taller. I think everyone secretly wishes they were as tall and willowy as a fashion model, so *when you wear something like this, that makes you look taller and slimmer, it makes you feel better as well.*

Maggie used a colour striper on the fabric, so I turned the front edge over just one centimetre to make a subtle striped binding that shadows the zebra design on the reverse side. I finished the front with press studs, since

buttons would have broken up the zebra pattern. We used both yarns in the collar and cuffs so they would blend with the tones in the zebra fabric, and we made the pile deeper than the patterned pile, so they feel as warm and luxurious as fur.

I used the cut-and-sew method because you don't want to worry about shaping when you knit a complicated fabric like this. The fabric is firm enough to be suitable for cutting, and the design benefits from the method as well, since a coat like this needs the extra support that machine-sewn seams give it.

Zebra Coat Instructions

For Knitmaster 324, 326, 328 and 329 machines with ribbing attachment and colour changer.

ABBREVIATIONS FOR ALL PATTERNS IN THIS SECTION

alt = alternate
beg = beginning
carr = carriage
cm = centimetres
col = colour
dec = decrease
ev = every
foll = following
gm = grams
inc = increase
N = needle
Ns = needles
patt = pattern
pos = position
rept = repeat
rem = remaining
st = stitch
sts = stitches
tog = together

Fabric Instructions

BACK AND FRONT (4 Pieces Alike)
Set up machine with ribber and colour changer. Arrange 176 Ns on each bed in double rib. Tension Dials at 4 on main machine and 3 on ribber. Using col 2 cast on as given in instruction book for double rib. Join all four cards in sequence, insert into machine and lock on row 1 card 1. Knit 2 rows rib. Row Counter 000. Transfer sts from main machine to ribber, leaving empty Ns in B pos. Push 3 Ns at each end of main machine back to A pos. Set machine for pile knitting as given in instruction book. Release card. With col 2 in colour changer, col 1 in main yarn feeder and cotton in auxiliary feeder, knit 792 rows (see Note) changing cols 1 and 2 ev 2 rows. Using waste yarn knit 8 rows and release from machine.

SLEEVES (2 Pieces Alike)
Follow instructions for Back and Front pieces but knit only 366 rows of pattern. Cast off. Remove card and colour changer.

CUFFS (2 Pieces Alike)
Arrange 125 Ns on main machine and 126 Ns on ribber in double rib. Tension Dials at 4 on main

DETAILS Shown here with earrings and bangles from Butler & Wilson, boots by Bally and sabre from Robert White & Sons.

MEASUREMENTS One size to fit 32–38 in (81–96 cm) bust.

MATERIALS 3 200 gm cones Kyoto in col 1 (cream); 3 200 gm cones Kyoto in col 2 (dark brown); 2 large and 1 small cone Thirties Cotton (camel); 3 large press fasteners; sewing thread; small amount lining for pockets. Yarn available from Silverknit, Dept K, The Old Mill, Epperstone-By-Pass, Woodborough, Nottingham.

TENSIONS 34 sts and 61 rows to 10 cm measured over 2 col pile knitting. Tension Dials at approx 4 on main bed and 3 on ribber. 30 sts and 53 rows to 10 cm measured over plain pile knitting. Tension Dials at approx 4 on main bed and 5 on ribber.

NOTE When working 2 col pile knitting it is important to pull cotton forwards to avoid it entering the colour changer.

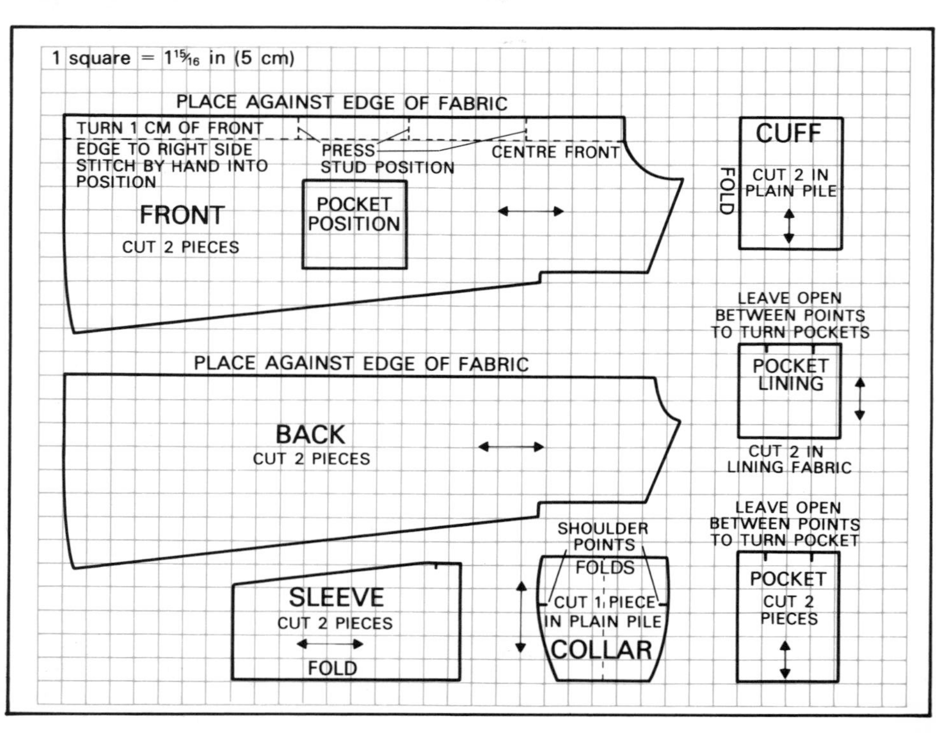

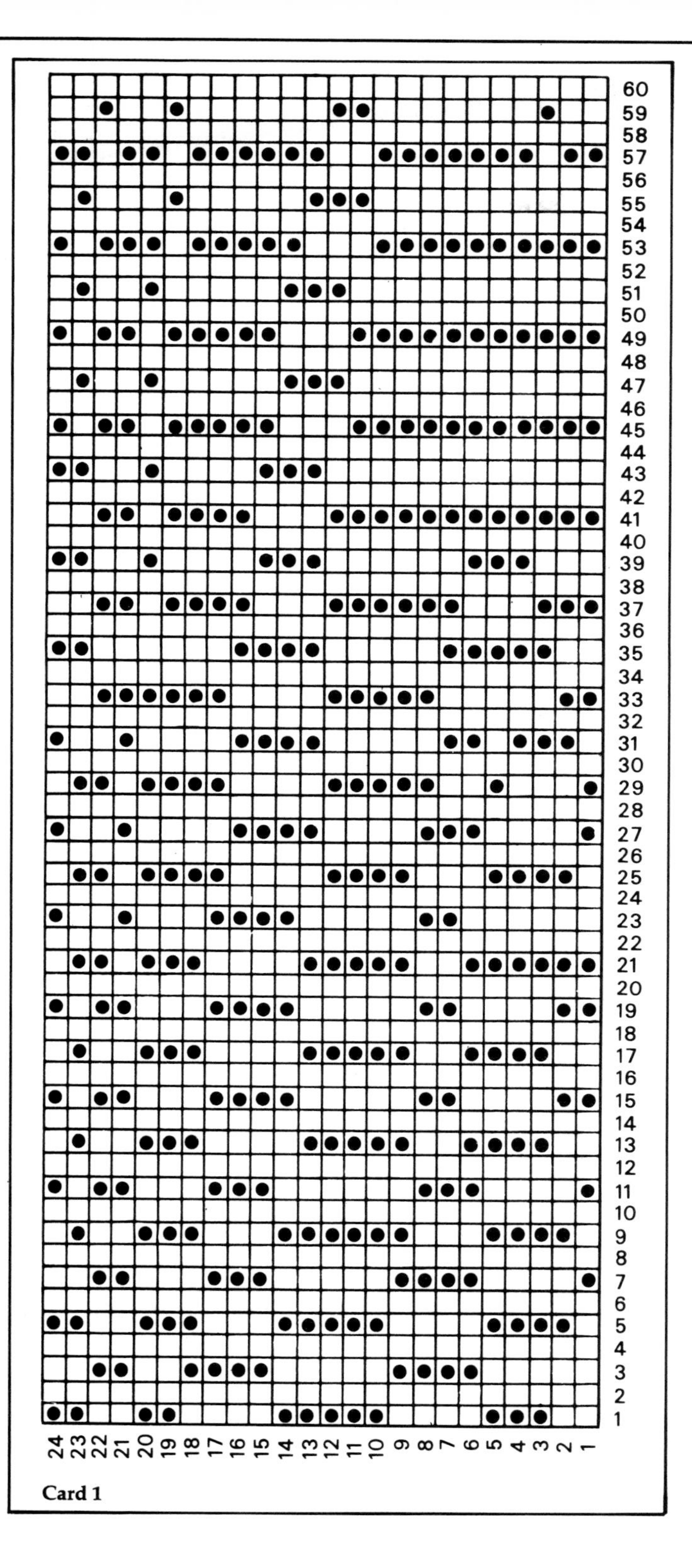

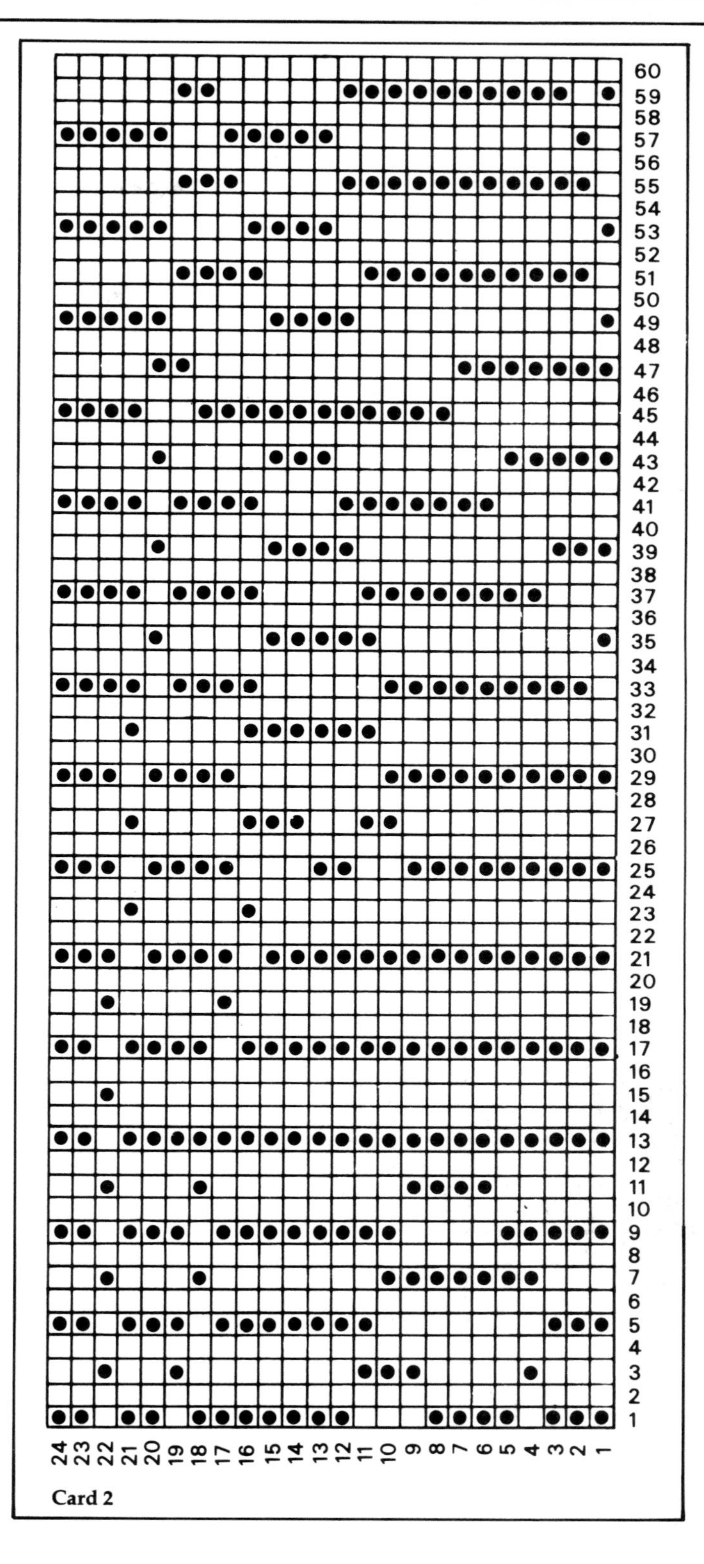

machine and 5 on ribber. Using col 2, cast on as given in instruction book for double rib. Row Counter 000. Transfer sts from main machine to ribber, leaving empty Ns in B pos. Set machine for pile knitting. With col 2 in main yarn feeder and col 1 in auxiliary feeder, knit 148 rows. Using waste yarn, knit 8 rows and release from machine.

COLLAR (1 Piece)

Follow instructions for cuffs but work over 84 sts and knit 276 rows. Using waste yarn, knit 8 rows and release from machine.

TO MAKE UP

Follow the chart to cut the pattern. Cut out pieces from the lengths of knitting. Cut pockets from wastage on back pieces. Seam allowances are ¾ in (2 cm) on all edges except collar, neckline, pockets and centre front, where they are ⅜ in (1 cm). Finish all edges with zig zag st. Line pockets and hand stitch into position shown on pattern. Stitch centre back seam, using 1 stitch from each edge only. Stitch shoulder seams. Sew sleeves to armholes. Sew ends of collar.

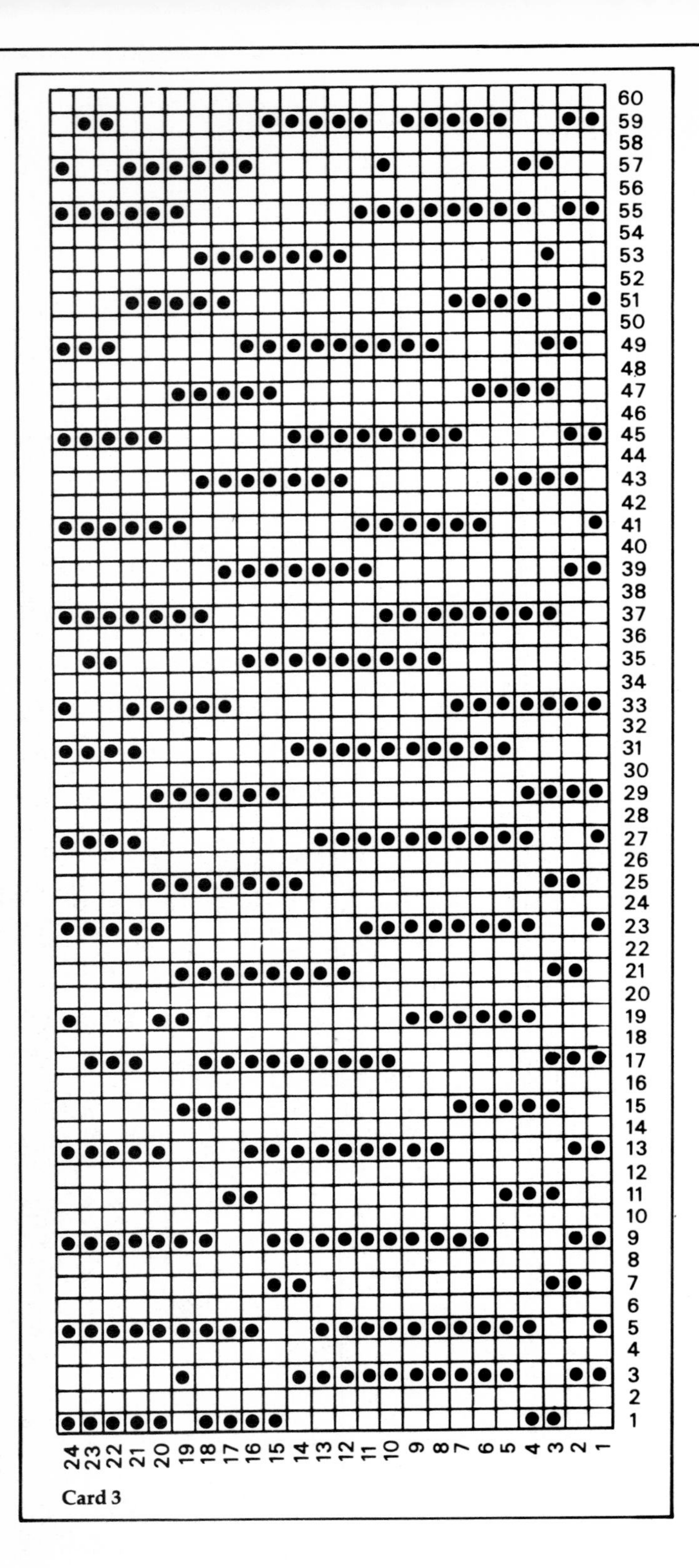

Card 3

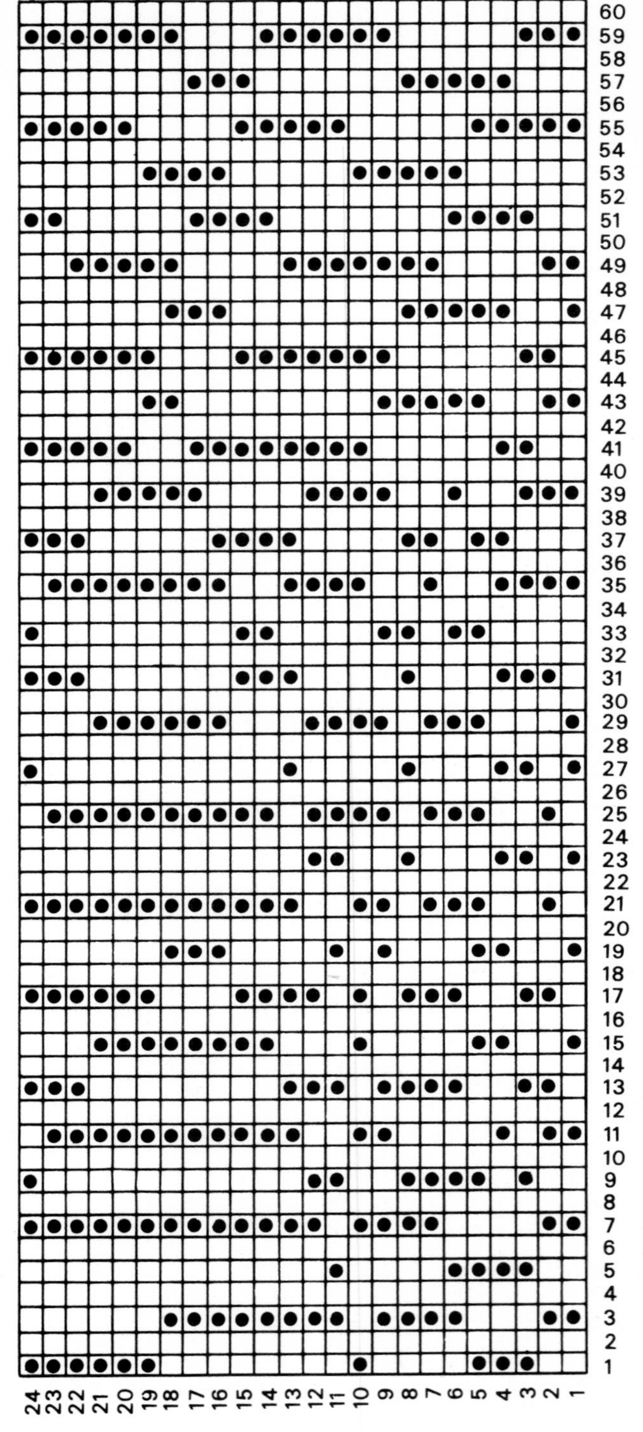

Card 4

Sew one side to neck edge, fell other side to enclose all seam allowances. Sew sleeve seams and side seams. Sew cuff edges together. Sew to sleeve and fell loose edge to inside cuff. Fold cuffs back and stitch to sleeve seam to hold cuffs in position. Turn front edge of coat to right side to give the appearance of a bound edge. Sew press studs where marked. Allow to drop before levelling and turning up hem.

Pink Suit

(Fabric by Maggie Dyke, design by Laraine McCarthy)

Laraine: Knitwear isn't often feminine, so this suit makes a nice change from the usual jumpers and cardigans. The fabric is done in *tuck stitch* and forms its own scallops, which I've used in the edging at collar, cuffs and hip.

A soft, gentle shape like this is very wearable, and I've used raglan sleeves because they look neat even when they're nice and full as they are here. Set-in sleeves usually look *too* big when they're full. The deep armholes and flounce at the hip make the waist look smaller, and the four-panel skirt is lightly gathered because I don't think knitted skirts should be close fitting.

The weight of the yarn we've used is very important to the overall design. The yarn is unusually light, just a 2-ply, but since tuck stitch makes a fabric more solid, it's given a bit of body to a yarn that might otherwise be too flat and flimsy. The result is a fabric that's light enough to hang well and wear comfortably, strong enough to carry the flounces, and soft enough to gather.

The making-up method is important, too. When you cut and sew you have to have a seam allowance, which might result in lumps and bumps on a fine fabric like this. So I used the knitting-to-shape method, which loses only one stitch on the edge and leaves a very small, neat seam. You can hardly see the raglan joins on the shoulders. Although it's extra work the shaping is not very complicated and the result is well worth it.

Pink is the most obviously feminine colour, but the suit would look just as nice in blue trimmed with pale pink. And you could change the impact of the design completely, turn it into something really dramatic, by changing the colours to black and white.

Pink Suit Instructions

For Knitmaster 324, 326, 328 and 329 machines.

DETAILS Shown here with shoes by Anello & Davide and cactus by Walter Damnes.

MATERIALS **Yarn for top**: 2(2, 2, 2, 3, 3) 200 gm cones Dakota Silverknit in colour 1, small amount in colour 2. **Yarn for skirt:** 2 200 gm cones Dakota Silverknit in colour 1. **Trimming:** length of elastic to fit waist.

MEASUREMENTS Six sizes, **top** to fit 32–42 in (81–106 cm) bust; **skirt** to fit 34–44 in (86–112 cm) hip. **Top, completed length** at centre back, 25¾(26⅛, 26½, 27, 27⅜, 27¾) in (65.5(66.5, 67.5, 68.5, 69.5, 70.5) cm). **Skirt, completed length** including waistband, 31 in (79 cm).

TENSION 34 sts and 64 rows to 10 cm measured over pattern, counting Ns in A pos as sts. Tension Dial at approx 3·.

NEEDLE ARRANGEMENT

```
 1  1   1  1   =A pos
11 11 101 11 11 =B pos
0= centre of machine
```

NOTES Figures in brackets refer to the larger sizes respectively. Purl side of pattern is used as right side. When dec-ing and inc-ing, if an A pos N becomes an end N, bring into B pos until a further st is dec or inc.

TOP BACK

** **Flounce** (2 pieces alike): counting from centre of machine push 64(67:73:76:79:82) Ns at right and 67(70:73:76:82:85) Ns at left to B pos. Arrange Ns as given in Notes. Using col 2 cast on by hand over rem 88(92:98:102:108:112) Ns in B pos. Row Counter 000. Tension Dial at 3·. Insert card and lock on row 1. Knit 2 rows. Release card and set machine for tuck stitch. Knit 6 rows. Change to colour 1, knit 145 rows. Lock card. Using waste yarn, knit 8 rows stockinet and release from machine. Join the 2 pieces together. Counting from centre of machine, push 82(85:91:94:97:100) Ns at left to B pos. Arrange Ns as before. With purl side of flounce facing, replace sts from below waste yarn on to Ns in B pos as follows. **1st, 2nd, 3rd, 4th and 5th sizes:** 1 st on each of the first 6(5:3:2:1) Ns and 2 sts tog on each of the next 2(2:1:1:2) Ns. **6th size:** 2 sts tog on each of the first 2 Ns. **All sizes*** 1 st on next N and 2 sts tog on each of next 2 Ns*. Rept from * to * all along the row ending with 1(1:1:1:1:2) sts on each of the last 6(5:3:2:1:2) Ns. Unravel waste yarn. Release card and set machine for tuck stitch. Row Counter 000. Knit 1 row. Mark both edges with waste yarn. Knit 12 rows. Mark both edges with waste yarn. Knit 89 rows. Row Counter 102. Carr at right. **Shape Raglan Armholes:** cast off 6(7:7:7:9:10) sts (see Notes) at beg of next 2 rows **. **1st, 2nd, 4th, 5th and 6th sizes:** dec 1 st both ends of next and ev foll 3rd row 54(56, 3, 5, 7) times in all. **3rd size:** dec 1 st both ends of next row. **3rd, 4th, 5th and 6th sizes:** knit 1 row. * Dec 1 st both ends of next and ev foll 3rd row 9 times in all, knit 1 row *. Rept the last 26 rows from * to * 5 times more. Dec 1 st both ends of next and ev foll 3rd row (5:6:6:6) times in all. **All sizes:** Knit 2(2:3:0:0:0) rows. Row Counter 266(272:278:284:290:296). Cast off rem 44(44:45:45:46:46) sts.

FRONT

Foll instructions for Back from ** to **. **1st, 2nd, 4th, 5th and 6th sizes:** dec 1 st both ends of next and ev foll 3rd row 40(42:3:5:7) times in all, knit 2(2:1:1:1) rows. **3rd size:** dec 1 st both ends of next row, knit 1 row. **3rd, 4th, 5th and 6th sizes:** * dec 1 st both ends of next and ev foll 3rd row 9 times in all, knit 1 row *. Rept the last 26 rows from * to * 4 times more. **All sizes:** 72(72:73:75:76:76) sts rem. Row Counter 224(230:236:242:248:254). Carr at right. **Shape Neck:** using a length of col 1 cast off centre 28(28:29:29:30:30) sts. Using nylon cord, knit 22(22:22:23:23:23) sts at left by hand taking Ns back to A pos. Note row number showing on pattern panel. **Knit Right Part** as follows: dec 1 st beg of next row. Cast off 3(3:3:4:4:4) sts beg of next row, knit 1 row. Cast off 2 sts at beg and dec 1 st at end of next row, knit 2 rows. Dec 1 st both ends of next and ev foll 3rd row 7 times in all, knit 1 row. Fasten off rem st. Carr at right. Set card at number previously noted and lock. Take carr to left, release card. Unravel nylon cord bringing Ns to correct pos. Knit Left Part as for Right Part.

LEFT SLEEVE

Lock card on row 1. Counting from centre of machine, push 58 Ns at right and 58 Ns at left to B pos. Arrange Ns as before. Using colour 2 cast on over rem Ns in B pos. Row Counter 000. Tension Dial at 3·. Knit 2 rows. Release card and set machine for tuck stitch. Knit 6 rows. Change to colour 1, knit 44 rows. Mark both edges with waste yarn. Knit 12 rows. Mark both edges with waste yarn. Row Counter 064. Inc 1 st (see Notes) both ends of next and ev foll 10th(9th:8th:8th:7th:7th) row 23(25:27:29:31:32) times in all. 162(166:170:174:178:180) sts. Knit 13(17:25:9:23:16) rows without shaping. Row Counter 298 (knit 1 row extra for Right Sleeve only). Carr at right. **Shape Raglan Top:** cast off 6(7:7:7:9:10) sts beg of next 2 rows. Knit 0(0:0:1:0:0) row. **2nd size:** dec 1 st both ends of next row. **3rd and 6th sizes:** dec 1 st both ends of next and ev foll alt row (5, 3) times in all, knit

(2, 1) rows. **1st, 2nd, 3rd, 4th and 5th sizes:** * dec 1 st both ends of next and ev foll alt row 10(4:3:3:2) times in all, knit 2 rows. * Rept the last 21(9:7:7:5) rows from * to * 5(14:18:21:32) more times. **1st and 2nd sizes:** dec 1 st both ends of next and foll alt row. **3rd size:** dec 1 st both ends of next and ev foll alt row 3 times in all. **4th and 5th sizes:** dec 1 st both ends of next row, knit 1 row. **6th size:** * dec 1 st both ends of next and ev foll 3rd row 3 times in all, knit 1 row. * Rept the last 8 rows from * to * 20 more times. Dec 1 st both ends of next row, knit 2 rows. **All sizes:** 26 sts rem. Row Counter 429(439:449:457:467:477). Carr at left. Cast off 7 sts at beg and dec 0(0:0:1:1:1) st at end of next row. **1st, 2nd and 3rd sizes:** dec 1 st beg of next row. Cast off 4 sts beg of next row. Dec 1 st beg of next row. Cast off 3 sts beg of next row. Dec 1 st beg of next row. Cast off 2 sts beg of next row. Dec 1 st beg of next 6 rows. **4th, 5th and 6th sizes:** knit 1 row. Cast off 4 sts at beg and dec 1 st at end of next row. Knit 1 row. Cast off 3 sts at beg and dec 1 st at end of next row, knit 1 row. Cast off 2 sts at beg and dec 1 st at end of next row, knit 1 row. Dec 1 st both ends of next and foll alt row, knit 1 row. Dec 1 st beg of next row. **All sizes:** fasten off rem st.

RIGHT SLEEVE

Foll instructions for Left Sleeve but reverse the shapings by noting alteration in number of rows worked and then reading left for right and vice versa.

COLLAR (4 Pieces)

Front: lock card on row 1. Counting from the centre of machine, push 76 Ns at right and 73 Ns at left to B pos. * Arrange Ns as before. Using col 2, cast on over rem Ns in B pos. Row Counter 000. Tension Dial at 3·. Knit 2 rows. Release card and set machine for tuck stitch, knit 6 rows. Change to col 1, knit 30 rows. Mark both edges with waste yarn. Knit 12 rows. Mark both edges with waste yarn. Knit 8 rows. Cast off. * **Back:** counting from centre of machine, push 58 Ns at right and 58 Ns at left to B pos. Foll instructions for Front piece from * to *. Join the 2 pieces together. Knit another 2 pieces in the same way. Remove card.

CORDS

(Knit 8.) Set machine as follows: carr at right. Right side lever to front, left side lever to back. Right front lever to I, left front lever to II. Cam lever to S. Tension Dial at 0. Push 4 Ns to D pos. Using col 1 cast on by hand. Knit 1 row towards the left (Ns are in B pos). Move carr to the right (the yarn is in front of the Sinker Gate). Continue to knit, pulling lightly on the cord at the same time. Knit 2 cords each approx 59 in (150 cm) long (for waist), 2 cords each approx 43½ in (110 cm) long (for collar) and 4 cords each approx 27½ in (70 cm) long (for sleeves).

TO MAKE UP TOP

Pin out each piece and press carefully with a cool iron over a dry cloth. Join raglan, side and sleeve seams. With plain sides of collar pieces together, join by working 1 row of double crochet around cast-on edges using col 2. With seams at shoulder points, sew cast-off edges into pos gathering to fit. Thread cords through holes made by Ns in A pos in line with marked points. Give final light press. Press edges in col 2 into scallops.

SKIRT (4 Panels)

Back and Front Panels (both alike): insert card and lock on row 1. Push 64(67:73:76:82:85) Ns at right and 64(69:70:78:81:85) Ns at left of centre 0 to B pos. 128(136:143:154:163:170) Ns. Using col 1, cast on by hand. Tension Dial at 3, knit 10 rows. Turn up a hem. Row Counter 000. Tension Dial at 3·. Release card. Set machine for tuck stitch, knit 4(7:4:7:7:4) rows. Dec 1 st (see Notes) both ends of next and ev foll 69th(60th:69th:60th:60th:69th) row 8(9:8:9:9:8) times in all. 112(118:127:136:145:154) sts. Row Counter 488. Set machine for stockinet, knit 1 row. Bring A pos Ns to B pos. Pick up a loop from the adjacent st and place on to empty N. **3rd and 5th sizes:** dec 1 st at right edge, (126:144) sts. **All sizes:** knit 1 row. Using waste yarn, knit 8 rows and release from machine. **Side Panels (both alike):** work as given for Back and Front panels but finish waste yarn at waist as follows. Push 55(59:63:68:72:77) Ns at left to D pos. Using waste yarn, knit 8 rows over rem 55(59:63:68:72:77) Ns at right and release from machine. Push 55(59:63:68:72:77) Ns at left from D pos to C pos. Using waste yarn, knit 8 rows and release from machine.

WAISTBAND (2 Pieces)

Join all panels. Push 110(118:126:136:144:154) Ns to B pos. With purl side facing pick up sts from below waste yarn on back panel and half of each side panels and place on to Ns with 2 sts tog on each N. Unravel waste yarn. * Tension Dial at 3. Using col 1, knit 15 rows. Tension Dial at 5, knit 1 row. Tension Dial at 3, knit 17 rows. Using waste yarn, knit 8 rows and release from machine. * Push 110(118:126:136:144:154) Ns to B pos. With purl side facing pick up sts from below waste yarn on front panel and rem half of each side panels and place on to Ns with 2 sts tog on each N. Unravel waste yarn. Work as given for 1st piece from * to *.

TO MAKE UP SKIRT

Pin out to size and press carefully with a cool iron over a dry cloth. Join waistband seams leaving a small opening in which to insert elastic. Fold waistband in half on to the outside. Pin into pos and backstitch through the open loops of last row knitted in col 1. Unravel waste yarn. Insert elastic, secure ends. Close opening. Press hem into scallops. Give final light press.

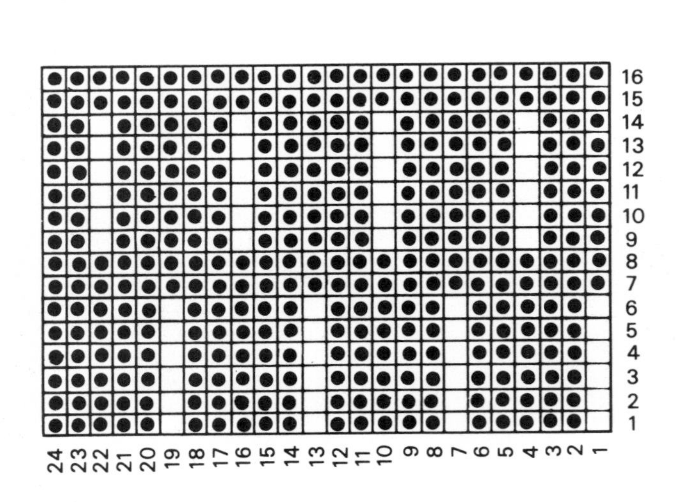

Fair Isle Quartet

(Fabric by Maggie Dyke, design by Susan Guess)

Susan: This design shows how much variety you can achieve with the same four yarns just by using different stitches and sides of the fabric. The waistcoat and jumper are done in the same combination of stripes, but the intermittent bands of tuck stitch on the waistcoat make the stripes look closer together. And I've used the purl side of the fabric for both pieces because the colours look much softer, and have more interesting textures.

I like sweaters to be long enough to keep you warm, and I like the way the sweater sets off the waistcoat by coming out at the collar, sleeves and hip. I've been careful to make the waistcoat long enough to wear on its own with a skirt or trousers, because if a waistcoat doesn't overlap the waistband, you look like you're 'coming apart at the seams'.

I chose a raglan sleeve for the sweater because it gives a lovely soft shape, and because it allows you to match up the stripes perfectly on the sleeve line. I used a raglan on the jacket, too, but I made the armholes even deeper, to give a fuller, cosier look.

The rib on the neck of the sweater is a *Continental rib*, which gives a double thickness of fabric, so the collar is soft and firm at the same time – it just *curves* around the neck. I used the same textured curve on the jacket as well, where the shawl collar and narrow double breast make the ribbing into a positive feature of the design. Very often people will knit or sew everything else, but they'll always buy their trousers ready-made. There's really no reason to be frightened of trousers, as long as you choose the right shape and fabric. These have a nice tailored line, easing gently in to the elastic at the waist. The yarn is pure wool, so it will stand up to a lot of wear, and I've used a Fair Isle stitch to give the fabric the extra body needed for trousers.

Fair Isle Quartet Instructions

For Knitmaster 324, 326, 328 and 329 machines.

MATERIALS **Yarn:** Jaeger Matchmaker 4-ply in cols 1, 2 and 3; Silverknit Saturn in col 4. **Jacket:** 11(11:12) 50 gm balls in col 1, 6(7:7) 50 gm balls in col 2. **Waistcoat:** 2 50 gm balls in col 1; 2 50 gm balls in col 2, 2 50 gm balls in col 3, 100 gm in col 4. **Jumper:** 3(3:4) 50 gm balls in col 1, 3 50 gm balls in col 2, 3 50 gm balls in col 3, 100 gm in col 4. **Trousers:** 12(12:13) 50 gm balls in col 1. **Trimming:** 10 buttons for jacket, 5 buttons for waistcoat, elastic to fit waist for trousers.

MEASUREMENTS Three sizes, to fit 34–38 (86–96 cm) bust and 36–40 (91–101 cm) hip. **Jacket:** length 30(30½:31) in (76(77.5:78.5) cm); sleeve seam 17(17½:18) in (43(44.5:45.5) cm). **Waistcoat:** length 21½(22:22½) in (54.5(56:57) cm). **Jumper:** length 24¾(25¼:25¾) in (63(64:65) cm), sleeve seam 18 in (46 cm). **Trousers:** length side seam 44(44½:45) in (113(113.5:114) cm), inside leg 33½ in (85 cm).

TENSIONS **Jacket:** 23 sts and 37 rows to 10 cm measured over weaving pattern. Tension Dial at approx 8·. **Waistcoat:** 24 sts and 48 rows to 10 cm measured over stripe pattern. Tension Dial at approx 8·. **Jumper:** 27 sts and 36 rows to 10 cm measured over stripe stockinet. Tension Dial at approx 8·. **Trousers:** 33 sts and 32 rows to 10 cm measured over pattern. Tension Dial at approx 6.

NOTES Figures in brackets refer to the larger sizes respectively. When using col 4 use 2 strands together. The purl side of jacket, waistcoat and jumper is used as right side. Plain side of trousers is used as right side. **Waistcoat Pattern:** Release card. * Set machine for tuck stitch. ** Knit 2 rows in each of the following cols: 1, 3, 4, 3, 1 and 2 **. Set machine for stockinet. Rept from ** to **. * The last 24 rows from * to * form one complete stripe pattern. **Jumper Pattern:** * Knit 2 rows in each of the following cols: 2, 1, 3, 4, 3 and 1 *. The last 12 rows from * to * form one complete col sequence.

JACKET

BACK

Insert punch card 7A into machine and lock on row 1. Push 131(137:143) Ns to B pos. * Arrange Ns for 2xl welt. Tension Dial at 6. Cast on using waste yarn and knit 7 rows (knit 1 row extra for Right Front only). Carr at left. Using col 1, knit 23 rows. Tension Dial at 8, knit 1 row. Tension Dial at 6, knit 23 rows. Turn up a hem. Carr at right *. Inc 2 sts. 133(139:145) sts. Row Counter 000. Tension Dial at 8. Release card. Set machine for weaving. With col 2 in feeder 1, weave with 2 ends of col 1 tog throughout. Knit 152 rows. Carr at right. **Shape Raglan Armholes:** cast off 7 sts at beg of next 2 rows, knit 2 rows. Dec 1 st at both ends of next and ev foll 4th row 6(6:5) times in all, knit 1 row. Dec 1 st at both ends of next and ev foll alt row 41(44:48) times in all, knit 1 row. Cast off rem 25 sts.

LEFT FRONT

Lock card on row 1. Push 59(62:65) Ns to B pos. Foll instructions for Back from * to *. Row Counter 000. Tension Dial at 8. Working in pattern as given for Back, knit 33 rows. Carr at left. **Pocket Opening:** using nylon cord, knit 21(22:23) sts at right by hand taking Ns back to A pos. Note row number showing on pattern panel. **Knit Left Part as follows:** knit 55 rows. Using nylon cord knit 38(40:42) sts by hand taking Ns down to A pos. With carr at right, set card at number previously noted and lock. Take carr to left. Release card. Unravel nylon cord over 21(22:23) Ns at right bringing Ns to correct pos. Reset Row Counter to 033. Knit 55 rows. Carr at right. Unravel nylon cord over 38(40:42) Ns at left bringing Ns to correct pos. Knit 59 rows. Row Counter at 147. Carr at left. **Shape Front Edge:** mark left edge with waste yarn. Dec 1 st at beg of next row, knit 4 rows. Row Counter 152. Carr at right. **Shape Raglan Armhole:** dec-ing 1 st at front edge of 3rd(4th:4th) and ev foll 7th(8th:8th) row 12 times in all, **at the same time** cast off 7 sts at beg of next row, knit 3 rows. Dec 1 st at beg of next and ev foll 4th row 6(6:5) times in all, knit 1 row. Dec 1 st at beg of next and ev foll alt row 31(34:38) times in all. Fasten off rem 2 sts. **Right Front:** follow instructions for Left Front but reverse the shapings by noting alteration in number of rows worked and reading left for right and vice versa.

LEFT SLEEVE

Lock card on row 1. Cast on using waste yarn 85(91:97) sts. Tension Dial at 8, knit 7 rows (knit 1 row extra for right sleeve only). Carr at left. Row Counter 000. Working in pattern as given for Back, knit 4 rows. Inc 1 st at both ends of next and ev foll 9th row 15 times in all. 115(121:127) sts. Knit 6(12:16) rows. Row Counter 137(143:147). Carr at right. **Shape Raglan Top:** cast off 7 sts at beg of next 2 rows, knit 2 rows. Dec 1 st at both ends of next and ev foll 4th row 6(6:5) times in all, knit 1 row. Dec 1 st at both ends of next and ev foll alt row 31(34:38) times in all. 27 sts rem. Carr at left. * Cast off 4 sts at beg of next row. Dec 1 st at beg of next row *. Rept from * to * twice more. ** Dec 1 st at beg of next 2 rows, knit 1 row. Dec 1 st at beg of next and foll alt row **. Rept from ** to ** once more. Dec 1 st at beg of next 2 rows, knit one row. Fasten off rem 2 sts. **Right Sleeve:** follow instructions for Left Sleeve but reverse the shapings by noting alteration in number of rows worked and reading left for right.

CUFFS (Both Alike)

Push 77(80:83) Ns to B pos. Arrange Ns for 2xl welt. 52(54:56) Ns. With purl side facing, pick up 85(91:97) sts from below waste yarn and place on to Ns as follows: 1 st on to each of first 10(9:8) Ns, 2 st tog on to each of next 33(37:41) Ns and 1 st on to each of last 9(8:7) Ns. Unravel waste yarn. Tension Dial at 6. Using col 1, knit 23 rows. Tension Dial at 8, knit 1 row. Tension Dial at 6, knit 23 rows. Push empty Ns to B pos. Knit 1 row. Using waste yarn, knit 8 rows and release from machine.

POCKET BANDS (Both Alike)

Push 56 Ns to B pos. Arrange Ns for 2xl. 38 Ns. With purl side facing, pick up 38 sts from front edge of pocket opening and place on to Ns. Tension Dial at 6. Using col 1, knit 7 rows. Tension Dial at 8, knit 1 row. Tension Dial at 6, knit 7 rows. Push empty Ns to B pos. Knit 1 row. Using waste yarn, knit 8 rows and release from machine.

POCKET LININGS

With carr at right and using col 1, cast on by hand 80 sts. Counting from right edge mark between sts 30 and 31. Tension Dial at 5, knit 2 rows. Always taking yarn round first inside N in D pos. Push 1 N at opposite end to carr into D pos, knit 4 rows. Push 1 N at opposite end to carr into D pos, knit 4 rows. Push 1 N at opposite end to carr into D pos, knit 4 rows. Push 1 N at opposite end to carr into D pos on next and ev foll alt row 8 times in all, knit 1 row. Push 2 Ns at opposite end to carr into D pos on next and ev foll alt row 8 times in all, knit 1 row. Push 3 Ns at opposite end to carr into D pos on next and ev foll alt row 3 times in all, knit 1 row. Push 4 Ns at opposite end to carr into D pos, knit 2 rows. Push 4 inside Ns at opposite end to carr from D pos down into C pos, knit 2 rows. Push 3 inside Ns at opposite end to carr from D pos down into C pos on next and ev foll alt row 3 times in all, knit 1 row. Push 2 inside Ns at opposite end to carr from D pos down into C pos on next and ev foll alt row 8 times in all, knit 1 row. Push 1 inside N at opposite end to carr from D pos down into C pos on next and ev foll alt row 7 times in all, knit 1 row. Push 1 inside N at opposite end to carr from D pos down into C pos, knit 4 rows. Push 1 inside N at opposite end to carr from D pos down into C pos, knit 4 rows. Push 1 inside N at opposite end to carr from D pos down into C pos, knit 4 rows. Push 1 N at opposite end to carr from D pos down into C pos, knit 2 rows. Counting from right edge, mark between sts 30 and 31. Cast off. Knit second pocket lining as for first pocket lining but reverse the shapings.

RIGHT COLLAR

Join raglan seams. Push 162(168:171) Ns to B pos. Push the 2nd and ev foll 3rd N back to A pos. 108(112:114) Ns. With carr at right and purl side facing, pick up 108(112:114) sts from marked point on right front to centre back of neck and place on to Ns. Inc 1 st at both edges. 110(114:116) sts. Tension Dial at 6. Using col 1, knit 1 row. Always taking yarn round first inside N in D pos, push 84(88:90) Ns at opposite end to carr from B pos to D pos, knit 2 rows. Push 9 inside Ns at opposite end to carr from D pos down into C pos on next and ev foll alt row 8 times in all, knit 1 row. Push rem 12(16:18) Ns at opposite end to carr from D pos down into C pos. Knit 37 rows. Tension Dial at 8, knit 1 row. Tension Dial at 6, knit 38 rows. Always taking yarn round first inside N in D pos, push 12(16:18) Ns at opposite end to carr from B pos to D pos, knit 2 rows. Push 9 Ns at opposite end to carr from B pos to D pos on next and ev foll alt row 8 times in all, knit 1 row. Push 84(88:90) Ns at opposite end to carr from D pos down into C pos. Push empty Ns to B pos, knit 1 row. Using waste yarn knit 8 rows and release from machine. **Left Collar:** follow instructions for Right Collar but reverse the shapings by reading left for right.

BUTTONHOLE BORDER

Push 131 Ns to B pos. Push the 3rd and ev foll 3rd N back to A pos. 88 Ns. With purl side facing, pick up 86 sts from marked point on right front to lower edge and place on to Ns. Inc 1 st at both edges. 88 sts. Tension Dial at 6. Using col 1, knit 8 rows. Counting from lower edge, make buttonholes over Ns 6 and 7 and ev foll 19th and 20th. Knit 23 rows. Make buttonholes over same Ns as before. Knit 7 rows. Tension Dial at 8, knit 1 row. Tension Dial at 6, knit 7 rows. Make buttonholes over same Ns as before. Knit 23 rows. Make buttonholes over same Ns as before. Knit 8 rows. Push empty Ns to B pos, knit 1 row. Using waste yarn knit 8 rows and release from machine. **Button Border:** follow instructions for Buttonhole Border but read left for right and omit buttonholes.

TO MAKE UP

Pin out each piece to size and press carefully with a warm iron over a damp cloth. Join side, sleeve and cuff seams. Join centre back seam of collar. Join ends of collar to front borders. Fold cuffs, pocket bands, collar and front borders in half on to outside. Pin into pos and backstitch through open loops of last row knitted in col 1. Unravel waste yarn. Sew pocket linings into pos. Catch down ends of pocket bands. Neaten ends of front borders. Finish buttonholes and sew on buttons to correspond.

WAISTCOAT

LEFT FRONT

Insert card 3A and lock on row 1. With carr at left and using col 2, cast on by hand 60(63:66) sts. Row Counter 000. Tension Dial at 8·, knit 2 rows. Working in patt as given in Notes and always taking yarn round first inside N in D pos, push 57(60:63) Ns at opposite end to carr into D pos, knit 2 rows. Using a transfer tool, push 3 inside Ns at opposite end to carr from D pos down into B pos on next and ev foll alt row 17 times in all, knit 1 row. Using a transfer tool, push rem 6(9:12) Ns at opposite end to carr from D pos down into B pos. Knit 32 rows. **Pocket Opening:** counting from left edge and using waste yarn, make pocket opening using the same method as for buttonholes over Ns 14–44(15–45,16–46) inclusive. Knit 89 rows. Row Counter 159. Carr at right. **Shape Armhole:** cast off 5 sts at beg of next row, knit 1 row. Cast off 2 sts at beg of next row, knit 1 row. Dec 1 st at beg of next and ev foll alt row 3 times in all, knit 1 row. Dec 1 st at beg of next and foll 4th row. 48(51:54) sts rem. Row Counter 174. Carr at left. **Shape Front Edge:** dec-ing 1 st at left edge of next and ev foll 4th row **at the same time** dec 1 st at armhole edge of 4th and ev foll 4th row 3 times in all, knit 5 rows. Dec 1st at arm-

hole edge of next and foll 6th row. 37(40:43) sts rem. Keeping armhole edge straight, dec 1 st at left edge of next and ev foll 4th row 18 times in all. 19(22:25) sts rem. Knit 7(13:19) rows without shaping. Row Counter 274(280:286). Carr at left. **Shape Shoulder:** always taking yarn round first inside N in D pos, push 3(3:4) Ns at opposite end to carr into D pos, knit 2 rows. Push 3 Ns at opposite end to carr into D pos on next and foll alt row, knit 1 row. Push 2(3,3) Ns at opposite end to carr into D pos on next and ev foll alt row 3 times in all, knit 1 row. Push 2(2,3) Ns at opposite end to carr into D pos, knit 2 rows. Using a transfer tool, push 17(20:22) Ns at opposite end to carr from D pos down into B pos, knit 1 row. Using waste yarn knit 8 rows in stockinet and release from machine. **Right Front:** follow instructions for Left Front but reverse the shapings by reading left for right and vice versa.

BACK

Lock card on row 37. Push 119(125:131) Ns to B pos. Arrange Ns for 2xl welt. Tension Dial at 6. Cast on using waste yarn and knit 8 rows. Carr at right. Using col 2, knit 7 rows. Tension Dial at 8, knit 1 row. Tension Dial at 6, knit 7 rows. Push empty Ns to B pos. Turn up a hem. Carr at left. Inc 1 st. 120(126:132) sts. Row Counter 000. Tension Dial at 8·, knit 2 rows. Working in pattern as given in Notes (starting with row 13), knit 121 rows. Row Counter 123. Carr at right. **Shape Armholes:** Cast off 5 sts at beg of next 2 rows and 2 sts at beg of next 2 rows. Dec 1 st at both ends of next and ev foll alt row 3 times in all, knit 1 row. Dec 1 st at both ends of next and ev foll 4th row 5 times in all, knit 5 rows. Dec 1 st at both ends of next and foll 6th row. 86(92:98) sts rem. Knit 76(82:88) rows. Row Counter 238(244:250). Carr at left. **Shape Shoulders:** always taking yarn round first inside N in D pos, push 3(3:4) Ns at opposite end to carr into D pos on next 2 rows. Push 3 Ns at opposite end to carr into D pos on next 2 rows. **Shape Neck:** using an odd length of col 2, cast off centre 14 sts. Using nylon cord knit 30(33:35) sts at right by hand taking Ns down into A pos. Note row number showing on pattern panel. **Knit Left Part as follows:** knit 1 row. * Push 3 Ns at opposite end to carr into D pos. Cast off 4 sts at beg of next row, knit 1 row. Push 2(3:3) Ns at opposite end to carr into D pos. Cast off 4 sts at beg of next row, knit 1 row. Push 2(3:3) Ns at opposite end to carr into D pos. Cast off 3 sts at beg of next row, knit 1 row. Push 2(3:3) Ns at opposite end to carr into D pos. Cast off 3 sts at beg of next row, knit 1 row. Push 2(2:3) Ns at opposite end to carr into D pos. Cast off 3 sts at beg of next row, knit 1 row. Using a transfer tool, push 17(20:22) Ns at opposite end to carr from D pos down into B pos, knit 1 row. Using waste yarn knit 8 rows stockinet and release from machine *. With carr at right, set card at number previously noted and lock. Take carr to left. Release card. Unravel nylon cord bringing Ns to correct pos. Knit right part as for left part from * to *, reversing shaping.

POCKET LININGS (Both Alike)

Place both sets of sts from pocket opening on to 2 st holders. Push 31 Ns to B pos. With purl side facing pick up 31 sts from top edge (back) of pocket opening and place on to Ns. Tension Dial at 6, using col 2 knit 33 rows. Tension Dial at 8, knit 1 row. Tension Dial at 6, knit 33 rows. Using waste yarn knit 8 rows and release from machine.

POCKET TOPS (Both Alike)

Push 44 Ns to B pos. Arrange Ns for 2xl welt. 30 Ns. Fold pocket lining in half. With plain side of pocket lining facing, place 2 sts tog on to 1 N and 1 st on to each of rem 29 Ns. Unravel waste yarn. With purl side facing, pick up 31 sts from lower edge (front) of pocket opening and place on to Ns as for pocket lining. Tension Dial at 6. Using col 2, knit 7 rows. Tension Dial at 8, knit 1 row. Tension Dial at 6, knit 7 rows. Push empty Ns to B pos. Pick up loops from row below and place on to Ns. Knit 1 row. Using waste yarn knit 8 rows and release from machine.

LOWER FRONT BANDS (Both Alike)

Push 65(68:71) Ns to B pos. Arrange Ns for 2xl welt. 44(46:48) Ns. With purl side facing, pick up 44(46:48) sts along lower edge of front and place on to Ns. Tension Dial at 6. Using col 2, knit 1 row. Inc 1 st at front edge on next and ev foll alt row 3 times in all, knit 1 row. Tension Dial at 8, knit 1 row. Tension Dial at 6, knit 1 row. Dec 1 st at front edge on next and ev foll alt row 3 times in all, knit 1 row. Push empty Ns to B pos. Pick up loops from row below and place on to Ns. Knit 1 row. Using waste yarn, knit 8 rows and release from machine.

BACK NECKBAND

Push 54 Ns to B pos. With purl side facing, pick up 54 sts around back of neck and place on to Ns. Inc 1 st at both edges. 56 sts. * Tension Dial at 8. Using col 2, knit 1 row. Transfer sts for 2xl welt. Push empty Ns down into A pos. Tension Dial at 6, knit 7 rows. Tension Dial at 8, knit 1 row. Tension Dial at 6, knit 7 rows. Push empty Ns to B pos. Pick up loops from row below and place on to Ns. Knit 1 row. Using waste yarn, knit 8 rows and release from machine *.

BUTTONHOLE BAND

Fold lower band in half on to outside. Pin into pos and backstitch through open loops of last row knitted in col 2. Unravel waste yarn. Push 186(189:192) Ns to B pos. With purl side facing, pick up 186(189:192) sts evenly along right front and place on to Ns. Inc 1 st at both edges. 188(191:194) sts. Tension Dial at 8. Using col 2, knit 1 row. Transfer sts for 2xl welt. Push empty Ns down into A pos. Tension Dial at 6, knit 3 rows. Counting from lower edge, make buttonholes over Ns 9–10; 33–34; 57–58; 81–82; 105 and 106. Knit 4 rows. Tension Dial at 8, knit 1 row. Tension Dial at 6, knit 4 rows. Make buttonholes over same Ns as before. Knit 3 rows. Push empty Ns to B pos. Pick up loops from row below and place on to Ns, knit 1 row. Using waste yarn, knit 8 rows and release from machine. **Button Band:** follow instructions for Buttonhole Band but read left for right and omit buttonholes.

ARMHOLE BANDS (Both Alike)

Graft shoulders. Unravel waste yarn. Push 168(177:183) Ns to B pos. With purl side facing, pick up 168(177:183) sts evenly around armhole edge and place on to Ns. Inc 1 st at both edges. 170(179:185) sts. Follow instructions for Back Neckband from * to *.

TO MAKE UP

Pin out each piece to size and press carefully with a cool iron over a damp cloth. Join side seams, press. Join ends of armhole bands and neckband to front bands. Fold front bands, neckband, armhole bands and pocket tops in half on to the outside. Pin into pos and backstitch through open loops of last row knitted in col 2. Unravel waste yarn. Neaten open ends of front bands. Join sides of pocket linings. Sew sides of pocket tops into pos. Finish buttonholes and sew on buttons to correspond. Give final pressing.

JUMPER

BACK

Push 128(134:143) Ns to B pos. Arrange Ns for 2xl welt. Using waste yarn, cast on and knit several rows ending with carr at left. Row Counter 000. Tension Dial at 6. Using col 2, knit 21 rows. Tension Dial at 8·, knit 1 row. Tension Dial at 6, knit 21 rows. Turn up a hem. **1st and 2nd sizes only:** inc 1 st at each end. **All sizes:** 130(136:143) sts. Row Counter 000. Tension Dial at 8·. Working in stripes as given in notes, knit 140 rows without shaping. **Shape Raglan Armholes:** cast off 7(7:8) sts beg of next 2 rows**. Dec 1 st both ends of next and ev foll alt row 5(7:2) times in all.* Dec 1 st both ends of next and ev foll alt row 6(6:7) times in all*. Rept the last 11(11:13) rows from * to * 6 more times. Cast off rem 22(24:25) sts.

FRONT

Follow instructions for Back to **. Dec 1 st both ends of next and ev foll alt row 5(7:2) times in all. * Dec 1 st both ends of next and ev foll alt row 6(6:7) times in all *. Rept the last 11(11:13) rows from * to * 2(2:3) more times. Dec 1 st both ends of next and ev foll alt row 5(5:3) times in all, knit 1(1:0) row. 60(62:61) sts rem. Row Counter 194(198:202). Carr at right. **Shape Neck:** using a length of appropriate col, cast off centre 8(8:9) sts. Using nylon cord, knit 26(27:26) sts at left by hand taking Ns back to A pos. **Knit Right Part as follows:** Dec 1(1:0) st beg of next row. Carr at left. Cast off 3 sts beg and dec 1 st at end of next row, knit 1 row. Cast off 2 sts at beg and dec 1 st at end of next row, knit 1 row. Rept the last 2 rows 1(2:2) more time. **1st and 2nd sizes only:** dec 1 st both ends of next and ev foll alt row 3(2) times in all. **All sizes:** dec 1 st beg of next 8(8:12) rows. Fasten off rem st. Take carr to left. Unravel nylon cord bringing Ns to B pos. Keeping stripes correct knit left part as for right part but read right for left.

RIGHT SLEEVE

Push 75(78:81) Ns to B pos. Using waste yarn, cast on and knit several rows ending with carr at right. Row Counter 000. Tension Dial at 8·. Using col 3, knit 2 rows. Using col 1, knit 2 rows. Continuing in stripes as given in notes, inc 1 st both ends of next and ev foll 6th row 22 times in all. 119(122:125) sts. Knit 13 rows without shaping. Row Counter 144. Carr at right. **Shape Raglan Top:** cast off 7(7:8) sts beg of next 2 rows. * Dec 1 st both ends of next and ev foll alt row 11(14:11) times in all, knit 0(0:2) rows *. Rept the last 21(27:23) rows from * to * 2(1:2) times more. Dec 1 st both ends of next and ev foll alt row 5(11:6) times in all. 29(30:31) sts rem. Knit 0(1:0) row. Row Counter 218(222:226). Carr at right. **1st and 3rd sizes only:** cast off 5 sts beg of next row. Dec 1 st beg of next row and cast off 3 sts beg of next row. Rept the last 2 rows 2(3) more times. **1st size only:** dec 1 st beg of next row and cast off 2 sts beg of next row. Rept the last 2 rows once more. Dec 1 st beg of next row. Cast off 2 sts at beg and dec 1 st at end of next row. **2nd size only:** cast off 5 sts at beg and dec 1 st at end of next row, knit 1 row. Cast off 3 sts at beg and dec 1 st at end of next and foll alt row. Dec 1 st beg of next row. Cast off 3 sts beg of next row. Dec 1 st beg of next row. Cast off 2 sts beg of next row and dec 1 st beg of next row. Rept the last 2 rows twice more. **3rd size only:** dec 1 st beg of next row. Cast off 3 sts beg of next row, knit 1 row. Cast off 3 sts at beg and dec 1 st at end of next row. **All sizes:** knit 1(0:1) row. Cast off rem 2 sts. **Left Sleeve:** follow instructions for Right Sleeve but reverse the shapings by reading left for right and vice versa throughout.

SLEEVE WELTS (Both Alike)

Push 47(50:53) Ns to B pos. With purl side of sleeve facing, pick up sts from below waste yarn and place on to Ns as follows. **1st and 2nd sizes only:** 2(1) sts on first N. * 2 sts tog on next N, 1 st on next N, 2 sts tog on each of the next 2 Ns, 1 st on next N, 2 sts tog on next N and 1 st on next N *. Rept from * to * all along the row. **3rd size only:** 2 sts tog on first N. * 2 sts tog on next N and 1 st on next N *. Rept from * to * all along the row ending with 2 sts tog on each of last 2 Ns. **All sizes:** unravel waste yarn. Tension Dial at 6. Using col 2, knit 1 row. Transfer the 3rd and ev foll 3rd st on to its adjacent N for 2xl. Knit 20 rows. Tension Dial at 8, knit 1 row. Tension Dial at 6, knit 20 rows. Push empty Ns to B pos. Pick up a loop from the row below and place on to empty N. Knit 1 row. Using waste yarn, knit 8 rows and release from machine.

POLO COLLAR

Join raglan seams leaving left back raglan open. Push 143(146:152) Ns to B pos. With purl side facing, pick up 143(146:152) sts evenly around neck edge and place on to Ns. Tension Dial at 6. Using col 2, knit 1 row. Transfer sts for 2xl, knit 20 rows. Tension Dial at 7, knit 21 rows. Tension Dial at 8, knit 21 rows. Tension Dial at 10, knit 1 row. Tension Dial at 8, knit 21 rows. Tension Dial at 7, knit 21 rows. Tension Dial at 6, knit 21 rows. Push empty Ns to B pos. Pick up a loop from the row below and place on to empty N. Knit 1 row. Using waste yarn knit 8 rows and release from machine.

TO MAKE UP

Pin out each piece to size and press carefully using a warm iron over a damp cloth. Join left back raglan, collar, side and sleeve seams. Fold collar and sleeve welts in half on to the outside. Pin into pos and backstitch through open loops of last row knitted in col 2. Unravel waste yarn. Give final light press.

TROUSERS

RIGHT FRONT

Cast on using waste yarn 65(69:73) sts. * Tension Dial at 5, knit 8 rows. Knit 1 row extra for left front and right back only. Carr at right. Using col 1, knit 13 rows. Tension Dial at 7, knit 1 row. Tension Dial at 5, knit 13 rows. Turn up a hem. Carr at left. Insert card 1A and lock on row 1. Row Counter 000. Tension Dial at 6, knit 1 row. Release card. With col 1 in feeders 1 and 2, work in one col Fair Isle throughout *. Knit 1 row. Inc 1 st at beg of next and ev foll alt row 4 times in all, knit 3 rows. Inc 1 st at beg of next and ev foll 4th row 5 times in all, knit 7 rows. Inc 1 st at beg of next and foll 14th row. 76(80:84) sts. Knit 8(6:4) rows without shaping. Carr at right. Inc 1 st at beg of next and ev foll 4th row 3 times in all, knit 3 rows. Inc 1 st at beg of next and ev foll alt row 3(5:7) times in all, knit 1 row. Cast on 2 sts at beg of next and foll alt row, knit 1 row. Cast on 4 sts at beg of next row. 90(96:102) sts. Row Counter 82(84:86). Carr at left. **Shape Leg:** knit 9(7:9) rows without shaping. Dec-ing 1 st at left edge of next and ev foll 36th(32nd:28th) row 8(9:10) times in all **at the same time** dec 1 st at right edge of next and ev foll 12th(10th:18th) row 3(5:7) times in all, knit 11(9:7) rows. Dec 1 st at right edge of next and ev foll 18th(16th:14th) row 13(14:15) times in all, knit 9(5:9) rows. 66(68:70) sts rem. Row Counter 353(355:357). Change to stockinet. Tension Dial at 5, knit 13 rows. Cast off. **Left Front:** follow instructions for Right Front but reverse the shapings by noting alteration in number of rows worked and reading left for right and vice versa.

LEFT BACK

Cast on using waste yarn 63(67:71) sts. Follow instructions for Right Front from * to *. **Shape Back:** always taking yarn round first inside N in D pos, push 54 Ns at opposite end to carr into D pos, knit 2 rows. Using a transfer tool, push 11 inside Ns at opposite end to carr from D pos down into B pos on next and ev foll alt row 4 times in all, knit 1 row. Using a transfer tool push rem 10 Ns at opposite end to carr from D pos down into B pos. Inc 1 st at beg of next row. Inc 1 st at beg of next and ev foll alt row 3 times in all, knit 2 rows. Inc 1 st at beg of next row, knit 2 rows. ** Inc 1 st at beg of next row, knit 4 rows. Inc 1 st at beg of next row, knit 2 rows **. Rept from ** to ** once more. Inc 1 st at beg of next row, knit 4 rows. Inc 1 st at beg of next and foll 8th row, knit 5 rows. Inc 1 st at beg of next and foll 6th row, knit 1 row. Inc 1 st at beg of next and ev foll alt row 6 times in all, knit 1 row. Cast on 2 sts at beg of next and ev foll alt row 5(6:7) times in all, knit 1 row. Cast on 4 sts at beg of next row, knit 1 row. Cast on 6 sts at beg of next and foll alt row. 109(115:121) sts. Row Counter 92(94:96). Carr at left. **Shape Leg:** knit 3 rows without shaping. Dec-ing 1 st at left edge of next and ev foll 12th row 7 times in all, **at the same time** dec 1 st at right edge of next and ev foll alt row 7(10:13) times in all, knit 3 rows. Dec 1 st at right edge of next and ev foll 4th row 3(2:3) times in all, knit 5 rows. Dec 1 st at right edge of next and ev foll 6th row 4(4:2) times in all, knit 7 rows. Dec 1 st at right edge of next and ev foll 8th row 3 times in all. 85(89:93) sts. Dec-ing 1 st at left edge of 12th(12th:11th) and ev foll 12th(12th:11th) row 15(16:17) times in all, **at the same time** dec 1 st at right edge of 8th and ev foll 60th(45th:36th) row 4(5:6) times in all, knit 7(5:7) rows. 66(68:70) sts rem. Row Counter 363(365:367). Change to stockinet. Tension Dial at 5, knit 13 rows. Cast off. **Right Back:** follow instructions for Left Back but reverse the shapings by noting alteration in number of rows worked and reading left for right and vice versa.

TO MAKE UP

Pin out each piece to size and press carefully with a warm iron over a damp cloth. Join outside leg seams, press. Join inside leg seams, press. Join back and front seams, press. Insert elastic into waistband and join ends. Turn up hems at lower edge and catch down on the inside. Give final pressing.

Bamboo and Fan Kimonos

(Fabric by Maggie Dyke, design by Laraine McCarthy)

Maggie: The bamboo fabric is done on an electronic machine, so you get a very large pattern that should only be used on a large garment like a poncho or kimono. The pattern is based on a seventeenth-century Japanese kimono and I used yarns with similar tones since too sharp a colour contrast would destroy the nice balance between the size of the overall pattern and the delicate shapes of the clouds and bamboo.

For the fan design I wanted an exciting contrast – striking colours and shapes against the black. An overall pattern would have been too busy and complicated, so we used a traditional Japanese technique and created an asymmetric pattern where the fans sweep up and over one shoulder. By isolating the colours and arranging them into a strong shape of their own you add a lot of drama to the design. We spent a lot of time working out the proportion of the coloured areas to the black – *if you want the pattern to be exciting, it mustn't be too balanced or regular*.

Laraine: The kimonos are based on the classic Japanese shape but I've slimmed them down through the body and added linings that link up to the colours in the fabrics. All the lines are straight except for the sleeves, because shaping would have chopped into the fabric patterns. You can see how different fabrics and colours change the character of the design. The bamboo fabric emphasizes the straight lines and simplicity of the kimono, while the asymmetric fan pattern emphasizes the waist and gives a strong impression of curves.

You can't always anticipate what knitted garments of this size are going to do, so you should hang the kimonos on a dummy or padded hanger for a week *before* you put up the hem, so you can see how much it's going to drop.

Fan Kimono Instructions

For Knitmaster Electronic machines.

DETAILS Shown with haircombs by Tailpieces and suede evening mules by Charles Jourdan.

MEASUREMENTS One size to fit 34–42 in (86–106 cm) bust. Length from top of shoulder 54 in (138 cm).

MATERIALS Sirdar Superwash and Wash 'n' Wear 4-ply. 51 25 gm balls Superwash in col 1; 3 25 gm balls Superwash in col 2; 3 25 gm balls Superwash in col 3; 3 20 gm balls Wash'n'Wear in col 4; 3 20 gm balls Wash'n'Wear in col 5; 3 25 gm balls Superwash in col 6; 2 press fasteners; lining if desired.

TENSION 30 sts and 39 rows to 10 cm measured over stockinet. Tension Dial at approx 7.

NOTE Colour Sequence. With col 1 in feeder 1 throughout, change col in feeder 2 as follows: 30 rows col 2; ** 30 rows col 3; *** 30 rows col 4; 30 rows col 5; 30 rows col 6. These 150 rows form 1 complete col sequence which should be repeated throughout.

BACK

Inspection button on. Insert card and set to row 1. Pattern width indicator at 60 throughout. Push 91 Ns at left and right of centre 0 to B pos (182 Ns altog). Using col 1, cast on by hand. Point cams to 91 and 75 both at left of centre 0. Needle 1 cam between 75 and 76 at left of centre 0. Buttons 1 (left) and 2 (left). Tension Dial at 7, knit 11 rows. Carr at left. Row Counter 000. Knit 120 rows. Inspection button off, insert yarn separators. Set carr for Fair Isle. Working in col sequence as given in note and moving right point cam 15 Ns to right after ev 30 rows, knit 240 rows. Row Counter 360. Keeping col sequence correct, move right point cam 15 Ns to left, knit 30 rows. Move right point cam 15 Ns to right, knit 30 rows. Row Counter 420. Mark both edges with waste yarn. Move right point cam 15 Ns to left. Keeping col sequence correct and moving right point cam 15 Ns to the left after ev 30 rows, knit 120 rows. Row Counter 540. Mark the 24th st at left and right of centre 0. Cast off.

LEFT FRONT

Inspection button on, set card to row 1. Push 91 Ns at left and 45 Ns at right of centre 0 to B pos (136 Ns altog). Using col 1, cast on by hand. Point cams to 45 at right and 75 at left of centre 0. Needle 1 cam between 45 and 46 at right of centre 0. Tension Dial at 7, knit 12 rows. Carr at right. Row Counter 000. Inspection button off, insert yarn separators. Set carr for Fair Isle. Working in col sequence as given in note (beg at **) knit 30 rows. Move left point cam 15 Ns to left, knit 30 rows. Move left point cam 15 Ns to right, knit 30 rows. Move left point cam 15 Ns to left. Keeping col sequence correct and moving left point cam 15 Ns to left after ev 30 rows, knit 270 rows. Remove card and yarn separators. Set carr for stockinet. Using col 1 knit 15 rows. Row Counter 375. Carr at left. **Shape Front Edge:** Dec 1 st at right edge of next and ev foll alt row 7 times in all, knit 2 rows. Rept the last 15 rows 9 more times, marking left edge when Row Counter shows 420. 66 sts rem. Knit 15 rows without shaping. Row Counter 540. Cast off.

RIGHT FRONT

Inspection button on. Insert card and set to row 1. Push 45 Ns at left and 91 Ns at right of centre 0 to B pos (136 Ns altog). Using col 1, cast on by hand. Point cams to 45 at left and 45 at right of centre 0. Needle 1 cam between 45 and 46 at right of centre 0. Tension Dial at 7, knit 11 rows. Carr at left. Row Counter 000. Inspection button off, insert yarn separators. Set carr for Fair Isle. Working in col sequence as given in note (beg at **) and moving right point cam 15 Ns to right after ev 30 rows, knit 120 rows. Keeping col sequence correct and right point cam remaining on 90, knit 89 rows. Row Counter 209. Carr at right. Move left point cam 15 Ns to right. Keeping col sequence correct and moving left point cam 15 Ns to right after ev 30 rows, knit 150 rows. Move left point cam 15 Ns to right, knit 15 rows. Row Counter 374. Carr at left. **Shape Front Edge:** * Knit 1 row. Dec 1 st at left edge on next and ev foll alt row 7 times in all, knit 1 row *. Repeating the last 15 rows from * to * continue in pattern as follows, move left point cam 15 Ns to the left. Knit 30 rows. Move left point cam 15 Ns to left, knit 1 row. Row Counter 420. Mark right edge with waste yarn, knit 29 rows. Move left point cam 15 Ns to left, knit 30 rows. Move left point cam 15 Ns to right, knit 30 rows. Move left point cam 15 Ns to right, knit 15 rows. 66 sts rem. Knit 16 rows without shaping. Row Counter 540. Cast off.

RIGHT SLEEVE (2 Pieces)

First Half (back)

Push 75 Ns at left of centre 0 to B pos. Inspection button on, set card to row 90. Point cams to 0 and 75 at left of centre 0. Needle 1 cam between 45 and 46 at left of centre 0. * Row Counter 000. Tension Dial at 7. Using col 1, cast on by hand and knit 2 rows (knit 1 row extra for second half). Carr at right. Inspection button off, insert yarn separators. Set carr for Fair Isle.

Working in col sequence as given in notes (beg at ***) and always taking yarn round the first inside N in D pos, push 35 Ns at opposite end to carr to D pos on next row, knit 1 row. Push 2 inside Ns at opposite end to carr from D pos back to C pos on next and ev foll alt row 9 times in all, knit 1 row. Push 1 inside N at opposite end to carr from D pos back to C pos on next and ev foll alt row 17 times in all (all Ns are back in B pos). Row Counter 055. Carr at left. Knit 37 rows. Mark both edges with waste yarn. Knit 120 rows. Row Counter 212. Cast off.

Second Half (front)

Push 75 Ns at right of centre 0 to B pos. Inspection button on, set card to row 90. Point cams to 0 and 75 at right of centre 0. Needle 1 cam between 75 and 76 at right of centre 0. Follow instructions for First Half from * to end but reverse the shapings by noting alteration in number of rows worked and reading left for right and vice versa. Remove card.

LEFT SLEEVE (2 Pieces)

First Half (front)

Follow instructions for First Half on Right Sleeve but work in stockinet throughout.

Second Half (back)

Follow instructions for First Half but reverse the shaping by noting alteration in number of rows worked and reading left for right and vice versa.

NECKBAND

Using col 1, cast on 22 sts by hand. Set carr for stockinet. Row Counter 000. Tension Dial at 6, knit 440 rows. Cast off.

BELT

Using col 1, cast on 46 sts by hand. Row Counter 000. Tension Dial at 7. Knit 1 row. Always taking yarn round the first inside N in D pos, push 22 Ns at opposite end to carr to D pos on next 2 rows (2 Ns rem in B pos). Push 2 inside Ns at opposite end to carr from D pos back to C pos on next 22 rows (all Ns are back in B pos). Row Counter 025. Knit 459 rows without shaping. Row Counter 484. Carr at right. Always taking yarn round the first inside N in D pos, push 23 Ns at left to D pos on next row, knit 1 row. * Push 2 inside Ns at opposite end to carr to D pos on the next and ev foll alt row 11 times in all, knit 1 row. Push rem N to D pos *. Break off yarn. Take carr to left. Push 22 Ns at left from D pos back to C pos. Knit the left half as for right half from * to *. Carr at left. Push all Ns from D pos back to C pos, knit 1 row. Cast off.

TO MAKE UP

Pin out each piece to size and press carefully with a cool iron over a damp cloth. Join shoulder seams. Join side seams to marked points. Join back and front pieces of sleeves at lower (shaped) edge from marked point to marked point, join top seam. Insert sleeves with shaped edge to outside. Sew one edge of neckband into pos, fold in half and catch down second edge on the inside. Neaten lower ends. Turn up 12 rows at lower edge and catch down on the inside. Sew on press fasteners where required. Line if desired. With plain sides of belt tog, join seam leaving an opening. Turn to right side. Close opening. Give final light press.

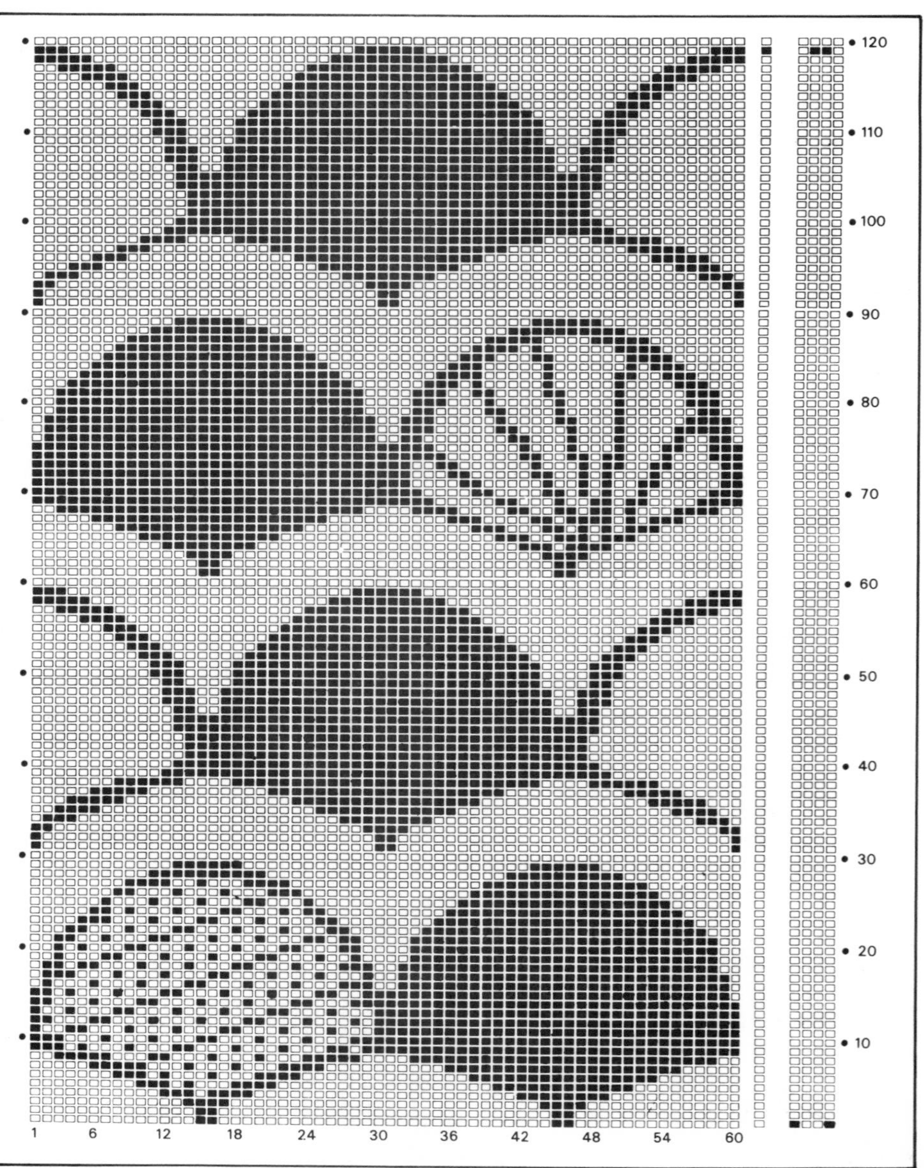

Bamboo Kimono Instructions

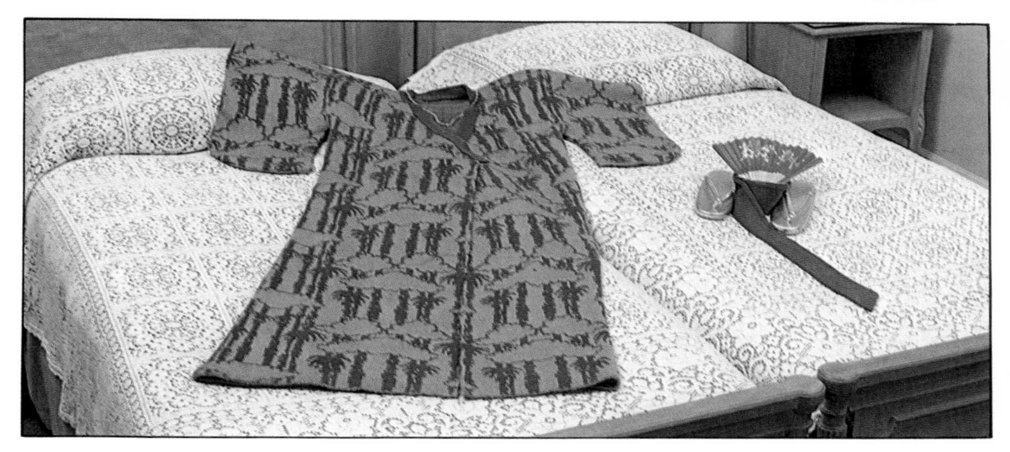

For Knitmaster Electronic machines.

DETAILS Photographed at the Savoy Hotel, London.

MEASUREMENTS One size to fit 34–40 in (86–101 cm) bust. Length from top of shoulder 57 in (144 cm). Finished length will drop to approx 59 in (150 cm).

MATERIALS Silverknit Dakota. 2 200 gm cones in col 1; 2 200 gm cones in col 2; 2 press fasteners; lining if desired. Yarn available from Silverknit, Dept K, The Old Mill, Epperstone-By-Pass, Woodborough, Nottingham.

TENSIONS 30 sts and 41 rows to 10 cm measured over pattern. Tension Dial at approx 4. 33 sts and 49 rows to 4 in (10 cm) measured over stockinet. Tension Dial at approx 3.

BACK

Inspection button on. Insert card and set to row 1. Pattern width indicator to 60. Using col 1, cast on by hand 162 sts. Point cams to all-over Fair Isle. Needle 1 cam at centre of machine. Buttons 1 (left), 2 (left) and No. 5. Insert yarn separators. Tension Dial at 3·, knit 14 rows stockinet. Row Counter 000. Tension Dial at 4. Inspection button off. Set carr for Fair Isle. With col 1 in feeder 1 and col 2 in feeder 2, knit 468 rows. Mark both edges with waste yarn, knit 116 rows. Row Counter 584. Carr at right. **Shape Back Neck:** Using a length of appropriate col, cast off centre 36 sts. Using nylon cord knit 63 sts at left by hand taking Ns back to A pos. Inspection button on. Take note of row number showing on pattern panel. Inspection button off. **Knit Right Part as follows:** Knit 1 row. Cast off 2 sts beg next and foll alt row, knit 1 row. Dec 1 st beg next and foll alt row. Row Counter 592. Cast off rem 57 sts. Inspection button on. Set card to number previously noted. Take carr to left. Unravel nylon cord bringing Ns to correct pos. Inspection button off. Knit left part as for right part. Remove yarn separators.

RIGHT FRONT

Inspection button on. Reset card to row 1. Push 81 Ns at right and 45 Ns at left of centre 0 to B pos (126 Ns altog). * Using col 1, cast on by hand. Point cams to all-over Fair Isle. Insert yarn separators. Tension Dial at 3·, knit 14 rows stockinet. (Knit 1 row extra for left front only.) Row Counter 000. Tension Dial at 4. Inspection button off. Set carr for Fair Isle. With col 1 in feeder 1 and col 2 in feeder 2, knit 420 rows. Carr at right. **Shape Front Edge:** Knit 2 rows. * Dec 1 st left edge next and foll alt row, knit 2 rows, dec 1 st left edge next row, knit 1 row *. Rept the last 7 rows from * to * 21 more times, **marking the right edge with waste yarn when row counter shows** 468. Dec 1 st left edge next and foll alt row, knit 2 rows. Dec 1 st left edge next row, knit 10 rows. Row Counter 592. Cast off rem 57 sts.

LEFT FRONT

Inspection button on. Reset card to row 1. Push 45 Ns at right and 81 Ns at left of centre 0 to B pos (126 Ns altog). Work as given for Right Front from * to end, but reverse the shapings by noting alteration in number of rows worked and reading left for right and vice versa.

SLEEVES (4 Pieces)

Right Front and Left Back (both alike)
Inspection button on. Reset card to row 27. Push 78 Ns at right and 1 N at left of centre 0 to B pos (79 Ns altog). Using col 1, cast on by hand. Point cams to all-over Fair Isle. Needle 1 cam between Ns 38 and 39 at right of centre 0. Insert yarn separators. Tension Dial at 4, knit 1 row (knit 1 row extra for left front and right back only). Carr at left. Row Counter 000. Inspection button off. Set carr for Fair Isle. With col 1 in feeder 1 and col 2 in feeder 2. Always taking yarn round first inside N in D pos. Push 36 Ns opposite end to carr to D pos on next row. 43 Ns rem in B pos. Knit 1 row. Push 2 inside Ns opposite end to carr from D pos to C pos on next and ev foll alt row 9 times in all, knit 1 row. Push 1 inside N opposite end to carr from D pos to C pos on next and ev foll alt row 18 times in all (all Ns are back in B pos). Row Counter 055, knit 31 rows. Row Counter 086. Mark both edges with waste yarn, knit 124 rows. Row Counter 210. Using waste yarn knit 8 rows and release from machine.

Left Front and Right Back (both alike)
Follow instructions for Right Front and Left Back but reverse the shapings by noting alteration in number of rows worked and reading right for left and vice versa throughout. Remove card.

NECKBAND

Inspection button on. Using waste yarn cast on 26 sts, knit several rows ending with carr at right. Row Counter 000. Tension Dial at 3. Set carr for stockinet. Using col 2, knit 636 rows. Using waste yarn knit 8 rows and release from machine.

BELT

Using col 2 cast on by hand 52 sts. Tension Dial at 3. Set carr for stockinet, knit 1 row. Row Counter 000. Always taking yarn round first inside N in D pos. Push 24 Ns opposite end to carr to D pos on next 2 rows. 4 Ns rem in B pos. Push 2 inside Ns opposite end to carr from D pos to C pos on next 24 rows. Row Counter 026. All Ns are back in B pos, knit 584 rows. Row Counter 610. Always taking yarn round first inside N in D pos, push 26 Ns opposite end to carr to D pos on next row, knit 1 row. * Push 2 inside Ns at opposite end to carr to D pos on next and ev foll alt row 12 times in all, knit 1 row *. Push rem 2 Ns to D pos. Break off yarn. Take carr to left. Push 26 Ns at left from D pos back to C pos. Knit the right part as for left part from * to *. Push all Ns from D pos back to C pos, knit 1 row. Cast off.

TO MAKE UP

Pin out each piece to size and press carefully with a cool iron over a dry cloth. Join shoulder seams. Join side seams to marked points. Join back and front pieces of sleeves at lower edge from marked point to marked point, graft top seam. Insert sleeves with shaped edge to outside. Sew one edge of neckband into pos. Fold in half and catch down second edge on the inside. Neaten lower edges. Turn up 14 rows at lower edge and catch down on the inside. Sew on press fasteners where required. Line if desired. With plain sides of belt tog, join seam leaving an opening. Turn to right side. Close opening. Give final light press.

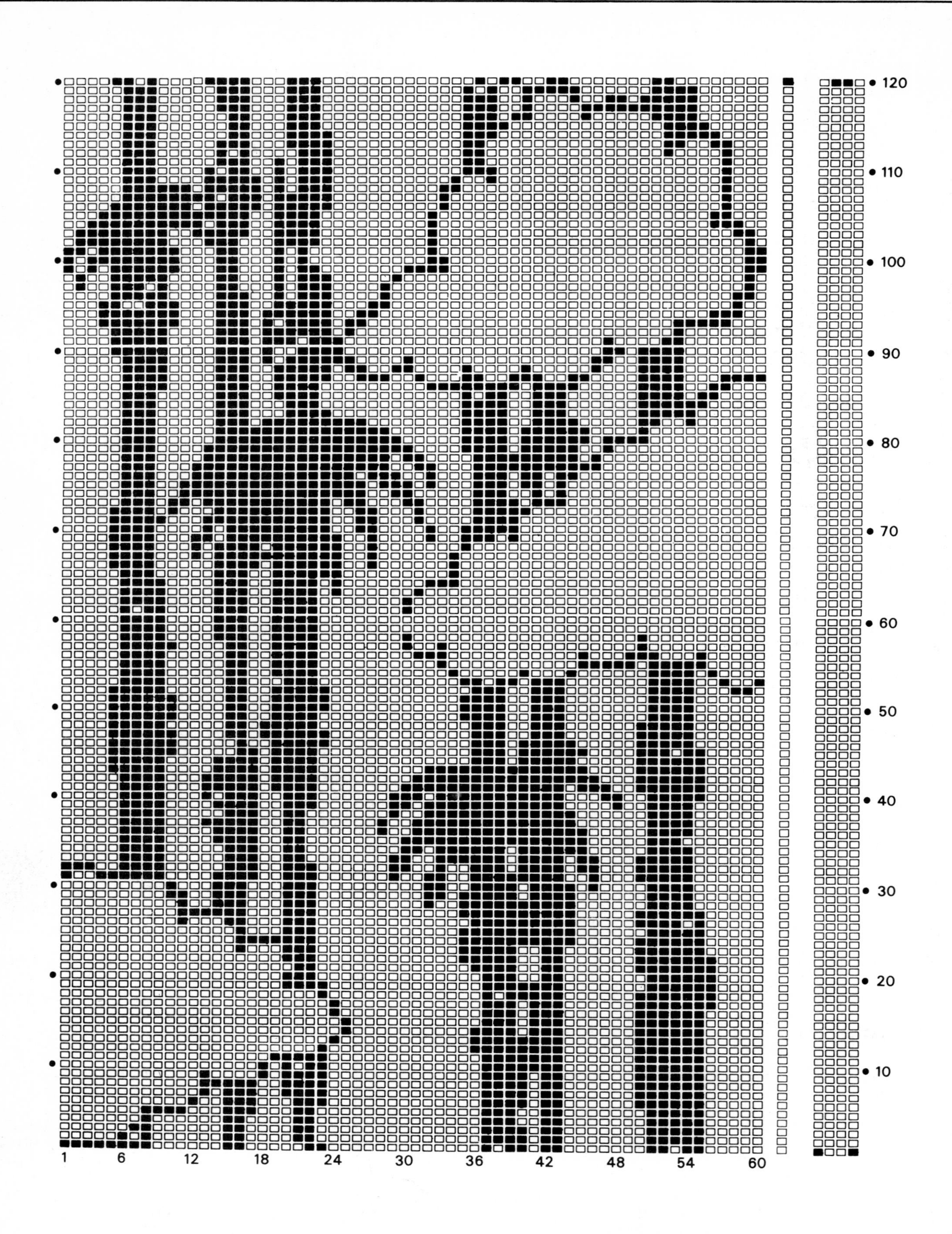
1
6
12
18
24
30
36
42
48
54
60
120
110
100
90
80
70
60
50
40
30
20
10

Hand Knitting

Introduction

In the Arabian city where the Queen of Sheba lived, they say that Eve knitted the pattern on the Serpent's back. She might well have done, for the knitters of ancient Arabia perfected silk knitting to such a degree that their multicoloured fabrics with patterns as lavish as a carpet and intricate as a snakeskin have not been surpassed in a thousand years. But whether or not it *is* older than sin, knitting is certainly about as old as man – and today it is still one of the most exciting and creative ways of dressing.

In Europe during the Middle Ages, wool knitting was one of the great craft arts like weaving or goldsmithing, and was mainly practised by men. An apprenticeship in one of the knitting guilds took six long years, at the end of which the fledgling Master Knitter had to knit a multicoloured carpet, a beret, a woollen shirt and a pair of hose – all in thirteen weeks! Durable rather than delicate, these wool knits were superseded in the sixteenth century when the introduction of silk resulted in the development of European silk knitting and the production of fabrics as rich and lustrous as brocade. Silk enjoyed a longer reign than Queen Elizabeth I, who refused to wear anything but knitted silk hose, and King Charles I who went to the scaffold in a tunic of fine knit silk, but it was finally supplanted in the eighteenth century when cotton and linen were introduced to Europe. This led to the vogue for *white knitting* – knitting white cotton and linen thread on wire-thin needles to create delicate open-patterned fabrics as fine as lace. In its turn *white knitting* lost ground to industrial knitting machines in the nineteenth century, and while traditional handknitting continued in rural areas and on a domestic basis, for a short while the art was in eclipse. Then fashion rediscovered handknits in the Twenties, with Chanel's sporty cardigans and Schiaparelli's witty knits, among them the famous cravat sweater shown here with a floppy bow knitted in at the neck, and the skeleton sweater reminiscent of ribs seen on an x-ray screen, both in the black and white that were her trademark.

Cravat sweater by Schiaparelli

There are many fascinating stories to be told about knitting in other parts of the world – in India and Tibet, for example, and in the Andes region of South America where colourful knits are produced using methods introduced by the Conquistadores. Today you have all the techniques and traditions of these earlier knits to draw on, a wealth of colours, patterns and yarns and, best of all, your own imagination. On the following pages, four very different designers talk about their work, and we present the Creative Dressing handknit collection.

Heinz Edgar Kiewe

Heinz Edgar Kiewe is a historian of the most rare and valuable kind – one who prefers action to dry analysis, and creativity to academic contemplation. Where other textile specialists are content to catalogue the technological details of the past, he has devoted his life to travelling, collecting, recording and preserving traditional designs and methods, in order to enrich and encourage the practice of the textile arts today. Over the years he has gained a tremendous reputation and following, and people from all over the world come to Art Needlework Industries, his shop in Oxford, to select from his unique collection of traditional wools, designs and materials for hand-knitting, tapestry, embroidery, crochet and rug-making. And to see Mr Kiewe himself, who has done more than any historian, anthropologist or textile scholar of our times to preserve and promote the most important things of all – *a belief in the value of the individual, and the means to make life into a work of art*.

'I feel that everyone has the urge to express themselves and to create things. And I feel that everyone has the *right* to be creative in their own way, express their *own* tastes and thoughts.

But ever since the Renaissance, art has suffered through the bureaucracy of the art schools who insist that work be done in this style or that style only, and the most terrible damage has been done by women's guilds. All these instructions, all this insistence on neatness. Over the last forty years, they've managed to kill *interest*!

It is better to be independent, and to develop your own aesthetic sense through observing, comparing, collecting and travelling. When I was young I had the good fortune to be surrounded by interesting things. My father had a large departmental store in Koenigsberg, a skyscraper in the Art Nouveau style, where he sold wonderful silks. In those days foreign goods were disapproved of, simply because they were foreign. But my father took great pride in his Chinese silks, and told me that whenever any exceptionally beautiful fabrics came into the store, fabrics with lovely patterns and colours, they invariably came from the East. He was decades ahead of his time in another textile field – that of knitted wool jersey – and he was one of the pioneers of the use of wool jersey for winter sports wear. So my interest in textiles and travel began early, and my life has revolved around them since I was sixteen.

The farther I travel, the more I see that there is no 'national art' – just regional variations. It is wrong to be chauvinistic and to pretend that art stops at the frontier. Art, craftsmanship and folk design have been part of the trade exchange between countries since the beginning of time. They provide an obvious channel of communication between peoples, because design is a truly international language, a language of the eye that can leap over the barrier of the spoken word. I think that by tracing the flow of design ideas in the past, by searching for similarities rather than differences and by looking at the history of folk designs, you gain many ideas that will enrich your own creative work.

The history of handknitting has received surprisingly little attention from others in the field, so as a result knitting is not generally appreciated for what it is – an art-form refined over many centuries, and one that provides a unique historical record of civilization. For example, you know that there are certain 'traditional' sweaters – Guernseys, Jerseys, Arans and so on – but have you ever wondered how and why they came into being?

The answer is very simple – the knitted wool sweater, which is over a thousand years old, is the best sea-faring garment that has ever been invented. As I say in *The History of Knitting*, sailors and fishermen from earliest times have found knitted wool sweaters a vital necessity. Knitted fabric stretches more than woven fabric or animal skins and, unlike synthetics, wool absorbs and sheds moisture naturally, allowing the body to breathe. Sea-faring is hard work – they were handling the heavy nets, rigging and sails day and night – so the breathing quality of the wool was essential to health and comfort. And so it is hardly surprising that all the traditional sweaters that have come down to our times are those of sea-faring folk – the people of Guernsey and Jersey, the Aran islands, and the lands of the Northern Seas.

The special relationship between sea-faring and knitted wool sweaters has been the starting-point of many fascinating researches I have made – Fair Isle knitting is a case in point. Everyone knows the Islands are famous for their knitting, but how did this knitting come into being? Did it just spring up out of nowhere? And why was it traditionally only found in the Islands?

Today, people think of Fair Isle knitting as something typically British – Scottish, to be precise – but history tells a very different story. The Orkneys and the Shetlands became part of Scotland only about three centuries ago, and the Fair Isle itself only became Scottish territory in 1954. Before that, the Islands were part of the Nordic world, the world of the Vikings – the great sea adventurers who dominated Northern Europe through the eleventh century, and who were responsible for establishing many contacts between East and West.

The Vikings were traders and merchants as well as warriors, and by AD 500 they had established regular trade routes that completely bypassed Britain. They sailed direct to Morocco and Rabat Salé where they exchanged fish and amber (the earliest European *jewel*) for all the goods of the East – spices, carpets, rugs and merino sheep, which give the best wool in the world.

Present evidence points to the ancient East as the home of knitting, possibly first with one hook or needle. By the time of the Viking trade visits to North Africa, the tribespeople had already developed intricate and colourful knitting by hand interlooping that repeated the patterns on their rugs and carpets. When the Vikings returned to their homeland, the knowledge of this craft travelled with them. The merino sheep adapted to the Northern climate and, using their wool, the Northern peoples began to knit patterned sea-faring sweaters, copying the motifs on the Berber rugs and carpets from North Africa. It was easy to adapt the trade goods to their own use, because they could count the stitches in the carpets and the patterns of the weaving and translate them into knitting stitches.

If you compare a traditional Northern sweater with a Berber carpet, you can see the relationship clearly. Of course, the Northerners made some changes – substituted a reindeer for a gazelle and so on – but basically they are much the same. As to why they should have done this at all, you have to remember that in North Africa and in the North, the patterns were *more* than decorative. In those days very few people could read and write, and the common people had to have some means of recording things, or recalling important things to mind, or identifying themselves to strangers and vice versa. So these patterns were made of symbols – symbols that had a story to tell, symbols that identified them as coming from a certain place or family. And of course, many of the symbols were magical emblems – charms to protect them from the evil eye, danger on the sea, and so on. This was very important, more than aesthetic feeling.

The sweaters that developed out of this contact – the very first Fair Isles if you like – were very primitive compared to those of today. The patterns were simpler and the colours were very limited indeed by modern standards. They were much like this Norwegian *Reindeer Slipover*, which has changed very little in over a thousand years, and which shows the Berber influence plainly.

From that time on, although Scandinavia gradually became cut off from the Eastern sphere, knitting continued to develop in the North. The next event of importance influenced the Islanders only. In 1558 some of the ships of the Spanish Armada were blown off course and wrecked off the Shetland Islands. The Armada sailors were found to be wearing patterned sweaters and scarves, the heritage of the centuries when Spain was ruled by the Moors of North Africa – the Spanish equivalent of the Northern Fair Isle knits. The Islanders copied the designs, and added them to their repertoire of traditional patterns.

The *Armada Scarf* you see here is one of those original designs. It is a copy of a sweater worn by an Armada sailor, kept in the family all this time for sentimental reasons. The island where this scarf was knitted is no longer inhabited, and except for fortunate circumstances the design might well have been lost for ever.

The pattern symbolizes the three cardinal virtues – the anchor of Faith, the cross of Hope and the heart of Charity. It is quite likely that the diamond motif is a charm – a protective device – like the eyes that are painted on boats in the East, in the Mediterranean and in Scandinavia. In the Islands, it is traditional for the person who is knitting the scarf to knit in one hair from her own head, to bring luck to the person who wears it. To me, the most interesting part of the design is the variety of the shades that run through the symbols. Faith, hope and charity are things that are shared, that involve someone else, and two colours are used to symbolize this.

After the Armada, the knitting of the Islands developed in a slightly different way to the rest of the old Nordic world. In time, every woman in the Islands had her own family pattern or motif, like a tartan or crest, and these were not used with others of the same kind. The mixture – one motif from this sweater, one motif from that, for purely decorative reasons – has only come about since the designs were mass-produced for export.

This mass production was the direct result of the Great Exhibition of 1851, held in London during Queen Victoria's reign. For the first time, ethnic textiles from all over the world were displayed, and that started a wave of popularity for British ethnic goods. Queen Victoria set the style with her fondness for tartans, but unfortunately there was very little else that was truly ethnic and British, so Fair Isle and Shetland knits were adopted instead, in brighter colours than had traditionally been used. By 1890, one 'became a tweedy Scotsman' – it was the smart thing to do – and Fair Isles were the fashion in society, very popular for shooting, and so on.

Fair Isle knitting remained popular during World War I because of all the knitting for the troops, and became even more popular after the war. The best lines were copied for royalty, as you see in the *Royal Fair Isle* sweater and the portrait (overleaf) of the late Duke of Windsor as Prince of Wales from which it was taken, painted in 1925. The striped ribbing on hem, cuffs and neck was a special feature of the Twenties – it was never used in Fair Isle knitting before, and has never been used since.

By now, all memory of the symbols in the pattern, and the origins of the technique itself, had been forgotten, and so it has continued to this day. So you see, this 'typically Scottish' Fair Isle knitting in the beginning had very little to do with Scotland at all, except for their Shetland wool and the perfection of Shetland knitters. It is a

regional variation of an art that had its beginnings in the East, and whose progress to the Islands can be traced in a scholarly way. To me, the history of the Fair Isle suggests many possibilities about the way the technique could be used today. You could transfer a beautiful Persian or Chinese carpet design into knitting, for example.

On a recent trip to Norway I was greatly saddened to see that the traditional knitting is very little worn these days – they wear nylon and plastic sports gear instead. If synthetic garments have become a part of everyday life, then I suppose one must learn to live with it. But I do not see why the two forms, old and new, cannot be combined. That is why I have included *Lapland Gauntlets* – they are for wearing *over* gloves or skiing mittens. It is a combination that gives you the best of the old and new.

Sadly, here we come to a problem. In the great enthusiasm for synthetics, fine traditional wools have almost disappeared. Real Shetland wool – wool from the northern merino sheep – is almost impossible to find now. That is why I allow no synthetics in my shop, and why I keep a unique stock of traditional yarns – homespun Harris Tweed yarn, Shetland 2-ply, fine lace weight Shetland 2-ply, Shetland 4-ply, Scottish homespun double-knit wool, mohair, Fisherman's Scottish homespun wool and many others. I feel it is vitally important to keep interest in these fine wools alive, because if the interest is not kept up my great fear is that in a few years time there will be no more Shetland wool, and with it will go a thousand years of history and beauty. I keep things like circular knitting needles and Tunisian knitting needles for the same reason.

Perfumes were once part of everyday life – they are forgotten today, and it's the same with colour. Today

people do not respect colour. But that cannot be entirely their fault, because what can they know of beautiful colours if they rely on imported mass-manufactured products – such *bad* colours, so few shades! That is why I keep hundreds of colours of yarns specially dyed for me, and specially dyed tapestry wools as well. The secret of tapestry is to shade well, so the tapestry looks three dimensional, and it is much the same with knitting. How can you design without a wealth of colours and shades!

Here you see what I call *chapelets* of shades – 'rosaries' of colour – tapestry yarns in the exact shades of the colours used in different historical periods that I have copied from original tapestries in museums, along with one of my own Florentine Flame Stitch tapestries, the pattern for which I obtained in 1947 from a private museum in Florence that is still closed to the public. It is not essential to use the original colours when you are doing a period tapestry but I feel it is important for them to be available because you can learn a great deal from them. The *chapelets* are an expression of an era – colours that express the feeling of the craft of a particular time and place. One *chapelet* is taken from the French fourteenth-century Angers tapestry – the deep indigo and scarlets were the first authentic dyes to come to Europe from India, a harbinger of the Eastern influences that were later to enrich the civilization of Europe. The other is an Art Nouveau *chapelet*, and in those bright greens and purples you see the mood of the time – the belief in the progress of man, in the greatness of chemistry, and the beauty of shades that owed nothing to Nature.

These *chapelets* relate to periods in time but, equally, colours express the character of an individual. You should collect wools in the shades that say something to you, and make up your own *chapelets*. It is like having the ingredients of a recipe that you intend to make later on. Look at the colours, think about harmonies and what you can do with them. As you walk through nature, take notes of the colours you really love – collect them, arrange them, and play with them.

You may approach things as I do or you may not – but the most important thing is that you should *do* something, create something that pleases *you*. The early weavers had no design books, the men and women who embroider in Kashmir today don't need a picture or pattern to tell them what to do – and neither do you. I believe in the natural artist, the Artist Craftsman. Beauty comes naturally, if you let it. Be creative. Be different. After all – *no two flowers are the same.'*

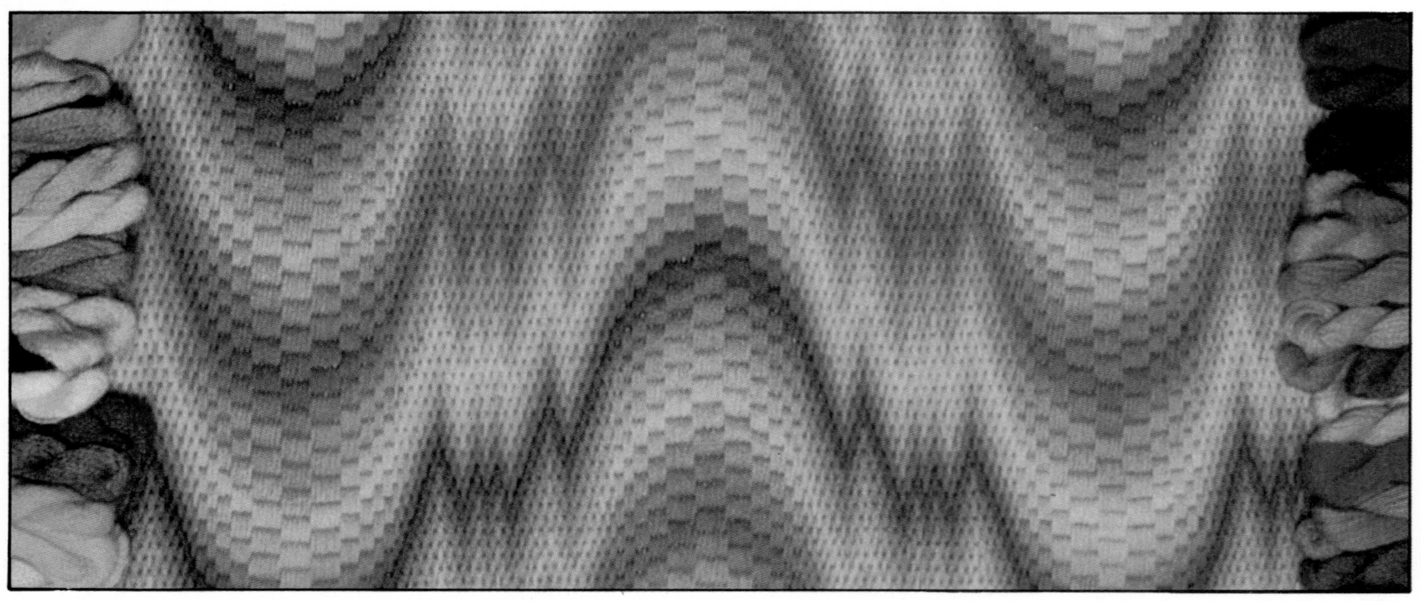

For further reading on the diffusion of design ideas from East to West, see *Civilization on Loan*, by Heinz Edgar Kiewe with Michael Biddulph and Victor Woods, Art Needlework Industries Publication, Oxford, 1973. Heinz Edgar Kiewe, is also the author of *The History of Knitting*, Art Needlework Industries Publication, Oxford, 1968.

NOTE All ANI yarns, tapestries and publications are available mail order. For yarn shade cards and current prices, send postal order or cheque for 30p (including brochures) to cover postage and packing if writing from within the UK, or the equivalent of £1 if writing from abroad, to Art Needlework Industries, 7 St Michael's Mansions, Ship Street, Oxford OX1 3DG.

Reindeer Fair Isle

MATERIALS **Yarn:** ANI Real Shetland Wool (2-ply, knits to 3-ply pattern), 5 oz in Natural, 2 oz in Brown, 1 oz in Gold, 1 oz in Rust. **Needles:** No. 12 British (No. 1 American, 2¾ mm).

MEASUREMENTS One size, to fit chest measurements 36–38 in (91–97 cm).

TENSION 7 sts and 7 rows to 1 in (2.5 cm).

ABBREVIATIONS **See** page 176.
Colours: N=Natural, B=Brown, Y=Gold, R=Rust.

BACK
With No. 12 needles cast on 145 sts in N. Work 3½ in (9 cm) in K1 P1 rib. Change to No. 9 needles and with N work 2 rows in st.st, dec 5 sts evenly on first row (140 sts). Work rows 1–21 of Chart 1: for K rows read the chart from right to left, and for P rows read the chart from left to right. Now work 3 rows in N. Now knit Fair Isle Mock Anchor band as follows. **1st row** K7B * 1N 4Y 1N 11B, rep from * to end but finish last rep with 8B. **2nd row** P9B * 1N 2Y 1N 13B, rep from * to end but finish last rep with 8B. **3rd row** K3B * 3N 2B 1N 2Y 1N 2B 3N 3B, rep from * to last st, 1B. **4th row** P2N * 1B 2N 4B 1N 2Y 1N 4B 2N, rep from * to last 2 sts, 1B 1N. **5th row** K1N * 1B 2N 3B 2N 2Y 2N 3B 2N, rep from * to last 3 sts, 1B 2N. **6th row** P4B * 5N 4Y 5N 3B, rep from * to end. **7th row** K7N * 6Y 11N, rep from * to end but finish last rep with 8N. **8th row** P7N * 8Y 9N, rep from * to end but finish last rep with 6N. **9th row** K9N * 2Y 15N, rep from * to end but finish last rep with 10N. Work 3 rows N, dec 1 st at beg of last row. Work Chart 2. Now work 3 rows in N, inc 1 st at beg of last row (140 sts). Work Chart 1 again and continue in this manner until required length to armhole. **Armhole shaping:** cast off 6 sts at beg of next 6 rows (104 sts). Continue to work straight in pattern until required length to shoulder, finishing at armhole edge. **Shoulder shaping:** cast off 10 sts at beg of next 6 rows. Cast off remaining sts.

FRONT
Work as Back until armhole has been reached. **Armhole shaping:** cast off 6 sts at beg of next 6 rows (104 sts). **Neck shaping:** K52 sts, turn and work on these sts only. P2tog at neck edge on next and every 3rd row until 30 sts remain. Continue straight until work is required length to shoulder, finishing at armhole edge. **Shoulder shaping:** cast off 10 sts at beg of next 3 alt rows. Return to remaining 52 sts and work to match first side.

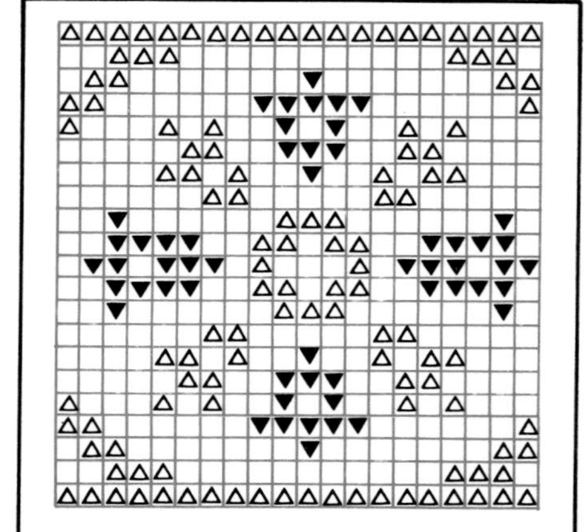
Chart 1

ARMBANDS
Join up shoulder seams. With No. 12 needles and right side of work facing, pick up and K 154 sts round arm. Work 7 rows in K1 P1 rib. Cast off in rib.

NECK RIBBING
With No. 12 needles and right side of work facing, pick up and K 65 sts down left front, 1 st from middle of V neck (mark this st with coloured thread), 65 sts from right side and 41 sts across back. **1st row** with K1 P1 rib, rib to within 2 sts of marked st, P2tog, K marked st, P2tog, K1 P1 to end. **2nd row** rib to within 2 sts of marked st, K2tog, K marked st, K2tog, rib to end. Continue in this manner for 5 more rows. Cast off in rib, then make up.

COLOUR KEY FOR CHARTS 1 and 2

Colour Key
△ = Rust (R)
○ = Gold (Y)
▲ = Brown (B)
□ = Natural (N)

Each square in these charts represents one stitch of knitting, and the colours to work the stitches are shown in the key. For K rows read the charts from right to left, and for P rows from left to right.

Chart 2

Armada Fair Isle Scarf

MATERIALS **Yarn:** ANI Real Shetland Wool (2-ply, knits to 3-ply pattern), 6 oz in Natural, 1 oz each in Wine Red, Light Blue, Old Gold, Dark Brown. **Needles:** No. 10 British (No. 3 American, 3¼ mm).

ABBREVIATIONS **See** page 176.
Colours: N=Natural.

SCARF

With No. 10 Needles cast on 112 sts. Work 6 rows st.st in N. Work rows 1–12 of Chart 1. Work 5 rows st.st in N. Work rows 1–14 of Chart 2 repeating pattern 4 times, then Chart 2A for last pattern, 5 repeats in all. Work 5 rows st.st in N. Work rows 1–17 of Chart 3. Work 5 rows of st.st in N. Work rows 1–14 of Chart 2 repeating pattern 4 times, then Chart 2A for last

Chart 1
14 sts 12 rows

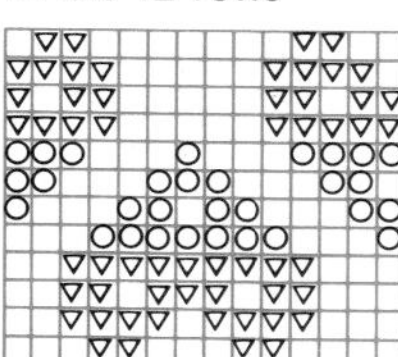

Chart 2
23 sts 14 rows

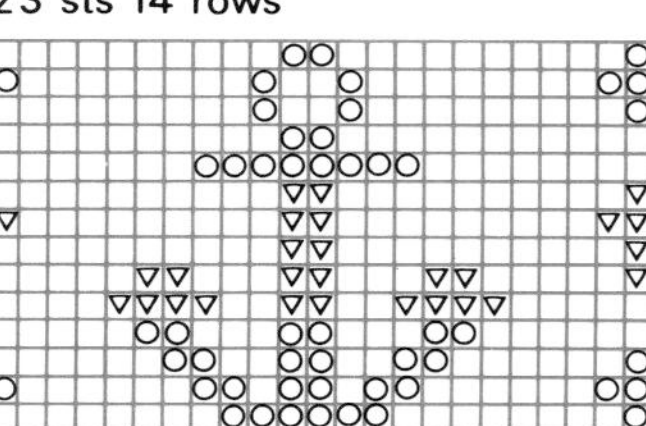

Chart 2a
20 sts 14 rows

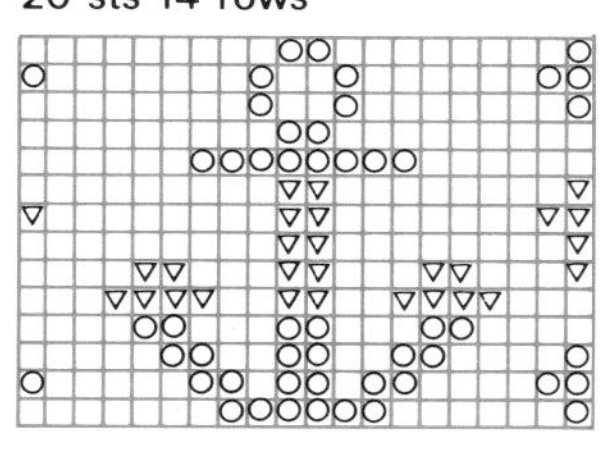

Chart 3
28 sts 17 rows

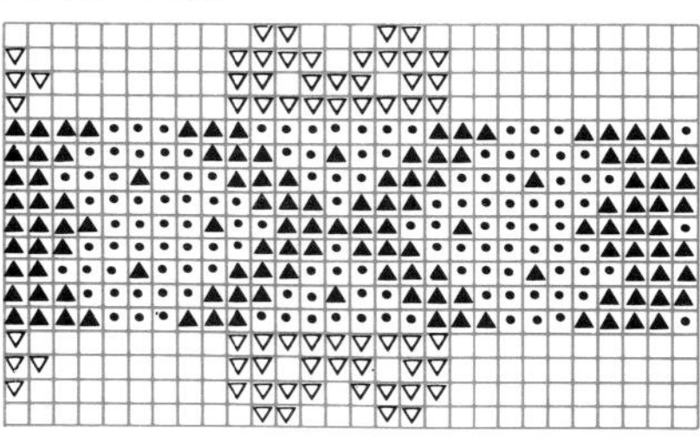

Chart 4
16 sts 10 rows

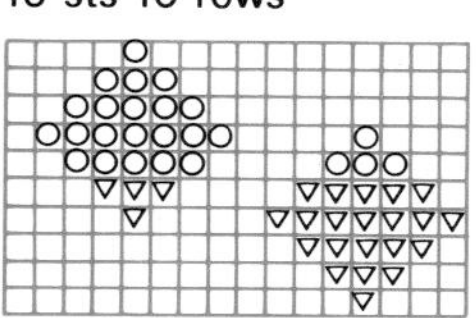

Colour Key
○ Light Blue
▽ Wine Red
• Dark Brown
▲ Old Gold
□ Natural

Chart 5
28 sts 17 rows

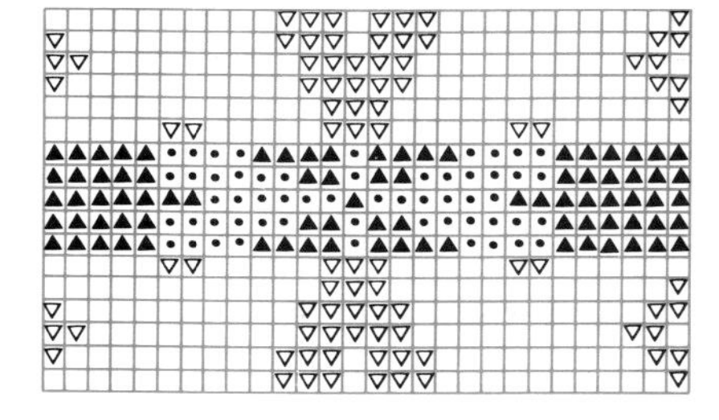

Chart 6
28 sts 17 rows

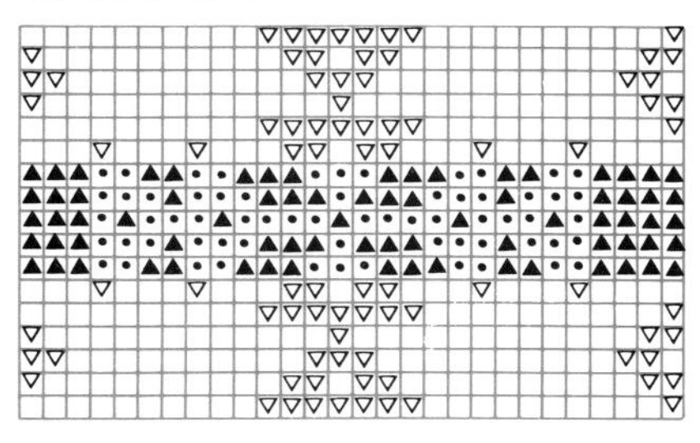

Chart 7
14 sts 12 rows

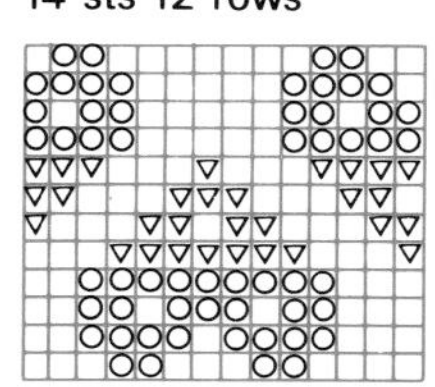

Chart 8
23 sts 14 rows

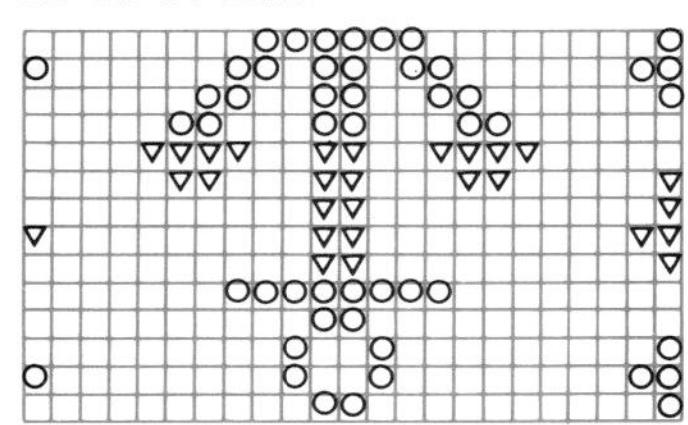

Chart 8a
20 sts 14 rows

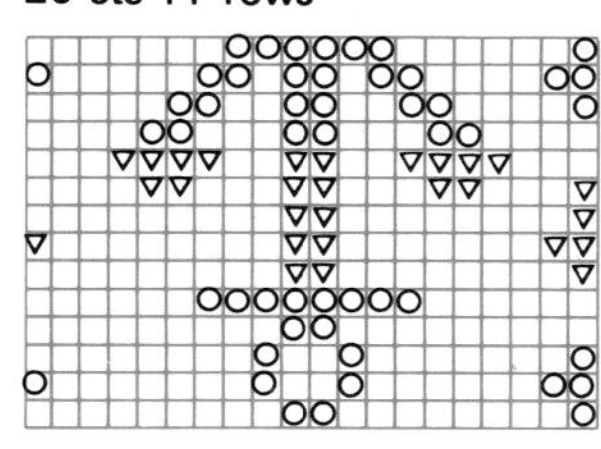

Chart 9
16 sts 10 rows

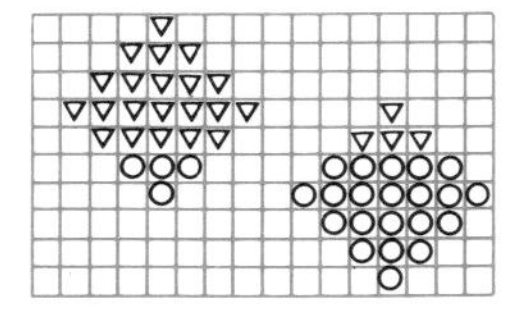

pattern, 5 repeats in all. Work 5 rows st.st in N. Work rows 1–10 of Chart 4. Work 5 rows st.st in N. Work rows 1–17 of Chart 5. Work 5 rows st.st in N. Work rows 1–14 of Chart 2 repeating pattern 4 times, then Chart 2A for last pattern, 5 repeats in all. Work 5 rows st.st in N. Work rows 1–12 of Chart 1. Work 5 rows st.st in N. Work rows 1–17 of Chart 6. Work 5 rows st.st in N. Work rows 1–10 of Chart 4. Work 5 rows st.st in N. Work rows 1–17 of Chart 6. Work 5 rows st.st in N. Work rows 1–12 of Chart 7. Work 5 rows st.st in N. Work rows 1–14 of Chart 8 repeating pattern 4 times, then Chart 8A for last pattern, 5 repeats in all. Work 5 rows st.st in N. Work rows 1–17 of Chart 5. Work 5 rows st.st in N. Work rows 1–10 of Chart 9. Work 5 rows st.st in N. Work rows 1–14 of Chart 8 repeating pattern 4 times, then Chart 8A for last pattern, 5 repeats in all. Work 5 rows st.st in N. Work rows 1–17 of Chart 3. Work 5 rows st.st in N. Work rows 1–14 of Chart 8, repeating pattern 4 times then Chart 8A for last pattern, 5 repeats in all. Work 5 rows st.st in N. Work rows 1–12 of Chart 7. Work 6 rows st.st in N. Cast off. Finish with fringe.

Lapland Gauntlets

DETAILS Photographed at the Savoy Hotel, London.

MATERIALS **Yarn:** ANI Homespun Wool, 4 oz in White, 2 oz each in Red, Green, Yellow and Blue. **Needles:** No. 10 British (No. 3 American, 3¼ mm).

MEASUREMENTS One size, to fit over glove size 7½.

ABBREVIATIONS **See** page 176. **Colours:** R=Red, G=Green, Y=Yellow, B=Blue, W=White.

BOTTOM WRIST

With W and No. 10 needles cast on 86 sts and K 1 row. **Row 1** K1W * in R K next st wrapping yarn round needle twice, K2W * rep from * to * to last st, K1W. **Row 2** K1 * K1 ML K1 bring yarn to front of work. Slip next red stitch purlwise, slipping both loops off needle and having one large red st on right-hand needle. Slip all 4 sts back on left-hand needle. Pull red stitch over 3 white sts and off needle and return 3 white sts to right-hand needle * rep from * to * to last st, K1W. Rep last 2 rows twice more first using G instead of R, then Y. Work 2 rows st.st in W. **Next row** K1W * 1R 3W * rep from * to * to last 2 sts, K1R 1W. **Next row** * P3R 1W * rep from * to * to last 2 sts, P2R. **Next row** K across in R. **Next rows** st.st 3 rows in G. **Next rows** st.st 3 rows in Y. **Next row** P2W * 1R 3W * rep from * to * to end. **Next row** K2W * 1R 1W 1R 5W * rep from * to * but end 1W. **Next row** * P1R 1W 1B 1W 1R 3W * rep from * to * but end 1W. **Next row** * K1R 1W 3B 1W 1R 1W * rep from * to * to end. **Next row** * P2B 1W 2B 1W 1R 1W * rep from * to * to last 6 sts, 2B 1W 2B 1W. **Next row** * K1R 1W 3B 1W 1R 1W * rep from * to * to last 6 sts, K1R 1W 3B 1W. **Next row** * P1R 1W 1B 1W 1R 3W * rep from * to * but end 1W. **Next row** K2W * 1R 1W 1R 5W * rep from * to * to last 4 sts, K1R 1W 1R 1W. **Next row** P2W * 1R 3W * rep from * to * to end. **Next rows** work 3 rows Y in st.st. **Next rows** work 3 rows G in st.st. **Next row** K 1 row R. **Next row** P2R * 1W 3R * rep from * to * to end. **Next row** K1W 1R * 3W 1R * rep from * to * to end. **Next row** (P2, P2tog W) 3 times (P1, P2tog W) to last 12 sts, P2, P2tog to end. 60 sts. **Next rows** work 2 rows st.st in W. **Next row** * in W K1, K2tog yfwd, * rep to last 3 sts, K3. **Next row** P 1 row in W.

TOP TO BOTTOM

Row 1 K23 sts in W inc in next st K1 inc in next st K15W K1R K18W. **Row 2** P17W 3R, P to end in W. **Row 3** K41W 2R 1G 2R, K to end in W. **Row 4** P15W 2R 3G 2R, P to end in W. **Row 5** K23W inc in next st K3W inc in next st K11W 2R 2G 1Y 2G 2R, K to end in white. **Row 6** P13W 2R 2G 3Y 2G 2R, P to end in W. **Row 7** K39W 2R 2G 2Y 1B 2Y 2G 2R, K to end in W. **Row 8** P11W 2R 2G 2Y 3B 2Y 2G 2R, P to end in W. **Row 9** K23W inc in next st K5W inc in next st 7W 2R 2G 2Y 2B 1W 2B 2Y 2G 2R, K to end in W. **Row 10** P9W 2R 2G 2Y 2B 3W 2B 2Y 2G 2R, P to end in W. **Row 11** K37W 2R 2G 2Y 2B 2W 1R 2W 2B 2Y 2G 2R, K to end in W. **Row 12** P7W 2R 2G 2Y 2B 2W 3R 2W 2B 2Y 2G 2R, P to end in W. **Row 13** K23W inc in next st K7W inc in next st K3W 2R 2G 2Y 2B 2W 5R 2W 2B 2Y 2G 2R, K to end in W. **Row 14** P7W 2R 2G 2Y 2W 3R 2W 2B 2Y 2G 2R, P to end in W. **Row 15** K39W 2R 2G 2B 2W 1R 2W 2B 2Y 2G 2R,K to end in W. **Row 16** P9W 2R 2G 2Y 2B 3W 2B 2Y 2G 2R, P to end in W. **Row 17** K23W inc in next st P9W inc in next st K7W 2R 2G 2Y 2B 1W 2B 2Y 2G 2R, K to end in W. **Row 18** P9W 2R 2G 2Y 2B 3W 2B 3W 2B 2G 2R, P to end in W.

DIVIDE FOR THUMB

K35W, turn and cast on 2 sts, P 14 sts including 2 cast-on sts, and cast on 2 sts at end of row (16 sts). Now work on these 16 sts for thumb. **Row 1** K8W 1R 7W. **Row 2** P6W 1R 1G 1R 7W. **Row 3** K6W 1R 1G 1Y 1G 1R 5W. **Row 4** P4W 1R 1G 3Y 1G 1R 5W. **Row 5** K6W 1R 1G 1Y 1G 1R 5W. **Row 6** P6W 1R 1G 1R 7W. **Row 7** K6W 1R 1G 1Y 1G 1R 5W. **Row 8** P4W 1R 1G 3Y 1G 1R 5W. **Row 9** K6W 1R 1G 1Y 1G 1R 5W. **Row 10** P6W 1R 1G 1R 7W. **Row 11** K8W 1R 7W. Now continue straight in W until thumb measures 2 in (5 cm). **Next row** * K1 K2tog, * rep from * to * to end. **Next row** P to end. **Next row** K2tog all across row. Break yarn. Draw through remaining sts and fasten off. With right side of work facing, rejoin wool at base of thumb and K up 4 sts from cast-on sts at base, K to end of row in pattern. **Pattern:** 2R 2G 2Y 2B 2W 1R 2W 2B 2Y 2G 2R. Continue straight but keeping pattern up the back. **Next pattern row** 2R 2G 2Y 2B 2W 3R 2W 2B 2Y 2G 2R. **Next row** 2R 2G 2Y 2B 2W 5R 2W 2B 2Y 2G 2R. **Next row** 2R 2G 2Y 2B 2W 3R 2W 2B 2Y 2G 2R. **Next row** 2R 2G 2Y 2B 2W 1R 2W 2B 2Y 2G 2R. **Next row** 2R 2G 2Y 2B 3W 2B 2Y 2G 2R. **Next row** 2R 2G 2Y 2B 1W 2B 2Y 2G 2R. **Next row** 2R 2G 2Y 2B 3W 2B 2Y 2G 2R. **Next row** 2R 2G 2Y 2B 2W 1R 2W 2B 2Y 2G 2R. Continue in this manner until 4½ in (11.5 cm) of patterns have been completed.

SHAPE TOP

K1 sl.1 K1 PSSO K22 K2tog sl.1 K1 PSSO, pattern to last 3 sts, K2tog K1. **Next row** P1 P2tog, pattern 28, P2tog tbl P2tog, P to last 3 sts, P2 tog tbl P1. Continue until 38 sts remain. **Next row** K1 sl.1 K1 PSSO K10 K2tog sl.1 K2tog PSSO, K to last 4 sts, K3tog K1. **Next row** P1 P2tog P14 P2tog tbl, P to end. **Next row** K1 sl.1 K1 PSSO K8 K2tog sl.1 K2tog PSSO K10 K3tog K1. **Next row** P1 P2tog P8 P2tog tbl, P to end. **Next row** Kl sl.1 K2tog PSSO K4 K3tog sl.1 K2tog PSSO K4 K3tog K1. **Next row** P1 P2tog P2 P2tog tbl P2tog P2 P2tog tbl P1. **Next row** K1 sl.1 K1 PSSO K2tog sl.1 K1 PSSO K2tog K1. **Next row** P to end. Break wool thread through sts and fasten off. Now take 6 strands of different coloured wool and plait enough to thread round wrists of gauntlets.

Royal Fair Isle

MATERIALS

Yarn: ANI Real Shetland Wool (2-ply, knits to 3-ply pattern), 6 oz in Natural, 2 oz each in Dark Green, Yellow, Beige and Brown, 1 oz each in Rust, Tan, Dark Rust and Light Green. **Needles:** one pair of needles **and** a set of 4 double pointed needles No. 12 British (No. 1 American, 2¾ mm), one pair of needles size No. 10 British (No. 3 American, 3¼ mm).

MEASUREMENTS

One size, to fit chest measurements 36–40 in (91–101 cm). **Length:** 28 in (71 cm). **Sleeve length:** 18 in (46 cm).

TENSION

7 sts and 7 rows to 1 in (2.5 cm).

ABBREVIATIONS

See page 176.
Colours: N=Natural, DG=Dark Green, T=Tan, LG=Light Green.

BACK

With N and No. 12 needles, cast on 138 sts. **Row 1** * K2DG P2N, * rep from * to * to last 2 sts, K2DG. **Row 2** * P2DG K2N, * rep from * to * to last 2 sts, P2DG. Rep last 2 rows for 2¾ in (7 cm). Change to No. 10 needles. K 4 rows N in st.st inc 6 sts on 1st row (144 sts). **Next row** * K3N K4T, * rep from * to * to last 4 sts, K4N. **Next row** P1N, * P3N P4T * rep from * to * to last 3 sts, P3N. **Next row** * K3N K4T, * rep from * to * to last 4 sts, K4N. These three rows complete the square pattern. Now work 3 rows st.st in N. Work rows 1–5 of Chart 1. Work 3 rows st.st in N. Work 3 rows of square pattern. Work 3 rows st.st in N. Work rows 1–15 of Chart 2. Work 3 rows st.st in N. Work 3 rows of square pattern. Work 3 rows st.st in N. Work rows 1–5 of Chart 1. Work 3 rows st.st in N. Work 3 rows of square pattern. Work 3 rows st.st in N. Work rows 1–17 of Chart 3. Work 3 rows st.st in N. Work 3 rows of square pattern. Work 3 rows st.st in N. Work rows 1–5 of Chart 1. Work 3 rows st.st in N. Work 3 rows of square pattern. Work 3 rows st.st in N. Work rows 1–14 of Chart 4. Work 4 rows st.st in N. This completes the pattern. Continue in pattern to required length to armhole, 18½ in or 47 cm. **Armhole shaping:** keeping pattern correct, cast off 8 sts at beg of next 2 rows. K2tog at beg of next 10 rows (118 sts). Work straight to shoulder, 27½ in or 70 cm. **Shoulder shaping:** cast off 12 sts at beg of next 6 rows. Cast off remaining sts.

FRONT

Work as for Back until armhole is reached. Cast off 8 sts at beg of next 2 rows (128 sts). **Next row** K2tog, pattern 60 sts, K2tog, turn and work on these 62 sts. Work one row. K2tog at armhole edge on next 4 alt rows then knit armhole straight. At the same time, K2tog at neck edge on every 3rd row until 36 sts remain. Work straight till required length finishing at armhole edge. Cast off 12 sts at beg of next 3 alt rows. Return to remaining 64 sts. With right side of work facing, join wool at neck edge. K2tog, pattern to end. **Next row** P2tog, pattern to end. Work to match first side.

SLEEVES

With No. 12 needles cast on 66 sts. **1st row** * K2DG P2N, * rep from * to * to last 2 sts, K2DG. **2nd row** * P2DG K2N, * rep from * to * to last 2 sts, P2DG. Rep last 2 rows for 2¾ in (7 cm). Change to No. 10 needles and N. Work 4 rows st.st in N. **Next row** * K3N K4T, * rep from * to * to last 3 sts, K3N. **Next row** * P3N P4T, * rep from * to * to last 3 sts, P3N. Rep first row again. Work 3 rows st.st in N, inc one st at each end of last row. **Row 1** K2N * K1LG K7 N, * rep from * to * to last 2 sts, K1LG K1N. **Row 2** * P1LG P1N P1LG P5N, * rep from * to * to last 4 sts, P1LG, P1N, P1LG, P1N. **Row 3** * K1LG K3N, * rep from * to * to end. **Row 4** P4N * P1LG P1N P1LG P5N, * rep from * to * to end. **Row 5** K6N * K1LG K7N, * rep from * to * to last 6 sts, K1LG K5N. Work 3 rows st.st in N, inc 1 st at each end of last row. **Next row** K5N * K4T K3N, * rep from * to * to last 2 sts, K2N. Rep row 1 once more. Work 3 rows st.st in N, inc 1 st at each end of last row. Work rows 1–15 of Chart 2, increasing 1 st at each end of every 6th row from last inc. Work 3 rows st.st in N. **Next row** K2N * K4T K3N, * rep from * to * to end but ending K2N. **Next row** P2N * P4T P3N, * rep from * to * to end but ending P2N. Rep row 1 once more. Work 3 rows st.st in N. Continue in this manner keeping pattern correct as on Back and increasing 1 st at each end of every 6th row until there are 100 sts. Work straight until required length to armhole. **Armhole shaping:** cast off 8 sts at beg of next 2 rows, then 1 st at beg only of every row until 48 sts remain. Cast off remaining 48 sts. Join all seams including shoulder seams.

NECKBAND

With set of 4 double-pointed No. 12 needles and N, pick up and knit 71 sts down left side, 71 sts up right side and 38 sts across back (160 sts). Starting with K2DG P2N, work all around neck once. **Next round** (K2DG P2N) 17 times K2tog DG P2N K2togDG, rib to end. **Next round** (K2DG P2N) 17 times K2togDG twice, rib to end. Continue in this manner for 8 rounds in all. Cast off in K2DG P2N rib. Make up.

Chart 1
5 rows 16 sts

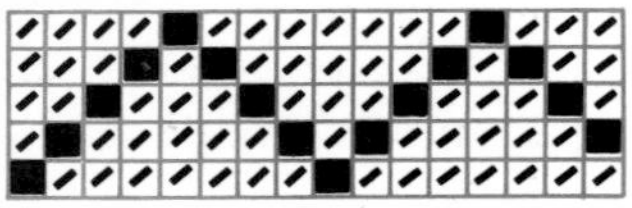

Chart 2
15 rows 24 sts

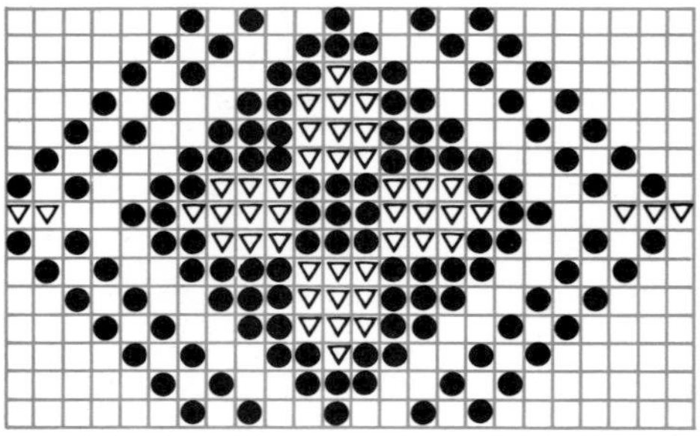

Chart 3
17 rows 24 sts

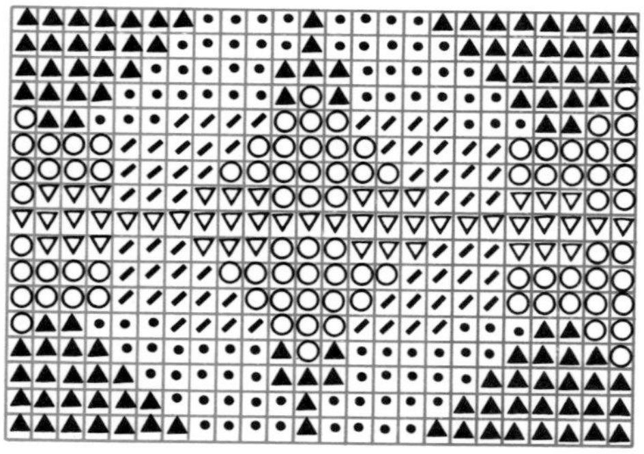

Chart 4
14 rows 24 sts

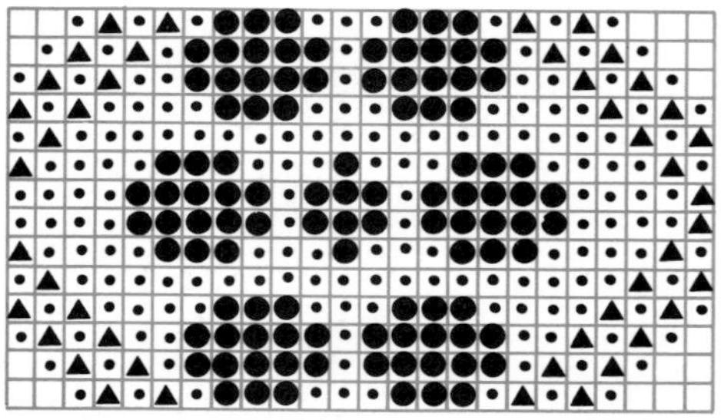

Colour Key
▽ = Rust
● = Yellow
□ = Brown
• = Dark Rust
○ = Dark Green (DG)
■ = Light Green (LG)
▲ = Beige
⁄ = Natural (N)

Kaffe Fassett

Kaffe Fassett's designs are collector's items from the moment they're finished. His use of colour and pattern turns knitting into an exciting visual medium, and his designs look like paintings, tapestries, mosaics, sunsets, anything but sweaters. After seeing what he does with yarn, your view of knitwear will never be the same again.

'Basically I'm a painter but I started getting interested in fabrics when I came to England from California, because of the yarns I discovered in Scotland. I learned to knit here, about ten years ago, and it's led on to designing tapestries and furnishing fabrics.

Today I see a continuous connection that flows between my paintings, tapestry work, fabric designs and my knitting. They all feed each other. For instance, I've learned a great deal about colour from knitting and needlepoint. You can mix paints until you get just the colour you want but you can't change the colour of yarn the same way. So you work with many, many shades of colour instead, learn to blend them through combinations and juxtapositions, and it's fed back into my painting.

I get a lot of ideas for patterns from carpets, decorative oriental and English china, and natural things like shells and leaves. I like to look at them, paint them, spend a lot of time absorbing the material that's there. I like to have them around me, so they're always talking to me.

I might start with a geranium leaf, paint it, turn it into a whole series of furnishing fabric designs. Or I might get an idea from a mosaic floor, and turn the idea into a jacket. And that's what you should do when you look at an object or a knitting pattern – get the *mood* of what's being put across, and apply it to your *own* personal needs. *Take one idea and translate it into another*.

I worked with the Missonis in Italy for a time – they're very creative people, and it was a very exciting collaboration. It was also my first encounter with high-powered industrial machines. I've designed machine knits for Bill Gibb since then but I always go back to handknitting because it's such a marvellous medium to work in. I get a great satisfaction out of knitting, so it's really a great joy to go on making things by hand, and it's not such a joy to just think up something theoretically for a machine.

I believe in making clothes that are wearable, that are beautiful and flattering to the body. But within those limitations I like to arrive at the most luscious colourings by using lots of colours and textures, taking advantage of the fact that I'm making it by hand, so I can put in the extras that a machine can't. I always like in my knitting to do the most difficult things, to try for really complicated effects. And I like my designs to be timeless. If they're beautiful today, I want them to be beautiful ten years from now. I want them to have qualities that will go on *saying* something – not just reflect what's accepted, what's 'in fashion', at the moment.

In my work I've concentrated on playing with colours and patterns because I think that there's such a lot to accomplish there. I like very brilliant and exciting colours, *dramatic* colours. I love very, very subtle colours that you can live with for years and not get tired of. And I love colours that have a conversation with each other, that speak to each other – where there's something happening, and there's a *movement* in the colours.

When I work with colours, I usually find that using several shades of the same colours helps the colours to move. For instance, magenta is helped by using shades of reds, maroons and dark pinks. The magenta can be the climax of the softer reds and pinks – everything *rises* to a magenta height. Another approach is to take many monotone colours – soft background colours – and then have one or two bright colours that lift out of that background. And sometimes I work with a lot of contrast too, where several of those approaches are combined, so you get an incredible tapestry effect.

Mixing patterns is like mixing colours, in that there's that same kind of movement. You get the relationship of the patterns, and the relationship and excitement of change of scale, where you have a small pattern relating to a bigger pattern. And you get a movement in that way, just the way you get a movement through shades of colours. It's something that's happening in nature all the time – in every garden, every forest, every series of rocks. A butterfly sitting on a flower is pattern on pattern. It's something that is always in our unconscious, if not in our conscious, mind.

The pattern influence has come to the West from the East, and I find it tremendously exciting. Look at Persian miniatures, Moroccan mosaics, all the Islamic use of pattern on pattern, the paintings of the Chinese, Japanese and Balinese. Outlandish combinations, subtle combinations – they've been playing with patterns for centuries. So when people say it's finished, it's out of fashion, I say, 'Rubbish, we're just starting!'

The slipover is a good example of the way I design. I spread out all the ANI wools, about fifty shades in all, and immediately saw these marvellous blues and high greens, and a lavender pink. They all worked together nicely, so I put them to one side. I decided to make something really jewel-like, so I picked out the maroon and rust to lift the colours even higher, and *inflame* them. I wanted to use the colours together but in a way that wasn't too overpowering. I wanted the fabric to be conservative but still have that inner light. So I picked out a beautiful neutral colour that was warm but soft for the background, and finished with the charcoal contrast.

Then I put all the yarns I'd chosen together, and sat there and looked at them until an idea emerged of doing a pattern like a formalized snakeskin or some textured thing found in nature. A pattern that could be knitted quite easily and written down, but was still exciting. So I took these simple geometric elements and arranged them in an unexpected way. Although the pattern looks like a 'repeat' at first glance, it isn't – every row is different. So you look and look and *look*, and try to figure out what sense it's making. It's an *adventure*. And the fabric is very rich, with little jewels of colour sparkling all over it. I particularly wanted the slipover to go with the beiges, neutrals, tweeds and browns that men like – all the soft 'nothing' colours that go so well together. When you put the slipover on over them, it adds a bit of light – it makes everything come together and *glow*.

I think there are two definite looks for men – the big chunky look, and things like this that really show and hug the body. The low cut makes it very comfortable on the shoulders, and exaggerates the shape of the arm and the flow of the sleeves. It's very svelte, and it's meant to be nippy, tight fitting and short – it should come down *just* to the belt. But since the length and size of the torso can vary a lot, there are separate instructions for a larger size in a more classic shape.

The slipover looks very handsome with a suit, and you can wear it just as happily with a T-shirt, a plain shirt in any of the colours, or a lightly patterned shirt in the darker shades, which would be my own choice. And you can change the look completely just by substituting a light cream for the charcoal. That one thing would make it very summery – creamy and glowing, something to wear with a crisp white shirt, summer trousers and a tan.

When I designed the cardigan, I looked through a lot of books on old textiles. I wanted a big motif, a nice big cloudlike shape that could be repeated. I came across the carnation on an old Turkish embroidery – it felt right, it had the advantage of being mostly two colours a row, and you could put the streaks right across horizontally.

I particularly wanted to use yarns that are flat with yarns that are more textured and fluffy. Different yarns carry the light in different ways so there are little surprises as the eye moves up the garment. The result is not a completely flat fabric – there's a slight in and out to it. You don't want to press things like this too violently, if at all. You *don't* want it to be like a sheet of plastic. A lot of knitters think that a 'proper' piece of knitting should be so flat and even that it looks like it's popped out of a machine. They don't understand that the slight unevenness is a *desirable* effect, something you're aiming for. You want the knitting to have a bit of *life* to it.

The shape of the cardigan is very generous, the opposite extreme to the slipover. It should be long and full, with a slight drape to it. The sleeves bell in gracefully – they're not too large, just a nice size. If there had been a seam on the shoulder it wouldn't have hung at all well, so it was very important to make it in one piece. I think the fact that the flowers are interrupted on the front adds excitement, and it's balanced by the fact that the flowers all join up on the side.

When you knit this you can choose similar colours, or make up your own combinations. For best results, use at *least* as many colours as we've used, so you have a nice *movement* in the colours. *You could change the nature of the design completely, make it very evening or daytime, depending on the colours you pick.* You should have a ball changing it. If the flowers were cream, I could see the cardigan being worn with cream trousers to an evening cocktail party on the deck of a cruise ship – it has that kind of elegance. You could do it in vivid, dramatic colours and wear it to the opening of a show with smart evening trousers. Or you could wear it just as happily everyday, with jeans.

I'd like to start you on a path of discovery for yourself, to find that personal kind of handwriting that you have *in* yourself, that nobody can teach or tell you. I don't want you just to sit down and make an exact 'Kaffe Fassett outfit' – what I'd like you to see is that this is *an* approach to design. Here's a geometric design, here's a great big motif design. This is the way I approach it, these are the techniques I use. Now take it away, and with what you've seen, you ought to be able to design it for yourself, make it personal, translate it into something that's *you*.

Think of it – this simple medium of handknitting gives you the chance to create exquisite, timeless objects to wear – and lets you build into them exactly what makes you feel on top of the world. It's work, but to make something really beautiful you've got to put in the work. And I know that once you start to do that, you'll see what a *joy* that work is. It's like reading a good book – *you won't want it to end too soon.*'

Postage Stamp Slipover, Small Size

MATERIALS **Yarn:** ANI Real Shetland Wool (2-ply, knits to 3-ply pattern), 3 oz in Dark Natural, 3 oz in Charcoal, 1 oz each of Pale Pink, Petunia, Royal Blue, Jade Green, Lilac, Morar Blue, Rust, Cyclamen, Old Gold and Turquoise. **Needles:** one pair No. 11 British (No. 2 American, 3 mm), one pair No. 9 British (No. 4 American, 3¾ mm). One circular needle size No. 9. One circular needle size No. 11.

MEASUREMENTS One size, to fit chest measurements 34–36 in (86.5–91.5 cm). **Length** from shoulder: 21 in (53.5 cm).

TENSION 7 sts and 7 rows to 1 in (2.5 cm).

ABBREVIATIONS See page 176. **Colours:** CH=Charcoal, N=Natural, PP=Pale Pink, PT=Petunia, RB=Royal Blue, J=Jade Green, L=Lilac, MB=Morar Blue, R=Rust, CY=Cyclamen, T=Turquoise, Z=Old Gold.

BACK

Using CH and No. 11 needles cast on 100 sts working K2 P2 rib for 16 rows in following colours. **Row 1** – CH. **Rows 2–3** – CY. **Row 4** – RB. **Row 5** – MB. **Rows 6–9** – N. **Row 10** – R. **Rows 11–12** – CH. **Row 13** – R. **Rows 14–16** – N. Change to No. 9 needles and CH. K6 inc in next stitch (K5 inc 1) to last 8 stitches K8 (118 sts). Purl one row. Now begin the pattern which is worked in st.st starting with a knit row. Only colour details are given.

Row 1 all CH. **Row 2** 8N * 1CH 13N rep from * to last 12 sts, 1CH 11N. **Row 3** 11N * 1CH 13N rep from * to last 9 sts, 1CH 8N. **Row 4** as 2nd row. **Row 5** 1N * 7PP 3N 1PP 3N rep from * to last 5 sts, 5PP. **Row 6** 4N 1PP * 3N 1CH 3N 1PP 5N 1PP rep from * to last st, 1N. **Row 7** 1N * 1PP 1N 3PT 1N 1PP 3N 1CH 3N rep from * to last 5 sts, 1PP 1N 3PT. **Row 8** * 1PT 1CH 1PT 1N 1PP 3N 1CH 3N 1PP 1N rep from * to last 6 sts, 1PT 1CH 1PT 1N 1PP 1N. **Row 9** as 7th row. **Row 10** as 6th row. **Row 11** as 5th row. **Row 12** as 2nd row. **Row 13** as 3rd row. **Row 14** as 2nd row. **Row 15** all CH. **Row 16** 3N * 1CH 7N rep from * to last 3 sts, 1CH 2N. **Row 17** 2N * 1CH 7N rep from * to last 4 sts, 1CH 3N. **Row 18** 1MB * 2N 1CH 2N 3MB rep from * to last 5 sts, 2N 1CH 2N. **Row 19** * 2N 1CH 2N 1MB 1CH 1MB rep from * to last 6 sts, 2N 1CH 2N 1MB. **Row 20** as 18th row. **Row 21** as 17th row. **Row 22** as 16th row. **Row 23** all CH. **Row 24** * 7N 1CH rep from * to last 6 sts, 6N. **Row 25** * 5J 1N 1CH 1N rep from * to last 6 sts, 5J 1N. **Row 26** as 24th row. **Row 27** all CH. **Rows 28–34** work as for rows 16–22, substituting colour L for colour MB. **Row 35** all CH. **Row 36** * 7N 1CH, rep from * to last 6 sts, 6N. **Row 37** 6N * 1CH 7N rep from * to end. **Row 38** * 2N 3RB 2N 1CH rep from * to last 6 sts, 2N 3RB 1N. **Row 39** 1N * 1RB 1CH 1RB 2N 1CH 2N rep from * to last 5 sts, 1RB 1CH 1RB 2N. **Row 40** as 38th row. **Row 41** as 37th row. **Row 42** as 36th row. **Row 43** all CH. **Row 44** 2N * 1CH 3N rep from * to end. **Row 45** * 1N 1R 1N 1CH rep from * to last 2 sts, 1N 1R. **Row 46** as 44th row. **Row 47** all CH. **Row 48** as 24th row. **Row 49** as 25th row, substituting CY for J. **Row 50** as 24th row. **Row 51** all CH. **Row 52** as 44th row. **Row 53** as 45th row, substituting O for R. **Row 54** as 52nd row. **Row 55** all CH. **Rows 56–62** as 28th to 34th rows, substituting PP for L. **Row 63** all CH. **Row 64** as 44th row. **Row 65** as 45th row, substituting J for R. **Row 66** as 44th row. **Row 67** working in CH, cast off 10 sts, work to end. **Row 68** cast off 10 sts in CH (1st on needle) 4N * 1CH 7N rep from * to last 5 sts, 1CH 2N work 2tog in N. **Row 69** work 2tog in N, 1N * 1CH 7N rep from * to last 6 sts, 1CH 3N work 2tog in N. **Row 70** work 2tog in T, * 2N 1CH 2N 3T rep from * to last 5 sts, 2N 1CH work 2tog in N. **Row 71** work 2tog in CH * 2N 1T 1CH 1T 2N 1CH rep from * to last 3 sts, 1N work 2tog in N. **Row 72** work 2tog in N * 1CH 2N 3T 2N rep from * to last 9 sts, 1CH 2N 3T 1N work 2tog in N. **Row 73** work 2tog in N * 5N 1CH 7N rep from * to last 2 sts, work 2tog in CH. **Row 74** work 2tog in N 6N * 1CH 7N rep from * to last 7 sts, 1CH 4N work 2tog in N. **Row 75** work 2tog in CH, work CH to last 2 sts, work 2tog. **Row 76** work 2tog in N * 1CH 3N rep from * to last 9 sts 1CH 6N work 2tog in N. **Row 77** work 2tog in N 5N * 1CH 1N 1MB 1N 1CH 7N rep from * to last 2 sts, work 2tog in CH. **Row 78** work 2tog in N 1N 3PT * 2N 1CH 3N 1MB 2N 3PT rep from * to last 13 sts, 2N 1CH 3N 1CH 2N 2PT work 2tog in PT. **Row 79** work 2tog in PT 1PT * 2N 5CH 2N 1PT 1CH 1PT rep from * to last 2 sts, work 2tog in N. **Row 80** work 2tog in PT 2PT * 2N 1CH 3N 1CH 2N 3PT rep from * to last 11 sts, 2N 1CH 3N 1CH 2N 2PT. **Row 81** 4N * 1CH 1N 1RB 1N 1CH 7N rep from * to last 10 sts, 1CH 1N 1RB 1N 1CH 5N. **Row 82** 5N * 1CH 3N 1CH 7N rep from * to last 9 sts, 1CH 3N 1CH 4N. **Row 83** all CH. **Row 84** 2N * 1CH 7N 1CH 5N rep from * to last 2 sts, 1CH 1N. **Row 85** * 1N 1CH 1N 3PP 1N 1CH 1N 5PP rep from * to last 4 sts, 1N 1CH 2N. **Row 86** as 84th row. **Row 87** all CH. **Row 88** 3N * 1CH 7N 1CH 11N rep from * to last 11 sts, 1CH 7N 1CH 2N. **Row 89** 2N * 1CH 7N 1CH 11N rep from * to last 12 sts, 1CH 7N 1CH 3N. **Row 90** 3N * 1CH 2N 3R 2N 1CH 11N rep from * to last 11 sts, 1CH 2N 3R 2N 1CH 2N. **Row 91** 2N * 1CH 2N 1R 1CH 1R 2N 1CH 3N 5R 3N rep from * to last 12 sts, 1CH 2N 1R 1CH 1R 2N 1CH 3N. **Row 92** * 3N 1CH 2N 3R 2N 1CH 3N 1R 3CH 1R rep from * to last 14 sts, 3N 1CH 2N 3R 2N 1CH 2N. **Row 93** 2N * 1CH 7N 1CH 3N 1R 1CH 1R 1CH 1R 3N rep from * to last 12 sts, 1CH 7N 1CH 3N. **Row 94** * 3N 1CH 7N 1CH 3N 1R 3CH 1R, rep from * to last 14 sts, 3N 1CH 7N 1CH 2N. **Row 95** 2N * 9CH 3N 5R 3N rep from * to last 12 sts, 9CH 3N. **Row 96** 3N * 1CH 3N 1CH 3N 1CH 11N rep from * to last 11 sts, 1CH 3N 1CH 3N 1CH 2N. **Row 97** 2N * 1CH 1N 1L 1N 1CH 1N 1L 1N 1CH 11N rep from * to last 12 sts, 1CH 1N 1L 1N 1CH 1N 1L 1N 1CH 3N. **Row 98** as 96th row. **Row 99** all CH. **Row 100** 8N * 1CH 13N rep from * to last 10 sts, 1CH 9N. **Row 101** 9N * 1CH 13N rep from * to last 9 sts, 1CH 8N. **Row 102** as 100th row. **Row 103** 6J * 3N 1CH 3N 7J rep from * to last 12 sts, 3N 1CH 3N 5J. **Row 104** 5J * 3N 1CH 3N 7J rep from * to last 13 sts, 3N 1CH 3N 6J. **Row 105** 1J * 3N 2J 3N 1CH 3N 2J rep from * to last 3 sts, 3N. **Row 106** * 1N 1CH 1N 2J 3N 1CH 3N 2J rep from * to last 4 sts, 1N 1CH 1N 1J. **Row 107** as 105th row. **Row 108** as 104th row. **Row 109** as 103rd row. **Row 110** as 100th row. **Row 111** as 101th row. **Row 112** as 100th row. **Row 113** all CH. **Row 114** 3N * 1CH 7N rep from * to last 7 sts, 1CH 6N. **Row 115** 6N * 1CH 7N rep from * to last 4 sts, 1CH 3N. **Row 116** 1Z * 2N 1CH 2N 3Z rep from * to last st, 1N. **Row 117** 1N * 1Z 1CH 1Z 2N 1CH 2N rep from * to last 9 sts, 1Z 1CH 1Z 2N 1CH 2N 1Z. **Row 118** as 116th row. **Row 119** as 115th row. **Row 120** as 114th row. **Row 121** all CH. **Row 122** 4N * 1CH 3N 1CH 3N rep from * to last 10 sts, 1CH 3N 1CH 5N. **Row 123** 4PT * 1N 1CH 1N 1PT 1N 1CH 1N 5PT rep from * to last 10 sts, 1N 1CH 1N 1PT 1N 1CH 1N 3PT. **Row 124** as 122nd row. **Row 125** all CH. **Row 126** as 44th row. **Row 127** as 45th row, substituting CY for R. **Row 128** as 44th row. **Row 129** all CH. **Row 130** 3N * 1CH 3N rep from * to last 5 sts, 1CH 4N. **Row 131** * 3RB 1N 1CH 1N rep from * to last 2 sts, 2RB. **Row 132** * 1CH 1RB 1N 1CH 1N 1RB rep from * to last 2 sts, 1CH 1RB. **Row 133** as 131st row. **Row 134** as 130th row. **Row 135** cast off 14 sts in CH, work to end. **Row 136** cast off 14 sts in CH, leave remaining sts on spare needle.

FRONT

Work as for Back till row 66. **Row 67** using CH cast off 10 sts, work 49 sts in CH (including stitches already on needle). Transfer remaining 59 sts to spare needle and continue to work on left side of neck. **Row 68** work 2tog in N 2N 1CH * 7N 1CH rep from * to last 4 sts, 2N work 2tog in N. **Row 69** work 2tog in N 1N * 1CH 7N rep from * to last 3 sts, 1N work 2tog in N. **Row 70** work 2tog in N * 1CH 2N 3T 2N rep from * to last 3 sts, 1CH work 2tog in N. **Row 71** work 2tog in CH * 2N 1T 1CH 1T 2N 1CH rep from * to last 9 sts, 2N 1T 1CH 1T 2N work 2tog in CH. **Row 72** work 2tog in N 1N * 3T 2N 1CH 2N rep from * to last 6 sts, 3T 1N work 2tog in N. **Row 73** work 2tog in N 5N * 1CH 7N rep from * to end of row. **Row 74** work 2tog in N 5N * 1CH 7N rep from * to last 6 sts, 4N work 2tog in N. **Row 75** work 2tog in CH, work in CH to end of row. **Row 76** work 2tog in N 1CH 7N 1CH 3N 1CH 7N 1CH 3N 1CH 6N work 2tog in N. **Row 77** work 2tog in N 5N * 1CH 1N 1MB 1N 1CH 7N rep from * to last 2 sts, 1CH 1N. **Row 78** work 2tog in CH * 2N 3PT 2N 1CH 3N 1CH rep from * to last 6 sts, 2N 2PT work 2tog in PT. **Row 79** work 2tog in PT 1PT 2N 5CH 2N 1PT 1CH 1PT 2N 5CH 2N 1PT 1CH 1PT 2N 1CH. **Row 80** work 2tog in N, 1N 3PT 2N 1CH 3N 1CH 2N 3PT 2N 1CH 3N 1CH 2N 2PT. **Row 81** 4N * 1CH 1N 1RB 1N 1CH 7N repeat from * to end. **Row 82** work 2tog in N 5N 1CH 3N 1CH 7N 1CH 3N 1CH 4N. **Row 83** all CH. **Row 84** work 2tog in N 3N 1CH 5N 1CH 7N 1CH 5N 1CH 1N. **Row 85** 1N 1CH 1N 3PP 1N 1CH 1N 5PP 1N 1CH 1N 3PP 1N 1CH 1N 3PP. **Row 86** work 2tog in N 2N 1CH 5N 1CH 7N 1CH 5N 1CH 1N. **Row 87** all CH. **Row 88** work 2tog in N 1CH 11N 1CH 7N 1CH 2N. **Row 89** 2N 1CH 7N 1CH 11N 1CH 1N. **Row 90** work 2tog in

CH 11N 1CH 2N 3R 2N 1CH 2N. **Row 91** 2N 1CH 2N 1R 1CH 1R 2N 1CH 3N 5R 3N 1CH. **Row 92** work 2tog in N 2N 1R 3CH 1R 3N 1CH 2N 3R 2N 1CH 2N. **Row 93** 2N 1CH 7N 1CH 3N 1R 1CH 1R 1CH 1R 3N. **Row 94** work 2tog in N 1N 1R 3CH 1R 3N 1CH 3N 1CH 2N. **Row 95** 2N 9CH 3N 5R 2N. **Row 96** work 2tog in N 8N 1CH 3N 1CH 3N 1CH 2N. **Row 97** 2N 1CH 1N 1L 1N 1CH 1N 1L 1N 1CH 9N. **Row 98** 9N 1CH 3N 1CH 3N 1CH 2N. **Row 99** all CH. **Row 100** work 2tog in N 2N 1CH 13N 1CH 1N. **Row 101** 1N 1CH 13N 1CH 3N. **Row 102** 3N 1CH 13N 1CH 1N. **Row 103** 1N 1CH 3N 7J 3N 1CH 3N. **Row 104** work 2tog in N 1CH 3N 7J 3N 1CH 1N. **Row 105** 1N 1CH 3N 2J 3N 2J 3N 1CH 2N. **Row 106** 2N 1CH 3N 2J 1N 1CH 1N 2J 3N 1CH 1N. **Row 107** as 105th row. **Row 108** work 2tog in N 1CH 3N 7J 1CH 1N. **Row 109** 1N 1CH 3N 7J 3N 1CH 1N. **Row 110** 1N 1CH 13N 1CH 1N. **Row 111** as 110th row. **Row 112** work 2tog in N 13N 1CH 1N. **Row 113** all CH. **Row 114** 1N 1CH 7N 1CH 6N. **Row 115** 6N 1CH 7N 1CH 1N. **Row 116** work 2tog in CH 2N 3Z 2N 1CH 2N 3Z 1N. **Row 117** 1N 1Z 1CH 1Z 2N 1CH 2N 1Z 1CH 2N 1CH. **Row 118** 1CH 2N 3Z 2N 1CH 2N 3Z 1N. **Row 119** 6N 1CH 7N 1CH. **Row 120** work 2tog N 6N 1CH 6N. **Row 121** all CH. **Row 122** 4N 1CH 3N 1CH 5N. **Row 123** 4PT 1N 1CH 1N 1PT 1N 1CH 1N 3PT. **Row 124** 4N 1CH 3N 1CH 5N. **Row 125** all CH. **Row 126** 2N 1CH 3N 1CH 3N 1CH 3N. **Row 127** 1N 1CY 1N 1CH 1N 1CY 1N 1CH 1N 1CY 1N 1CH 1N 1CY. **Row 128** 2N 1CH 3N 1CH 3N 1CH 3N. **Row 129** all CH. **Row 130** 3N 1CH 5N 1CH 4N. **Row 131** 3RB 1N 1CH 1N 3RB 1N 1CH 1N 2RB. **Row 132** 1CH 1RB 1N 1CH 1N 1RB 1CH 1RB 1N 1CH 1N 1RB 1CH 1RB. **Row 133** as 131st row. **Row 134** as 130th row. **Row 135** all CH, cast off all stitches. Join CH wool to centre of neck to begin right front.

RIGHT FRONT

Row 67 all CH. **Row 68** cast off 10 sts (1st on needle) 4N 1CH * 7N 1CH rep from * to last 3 sts, 1N work 2tog in N. **Row 69** work 2tog in N * 1CH 7N rep from * to last 6 sts, 1CH 3N work 2tog N. **Row 70** work 2tog in T 2N * 1CH 2N 3T 2N rep from * to last 2 sts, work 2tog CH. **Row 71** work 2tog in N 1N * 1T 1CH 1T 2N 1CH 2N rep from * to last 8 sts, 1N work 2tog N 2N 1CH 2N. **Row 72** work 2tog in N * 1CH 2N 3T 2N rep from * to last 8 sts, 1CH 2N 3T work 2tog in N. **Row 73** 6N * 1CH 7N rep from * to last 2 sts, work 2tog in CH. **Row 74** work 2tog in N 6N * 1CH 7N rep from * to last 6 sts, 4N work 2tog in N. **Row 75** all CH to last 2 sts, work 2tog CH. **Row 76** work 2tog in N * 1CH 7N 1CH 3N rep from * to last 10 sts, 1CH 7N work 2tog in CH. **Row 77** * 1CH 7N 1CH 1N 1MB 1N rep from * to last 10 sts, 1CH 7N work 2tog in CH. **Row 78** work 2tog in N 1N 3PT 2N 1CH 3N 1CH 2N 3PT 2N 1CH 3N 1CH 2N 3PT 1N work 2tog in N. **Row 79** 2N 1PT 1CH 1PT 2N 5CH 2N 1PT 1CH 1PT 2N 5CH 2N 1PT 1CH 1PT work 2tog in N. **Row 80** work 2tog in PT 2PT 2N 1CH 3N 1CH 2N 3PT 2N 1CH 3N 1CH 2N 3PT work 2tog in N. **Row 81** 6N 1CH 1N 1RB 1N 1CH 7N 1CH 1N 1RB 1N 1CH 5N. **Row 82** 5N 1CH 3N 1CH 7N 1CH 3N 1CH 4N work 2tog in N. **Row 83** all CH. **Row 84** 2N 1CH 7N 1CH 5N 1CH 7N 1CH

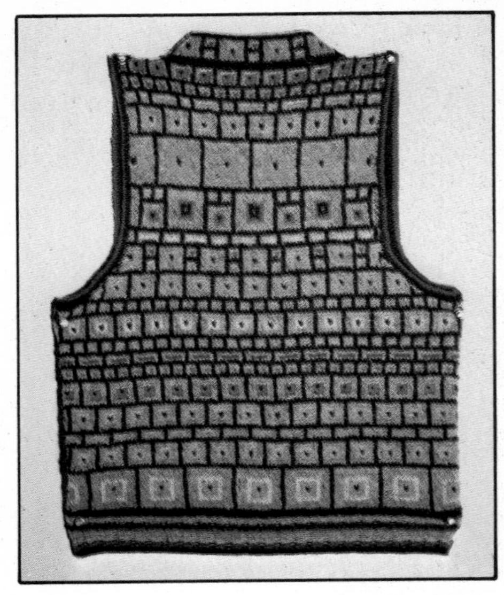

work 2tog in N. **Row 85** 1N 1CH 1N 5PP 1N 1CH 1N 3PP 1N 1CH 1N 5PP 1N 1CH 2N. **Row 86** 2N 1CH 7N 1CH 5N 1CH 7N work 2tog in CH. **Row 87** all CH. **Row 88** 3N 1CH 7N 1CH 11N work 2tog in CH. **Row 89** 1CH 11N 1CH 7N 1CH 3N. **Row 90** 3N 1CH 2N 3R 2N 1CH 10N work 2tog in N. **Row 91** 3N 5R 3N 1CH 2N 1R 1CH 1R 2N 1CH 3N. **Row 92** 3N 1CH 2N 3R 2N 1CH 3N 1R 3CH 1R 1N work 2tog in N. **Row 93** 2N 1R 1CH 1R 1CH 1R 3N 1CH 7N 1CH 3N. **Row 94** 3N 1CH 7N 1CH 3N 1R 3CH 1R work 2tog in N. **Row 95** 1N 5R 3N 9CH 3N. **Row 96** 3N 1CH 3N 1CH 3N 1CH 7N work 2tog in N. **Row 97** 8N 1CH 1N 1L 1N 1CH 1N 1L 1N 1CH 3N. **Row 98** 3N 1CH 3N 1CH 3N 1CH 8N. **Row 99** all CH. **Row 100** 1N 1CH 13N 1CH 2N work 2tog in N. **Row 101** 3N 1CH 13N 1CH 1N. **Row 102** 1N 1CH 13N 1CH 3N. **Row 103** 3N 1CH 3N 7J 3N 1CH 1N. **Row 104** 1N 1CH 3N 7J 3N 1CH 1N work 2tog in N. **Row 105** 2N 1CH 3N 2J 3N 2J 3N 1CH 1N. **Row 106** 1N 1CH 3N 2J 1N 1CH 1N 2J 3N 1CH 2N. **Row 107** as 105th row. **Row 108** 1N 1CH 3N 7J 3N 1CH work 2tog in N. **Row 109** 1N 1CH 3N 7J 3N 1CH 1N. **Row 110** 1N 1CH 13N 1CH 1N. **Row 111** 1N 1CH 13N 1CH 1N. **Row 112** 1N 1CH 13N work 2tog in CH. **Row 113** all CH. **Row 114** 6N 1CH 7N 1CH 1N. **Row 115** 1N 1CH 7N 1CH 6N. **Row 116** 1N 3Z 2N 1CH 2N 3Z 2N work 2tog in CH. **Row 117** 1CH 2N 1Z 1CH 1Z 2N 1CH 2N 1Z 1CH 1Z 1N. **Row 118** 1N 3Z 2N 1CH 2N 3Z 2N 1CH. **Row 119** 1CH 7N 1CH 6N. **Row 120** 6N 1CH 6N work 2tog in N. **Row 121** all CH. **Row 122** 4N 1CH 3N 1CH 5N. **Row 123** 4PT 1N 1CH 1N 1PT 1N 1CH 1N 3PT. **Row 124** 4N 1CH 3N 1CH 5N. **Row 125** all CH. **Row 126** 2N 1CH 3N 1CH 3N 1CH 3N. **Row 127** 1N 1CY 1N 1CH 1N 1CY 1N 1CH 1N 1CY 1N 1CH 1N 1CY. **Row 128** as 126th row. **Row 129** all CH. **Row 130** 3N 1CH 5N 1CH 4N. **Row 131** 3RB 1N 1CH 1N 3RB 1N 1CH 1N 2RB. **Row 132** 1CH 1RB 1N 1CH 1N 1RB 1CH 1RB 1N 1CH 1N 1RB 1CH 1RB. **Row 133** as 131st row. **Row 134** as 130th row. **Row 135** cast off in CH.

NECKBAND

Join shoulder seams and press. Using size 9 circular needle (or 4 double-pointed needles) and with right side of work facing, pick up in CH wool 70 sts on neck edge of right front, 71 sts on neck edge of left front and 46 sts from spare needle for back of neck. Mark this point with a coloured thread. **Round 1** K in CH. **Round 2** 2N * 1CH 7N 1CH 3N rep from * 4 times more, 1CH 5N sl.1 K1N PSSO K2tog in N, 5N 1CH 3N rep from * to last 10 sts of round, 1CH 7N 1CH 1N. **Round 3** K1MB 1N * 1CH 7N 1CH 1N 1MB 1N rep from * 4 times more, 1CH 4N sl.1 K1N PSSO K2tog in N 4N 1CH 1N 1MB 1N rep from * to last 10 sts of round, 1CH 7N 1CH 1N. **Round 4** K2N * 1CH 2N 3PT 2N 1CH 3N rep from * 4 times more, 1CH 2N 1PT sl.1 K1 in PT PSSO K2tog in PT 1PT 2N 1CH 3N rep from * to last 10 sts of round, 1CH 2N 3PT 2N 1CH 1N. **Round 5** K3CH * 2N 1PT 1CH 1PT 2N 5CH rep from * 4 times more, 2N sl.1 K1 in PT PSSO K2tog in PT 2N 5CH rep from * to last 9 sts of round, 2N 1PT 1CH 1PT 2N 2CH. **Round 6** K2N * 1CH 2N 3PT 2N 1CH 3N rep 4 times more from *, 1CH 1N sl.1 K1 in N PSSO K2tog in N 1N 1CH 3N rep from * till last 10 sts of round, 1CH 2N 3PT 2N 1CH 1N. **Round 7** K1RB 1N * 1CH 7N 1CH 1N 1RB 1N rep from * 4 times more, 1CH sl.1 K1 in N PSSO K2tog in N 1CH 1N 1RB 1N rep from * to last 10 sts, 1CH 7N 1CH 1N. **Round 8** K2N * 1CH 7N 1CH 3N rep from * 4 times more, sl.1 K1 in CH PSSO K2tog in CH 3N, rep from * to last 10 sts, 1CH 7N 1CH 1N. **Round 9** K61CH sl.1 K1 in CH PSSO K2tog in CH, K CH to end of round. **Round 10** change to No. 11 needle. K60 sts in CY, sl.1 K1 in CY PSSO K2tog in CY, K to end of round. **Round 11** P in CY. **Round 12** K59 in CY inc in next st K1, inc in next st, K to end in CY. **Round 13** complete neck in CH. K60, inc in next st K1, inc in next stitch, K to end. **Round 14** K61 inc in next st K1, inc in next st K to end. **Round 15** K62 inc in next st K1, inc in next st K to end. **Round 16** K63 inc in next st K1, inc in next st K to end. **Round 17** K64 inc in next st K1, inc in next st K to end. **Round 18** K65 inc in next st K1, inc in next st K to end. **Round 19** K66 inc in next st K1, inc in next st K to end. **Round 20** K67 inc in next st K1, inc in next st K to end. **Round 21** K68 inc in next st K1, inc in next st K to end. **Round 22** K69 inc in next st K1, inc in next st K to end. **Round 23** K70 inc in next st K1, inc in next st K to end. **Round 24** K71 inc in next st K1, inc in next st K to end. **Round 25** K72 inc in next st K1, inc in next st K to end. Cast off.

ARMHOLES

Both worked the same. Using size 11 needle and CH yarn, with right side facing pick up 155 sts. **Row 1** P in CH. **Row 2** K in CY. **Row 3** K in CY. **Row 4** K in CY. **Row 5** P in CH. **Row 6** K in CH. **Row 7** P in CH. **Row 8** K 10CH inc in next st K4 inc in next st K4 inc in next st till the last 21 sts inc in next st K4 inc in next st K4 inc in next st K10. **Row 9** P in CH. Cast off. Sew up side seams and lightly press finished garment.

Classic Postage Stamp Slipover

MATERIALS

Yarn: ANI Real Shetland Wool (2-ply, knits to 3-ply pattern), 4 oz in Light Natural, 3 oz in Charcoal, 1 oz each in Pale Pink, Medium Pink, Saxe Blue, Emerald Green, Lilac, Morar Blue, Rust, Cyclamen, Light Yellow and Turquoise. **Needles:** one pair No. 11 British (No. 2 American, 3 mm), one pair No. 9 British (No. 4 American, 3¾ mm), one circular needle size No. 9 British or equivalent, one circular needle size No. 11 British or equivalent.

MEASUREMENTS

One size to fit chest measurements 38–40 in (96.5–101.5 cm). **Length** from shoulder: 25 in (63.5 cm).

TENSION

7 sts and 7 rows to 1 in (2.5 cm).

ABBREVIATIONS

See page 176.
Colours: CH=Charcoal, N=Light Natural, PP=Pale Pink, PT=Medium Pink, RB=Saxe Blue, J=Emerald Green, L=Lilac, MB=Morar Blue, R=Rust, CY=Cyclamen, Z=Light Yellow, T=Turquoise.

BACK

Using CH and No. 11 needles cast on 108 sts working 16 rows in K2P2 rib as follows. **Row 1** CH. **Rows 2–3** CY. **Row 4** RB. **Row 5** MB. **Rows 6–9** N. **Row 10** R. **Rows 11–12** CH. **Row 13** R. **Rows 14–16** – N. Change to No. 9 needles and CH yarn. K 1 row inc 33 sts evenly across row (141 sts). Then P 1 row. Now begin the pattern, which is worked in stocking stitch starting with a K row. Only colour details are given.

Row 1 all CH. **Row 2** * 1CH 13N * rep from * to * to last st, 1CH. **Rows 3–4** as 2nd row. **Row 5** * 1CH 3N 7PP 3N * rep from * to * to last st, 1CH. **Row 6** *1CH 3N 1PP 5N 1PP 3N * rep from * to * to last st, 1CH. **Row 7** * 1CH 3N 1PP 1N 3PT 1N 1PP 3N * rep from * to * to last st, 1CH. **Row 8** * 1CH 3N 1PP 1N 1PT 1CH 1PT 1N 1PP 3N * rep from * to * to last st, 1CH. **Row 9** as 7th row. **Row 10** as 6th row. **Row 11** as 5th row. **Row 12** as 2nd row. **Row 13** as 3rd row. **Row 14** as 2nd row. **Row 15** all CH. **Row 16** 2N * 1CH 7N * rep from * to * to last 3 sts, 1CH 2N. **Row 17** as 16th row. **Row 18** 2N * 1CH 2N 3MB 2N * rep from * to * to last 3 sts, 1CH 2N. **Row 19** 2N * 1CH 2N 1MB 1CH 1MB 2N * rep from * to * to last 3 sts, 1CH 2N. **Row 20** as 18th row. **Rows 21–22** as 16th row. **Row 23** all CH. **Row 24** 6N * 1CH 7N * rep from * to * to last 6 sts, 6N. **Row 25** * 5J 1N 1CH 1N * rep from * to * to last 5 sts, 5J. **Row 26** as 24th row. **Row 27** all CH. **Rows 28–34** as for rows 16–22, substituting colour L for MB. **Row 35** all CH. **Row 36** 6N * 1CH 7N * rep from * to * but end 6N. **Row 37** as 36th row. **Row 38** 1N * 3RB 2N 1CH 2N * rep from * to * to last 4 sts, 3RB 1N. **Row 39** 1N * 1RB 1CH 1RB 1N 1CH 2N * rep from * to * to last 4 sts, 1RB 1CH 1RB 1CH. **Row 40** as 38th row. **Rows 41–42** as 36th row. **Row 43** all CH. **Row 44** 1CH * 3N 1CH * rep from * to * to end. **Row 45** 1CH * 1N 1R 1N 1CH * rep from * to * to end. **Row 46** as 44th row. **Row 47** all CH. **Row 48** as 24th row. **Row 49** 5CY * 1N 1CH 1N 5CY * rep from * to * to end. **Row 50** as 24th row. **Row 51** all CH. **Row 52** as 44th row. **Row 53** as 45th row, substituting colour Z for R. **Row 54** as 44th row. **Row 55** all CH. **Rows 56–62** as for rows 16–22, substituting colour PP for MB. **Row 63** all CH. **Row 64** 1CH * 3N 1CH * rep from * to * to end. **Row 65** 1CH * 1N 1J 1N 1CH * rep from * to * to end. **Row 66** as 64th row. **Row 67** all CH. **Rows 68–69** 6N 1CH * 7N 1CH * rep from * to * to last 6 sts, 6N. **Row 70** 1N * 3T 2N 1CH 2N * rep from * to * to last 4 sts, 3T 1N. **Row 71** 1N * 1T 1CH 1T 2N 1CH 2N * rep from * to * to last 4 sts, 1T 1CH 1T 1N. **Row 72** as 70th row. **Rows 73–74** as 68th and 69th rows. **Row 75** all CH. **Row 76** 2N * 1CH 3N 1CH 7N * rep from * to * to last 7 sts, 1CH 3N 1CH 2N. **Row 77** 2N * 1CH 1N 1MB 1N 1CH 7N * rep from * to * to last 7 sts, 1CH 1N 1MB 1N 1CH 2N. **Row 78** 2N * 1CH 3N 1CH 2N 3PT 2N * rep from * to * to last 7 sts, 1CH 3N 1CH 2N. **Row 79** 2N * 5CH 2N 1PT 1CH 1PT 2N * rep from * to * to last 7 sts, 5CH 2N. **Row 80** as 78th row. **Row 81** 2N * 1CH 1N 1RB 1N 1CH 7N * rep from * to * to last 7 sts, 1CH 1N 1RB 1N 1CH 2N. **Row 82** as 76th row. **Row 83** all CH. **Row 84** 1CH * 7N 1CH 5N 1CH * rep from * to * to end. **Row 85** * 1CH 1N 3PP 1N 1CH 1N 5PP 1N * rep from * to * to last st, 1CH. **Row 86** as 84th row. **Row 87** all CH. **Row 88** 1CH * 7N 1CH 11N 1CH * rep from * to * to end. **Row 89** as 88th row. **Row 90** 1CH * 2N 3R 2N 1CH 11N 1CH * rep from * to * to end. **Row 91** 1CH * 3N 5R 3N 1CH 2N 1R 1CH 1R 2N 1CH rep from * to * to end. **Row 92** 1CH * 2N 3R 2N 1CH 3N 1R 3CH 1R 3N 1CH rep from * to * to end. **Row 93** cast off 10 sts then continue 1N 1CH 7N 1CH * 3N 1R 1CH 1R 1CH 1R 3N 1CH 7N 1CH * rep from * to * to end. **Row 94** cast off 10 sts then continue 1N 1R 3CH 1R 3N 1CH * 7N 1CH 3N 1R 3CH 1R 3N 1CH * rep from * to * to last 10 sts, 7N 1CH 2N. **Row 95** work 2tog CH * 9CH 3N 5R 3N * rep from * to * to last 10 sts, 3N 5R work 2tog. **Row 96** work 2tog in N 7N * 1CH 3N 1CH 3N 1CH 11N * rep from * to * to last 10 sts, 1CH 3N 1CH 3N work 2tog. **Row 97** work 2tog in N * 1L 1N 1CH 1N 1L 1N 1CH 11N 1CH 1N * rep from * to * to last 15 sts, 1L 1N 1CH 1N 1L 1N 1CH 6N work 2tog. **Row 98** work 2tog in N 5N * 1CH 3N 1CH 3N 1CH 11N * rep from * to * to last 8 sts, 1CH 3N 1CH 1N work 2tog. **Row 99** work 2tog, work across in CH to last 2 sts, work 2tog. **Row 100** work 2tog 11N * 1CH 13N * rep from * to * to last 14 sts, 1CH 11N work 2tog. **Row 101** work 2tog in N 10N * 1CH 13N * rep from * to * to last 13 sts, 1CH 10N work 2tog. **Row 102** work 2tog in N 9N * 1CH 13N * rep from * to * to last 12 sts, 1CH 9N work 2tog. **Row 103** work 2tog in J 5J * 3N 1CH 3N 7J * rep from * to * to last 14 sts, 3N 1CH 3N 5J work 2tog. **Row 104** work 2tog 4J * 3N 1CH 3N 7J * rep from * to * to last 13 sts, 3N 1CH 3N 4J work 2tog. **Row 105** work 2tog in N 1N 2J 3N 1CH * 3N 2J 3N 2J 3N 1CH * rep from * to * to last 8 sts, 3N 2J 1N work 2tog. **Row 106** work 2tog in N 2J 3N 1CH * 3N 2J 1N 1CH 1N 2J 3N 1CH * rep from * to * to last 7 sts, 3N 2J work 2tog. **Row 107** 1N 2J 3N 1CH * 3N 2J 3N 2J 3N 1CH * rep from * to * to last 6 sts, 3N 2J 1N. **Row 108** 3J 3N 1CH * 3N 7J 3N 1CH * rep from * to * to last 6 sts, 3N 3J. **Row 109** as 108th row. **Row 110** 6N 1CH * 13N 1CH * rep from * to * to last 6 sts, 6N. **Row 111** as 110th row. **Row 112** as 110th row. **Row 113** all CH. **Row 114** 1CH * 7N 1CH * rep from * to * to end. **Row 115** as 114th row. **Row 116** 1CH * 2N 3Z 2N 1CH * rep from * to * to end. **Row 117** 1CH * 2N 1Z 1CH 1Z 2N 1CH * rep from * to * to end. **Row 118** as 116th row. **Row 119** as 114th row. **Row 120** as 114th row. **Row 121** all CH. **Row 122** 1CH * 7N 1CH 3N 1CH * rep from * to * to end. **Row 123** 1CH * 1N 1PT 1N 1CH 1N 5PT 1N 1CH * rep from * to * to end. **Row 124** as 122nd row. **Row 125** all CH. **Row 126** 1CH * 3N 1CH * rep from * to * to end. **Row 127** 1CH * 1N 1CY 1N 1CH * rep from * to * to end. **Row 128** as 126th row. **Row 129** all CH. **Row 130** 1CH * 5N 1CH * rep from * to * to end. **Row 131** 1CH * 1N 3RB 1N 1CH * rep from * to * to end. **Row 132** 1CH * 1N 1RB 1CH 1RB 1N 1CH * rep from * to * to end. **Row 133** as 131st row. **Row 134** as 130th row. **Row 135** all CH. **Rows 136–142** as for rows 114–120 incl substituting colour L for Z. **Row 143** all CH. **Row 144** 4N 1CH * 7N 1CH * rep from * to * to last 4 sts, 4N. **Row 145** as 144th row. **Row 146** 2T 2N 1CH * 2N 3T 2N 1CH * rep from * to * to last 4 sts, 2N 2T. **Row 147** 1CH 1T 2N 1CH * 2N 1T 1CH 1T 2N 1CH * rep from * to * to last 4 sts, 2N 1T 1CH. **Row 148** as 145th row. **Row 149** as 144th row. **Row 150** as 144th row. **Row 151** all CH. **Row 152** 2N 1CH * 3N 1CH * rep from * to * to last 2 sts, 2CH. **Row 153** 1R 1N 1CH * 1N 1R 1N 1CH * rep from * to * to last 2 sts, 1N 1R. **Row 154** as 152nd row. **Row 155** all CH. **Row 156** 1CH * 7N 1CH * rep from * to * to end. **Row 157** 1CH * 1N 5J 1N 1CH * rep from * to * to end. **Row 158** as 156th row. **Row 159** in CH cast off 24 sts K49 sts (including one st on right needle) cast off remaining 24 sts.

LEFT FRONT

Work as for Back until row 92. **Row 93** cast off 10 sts 1N 1CH 7N 1CH * 3N 1R 1CH 1R 1CH 1R 3N 1CH * rep from * to * once, 7N 1CH 3N 1R 1CH 1R 1CH 1R 1N turn. Work on these sts for left front. **Row 94** work 2tog in R 3CH 1R 3N 1CH * 7N 1CH 3N 1R 3CH 1R 3N 1CH * rep from * to * once more, 7N 1CH work 2tog in N. **Row 95** work 2tog in CH 8CH * 3N 5R 3N 9CH * rep from * to * once, 3N 3R work 2tog in R. **Row 96** work 2tog in N 5N 1CH * 3N 1CH 3N 1CH 11N 1CH * rep from * to * once, 3N 1CH 2N work 2tog in N. **Row 97** work 2tog in L * 1N 1CH 1N 1L 1N 1CH 11N 1CH 1N 1L * rep from * to * once, 1N 1CH 1N 1L 1N 1CH 4N work 2tog in N. **Row 98** work 2tog in N 3N 1CH * 3N 1CH 3N 1CH 11N 1CH * rep from * to * once, 3N 1CH work 2tog in N. **Row 99** work 2tog in CH, all CH to end work 2tog. **Row 100** work 2tog in N, 6N 1CH 13N 1CH 13N 1CH 10N work 2tog in N. **Row 101** work 2tog in N, 9N 1CH 13N 1CH 13N 1CH 6N work 2tog. **Row 102** work 2tog in N 5N 1CH 13N 1CH 13N 1CH 8N work 2tog. **Row 103** work 2tog in J, 3N 1CH * 3N 7J 3N 1CH * rep from * to * once, 3N 2J work 2tog in J. **Row 104** work 2tog in J, 1J 3N 1CH * 3N 7J 3N 1CH * rep from * to * once, 3N 3J work 2tog in J. **Row 105** work 2tog in N 2J 3N 1CH * 3N 2J 3N 2J 3N 1CH

* rep from * to * once, 3N 2J. **Row 106** work 2tog in J, 3N 1CH 3N 2J 1N 1CH 1N 2J 3N 1CH * rep from * to * once, 3N 2J 1N. **Row 107** 1N 2J 3N 1CH * 3N 2J 3N 2J 3N 1CH * rep from * to * once, 3N 1J. **Row 108** work 2tog in N 2N 1CH * 3N 7J 3N 1CH * rep from * to * once, 3N 3J. **Row 109** 3J 3N 1CH * 3N 7J 3N 1CH * rep from * to * once, 1CH 3N. **Row 110** work 2tog in N 1N 1CH * 13N 1CH * rep from * to * once, 6N. **Row 111** 6N 1CH 13N 1CH 13N 1CH 2N. **Row 112** work 2tog in N 1CH 13N 1CH 13N 1CH 6N. **Row 113** all CH. **Row 114** work 2tog in N 1N 1CH * 7N 1CH * rep from * to * end. **Row 115** 1CH * 7N 1CH * rep from * to * to last 2 sts, 2N. **Row 116** work 2tog in N 1CH * 2N 3Z 2N 1CH * rep from * to * to end. **Row 117** 1CH * 2N 1Z 1CH 1Z 2N 1CH * rep from * to * to last st, 1N. **Row 118** work 2tog in CH * 2N 3Z 2N 1CH * rep from * to * to end. **Row 119** 1CH * 7N 1CH * rep from * to * to end. **Row 120** work 2tog in N 6N 1CH * 7N 1CH * rep from * to * to end. **Row 121** all CH. **Row 122** work 2tog in N 1N 1CH * 3N 1CH 7N 1CH * rep from * to * to last 4 sts, 3N 1CH. **Row 123** 1CH * 1N 1PT 1N 1CH 1N 5PT 1N 1CH * rep from * to * to last 6 sts, 1N 1PT 1N 1CH 2N. **Row 124** work 2tog in N 1CH 3N 1CH 7N 1CH rep from * to * to last 4 sts, 3N 1CH. **Row 125** all CH. **Row 126** 1N 1CH * 3N 1CH * rep from * to * to end. **Row 127** 1CH * 1N 1CY 1N 1CH * rep from * to * to last st, 1N. **Row 128** work 2tog in CH * 3N 1CH * rep from * to * to end. **Row 129** all CH. **Row 130** 4N 1CH * 5N 1CH * rep from * to * to end. **Row 131** 1CH * 1N 3RB 1N 1CH * rep from * to * to last 4 sts, 1N 3RB. **Row 132** work 2tog in CH 1RB 1N 1CH * 1N 1RB 1CH 1RB 1N 1CH * rep from * to * to end. **Row 133** 1CH * 1N 3RB 1N 1CH * rep from * to * to last 3 sts, 1N 2RB. **Row 134** 3N 1CH * 5N 1CH * rep from * to * to end. **Row 135** all CH. **Row 136** work 2tog in N 1N 1CH * 7N 1CH * rep from * to * to end. **Row 137** 1CH * 7N 1CH * rep from * to * to last 2 sts, 2N. **Row 138** 2N 1CH 2N 3L 2N 1CH * rep from * to * to end. **Row 139** 1CH * 2N 1L 1CH 1L 2N 1CH * rep from * to * to last 2 sts, 2N. **Row 140** work 2tog in N 1CH * 2N 3L 2N 1CH * rep from * to * to end. **Row 141** 1CH * 7N 1CH * rep from * to * to last st, 1N. **Row 142** 1N 1CH * 7N 1CH * rep from * to * to end. **Row 143** all CH. **Row 144** work 2tog in N 3N 1CH * 7N 1CH * rep from * to * to last 4 sts, 4N. **Row 145** 4N 1CH * 7N 1CH * rep from * to * to last 4 sts, 4N. **Row 146** 2T 2N 1CH * 2N 3T 2N 1CH * rep from * to * to last 4 sts, 2N 2T. **Row 147** 1CH 1T 2N 1CH * 2N 1T 1CH 1T 2N 1CH * rep from * to * to last 4 sts, 2N 1T 1CH. **Row 148** work 2tog in T 2N 1CH * 2N 3T 2N 1CH * rep from * to * to last 4 sts, 2N 2T. **Row 149** 4N 1CH * 7N 1CH * rep from * to * to last 3 sts, 3N. **Row 150** 3N 1CH * 7N 1CH * rep from * to * to last 4 sts, 4N. **Row 151** all CH. **Row 152** 1N 1CH * 3N 1CH * rep from * to * to last 2 sts, 2N. **Row 153** 1R 1N 1CH * 1N 1R 1N 1CH * rep from * to * to last st, 1N. **Row 154** as 152nd row. **Row 155** all CH. **Row 156** * 7N 1CH * rep from * to * to end. **Row 157** 1CH 1N 5J 1N 1CH * rep from * to * to last 7 sts, 1N 5J 1N. **Row 158** as 156th row.

RIGHT FRONT

Rejoin wool at centre of work. **Row 93** 2N 1CH * 7N 1CH 3N 1R 1CH 1R 1CH 1R 3N 1CH * rep from * to * to end. **Row 94** cast off 10 sts 1N 1R

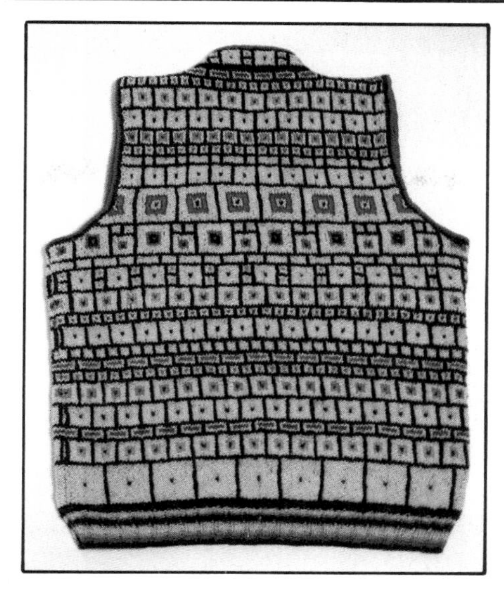

3CH 1R 3N 1CH * 7N 1CH 3N 1R 3CH 1R 3N 1CH * rep from * to * once, 7N 1CH work 2tog. **Row 95** work 2tog in CH 8CH * 3N 5R 3N 9CH * rep from * to * once, 3N 5R work 2tog in N. **Row 96** work 2tog in N 7N 1CH * 3N 1CH 3N 1CH 11N 1CH * rep from * to * once, 3N 1CH 2N work 2tog in N. **Row 97** work 2tog in L * 1N 1CH 1N 1L 1N 1CH 11N 1CH 1N 1L * rep from * to * once, 1N 1CH 1N 1L 1N 1CH 6N work 2tog. **Row 98** work 2tog 5N 1CH * 3N 1CH 3N 1CH 11N 1CH * 3N 1CH work 2tog in N. **Row 99** work 2tog in CH, work all across in CH, work 2tog. **Row 100** work 2tog in N 11N * 1CH 13N * rep from * to * once, 1CH 6N, work 2tog in N. **Row 101** work 2tog in N 5N 1CH * 13N 1CH * rep from * to * once, 10N work 2tog. **Row 102** work 2tog in N 9N 1CH * 13N 1CH * rep from * to * once, 4N work 2tog in N. **Row 103** work 2tog in J 3N 1CH * 3N 7J 3N 1CH * rep from * to * once, 3N 5J work 2tog in J. **Row 104** work 2tog in J 4J 3N 1CH * 3N 7J 3N 1CH * rep from * to * once, 2N work 2tog in N. **Row 105** 3N 1CH * 3N 2J 3N 2J 3N 1CH * rep from * to * once, 3N 2J 1N work 2tog in N. **Row 106** work 2tog in N 2J 3N * 1CH 3N 2J 1N 1CH 1N 2J 3N * rep from * to * once, 1CH 1N work 2tog. **Row 107** 2N 1CH * 3N 2J 3N 2J 3N 1CH * rep from * to * once, 3N 2J 1N. **Row 108** 3J 3N 1CH * 3N 7J 3N 1CH * rep from * to * once, 2N. **Row 109** 2N 1CH * 3N 7J 3N 1CH * rep from * to * once, 3N 3J. **Row 110** 6N 1CH * 13N 1CH * rep from * to * once, work 2tog. **Row 111** 1N 1CH * 13N 1CH * rep from * to * once, 6N. **Row 112** 6N 1CH * 13N 1CH * rep from * to * once, 6N. **Row 113** all CH. **Row 114** 1CH * 7N 1CH * rep from * to * to last 3 sts, 1N work 2tog. **Row 115** 2N 1CH * 7N 1CH * rep from * to * to end. **Row 116** 1CH * 2N 3Z 2N 1CH * rep from * to * to last 2 sts, 2N. **Row 117** 2N 1CH * 2N 1Z 1CH 1Z 2N 1CH * rep from * to * to end. **Row 118** 1CH * 2N 3Z 2N 1CH * rep from * to * to last 2 sts, work 2tog. **Row 119** 1N 1CH * 7N 1CH * rep from * to * to end. **Row 120** 1CH * 7N 1CH * rep from * to * to last st, 1N. **Row 121** all CH. **Row 122** 1CH * 7N 1CH 3N 1CH * rep from * to * once, 7N work 2tog in CH. **Row 123** 1CH 1N 5PT 1N 1CH * 1N 1PT 1N 1CH 1N 5PT 1N 1CH * rep from * to * to end. **Row 124** 1CH * 7N 1CH 3N 1CH * rep from * to * to last 8 sts, 7N 1CH. **Row 125** all CH. **Row 126** * 1CH 3N * rep from * to * to last 5 sts, 1CH 2N work 2tog. **Row 127** * 1N 1CY 1N 1CH * rep from * to * to end. **Row 128** * 1CH 3N * rep from * to * to end. **Row 129** all CH. **Row 130** * 1CH 5N * rep from * to * to last 2 sts, work 2tog. **Row 131** 1CH * 1N 3RB 1N 1CH * rep from * to * to end. **Row 132** 1CH * 1N 1RB 1CH 1RB 1N 1CH * rep from * to * to end. **Row 133** as 131st row. **Row 134** 1CH * 5N 1CH * rep from * to * to last 6 sts, 4N work 2tog. **Row 135** all CH. **Row 136** * 1CH 7N * rep from * to * to last 6 sts, 1CH 5N. **Row 137** 5N 1CH * 7N 1CH * rep from * to * to end. **Row 138** 1CH * 2N 3L 2N 1CH * rep from * to * to last 5 sts, 2N 1L work 2tog in L. **Row 139** 1CH 1L 2N 1CH * 2N 1L 1CH 1L 2N 1CH * rep from * to * to end. **Row 140** 1CH * 2N 3L 2N 1CH * rep from * to * to last 4 sts, 2N 2L. **Row 141** 4N 1CH * 7N 1CH * rep from * to * to end. **Row 142** 1CH * 7N 1CH * rep from * to * to last 4 sts, 2N work 2tog in N. **Row 143** all CH. **Row 144** 4N * 1CH 7N * rep from * to * to end. **Row 145** * 7N 1CH * rep from * to * to last 4 sts, 4N. **Row 146** 2T 2N 1CH * 2N 3T 2N 1CH * rep from * to * once, 2N 3T work 2tog. **Row 147** 1N 1T 1CH 1T 2N 1CH * 2N 1T 1CH 1T 2N 1CH * rep from * to * once, 2N 1T 1CH. **Row 148** 2T 2N 1CH * 2N 3T 2N 1CH * rep from * to * once to last 6 sts, 2N 3T 1N. **Row 149** 6N 1CH 7N 1CH 7N 1CH 4N. **Row 150** 4N 1CH 7N 1CH 7N 1CH 4N work 2tog. **Row 151** all CH. **Row 152** 2N 1CH * 3N 1CH * rep from * to * to last 3 sts, 3N. **Row 153** * 1N 1R 1N 1CH * rep from * to * to last 2 sts, 1N 1R. **Row 154** 2N 1CH * 3N 1CH * rep from * to * to last 3 sts, 1N work 2tog. **Row 155** all CH. **Row 156** 1CH * 7N 1CH * rep from * to * to end. **Row 157** 1CH * 1N 5J 1N 1CH * rep from * to * to end. **Row 158** * 1CH 7N * rep from * to * to last 9 sts, 1CH 6N work 2tog. Cast off remaining 24 stitches. Join shoulder seams.

NECKBAND

Using No. 9 circular needle or 4 size 9 double pointed needles and CH yarn, with right side of work facing pick up 70 sts down right front, 69 sts up left front. K sts from back dec 1 stitch at centre of back (48 sts). **Round 1** K in CH. **Round 2** 2N * 1CH 7N 1CH 3N * repeat from * to * 4 times more, 1CH 5N sl.1 K1 in N PSSO K2tog in N 5N 1CH 3N rep from * to * to last 10 sts, 1CH 7N 1CH 1N. **Round 3** 1MB 1N * 1CH 7N 1CH 1N 1MB 1N * rep from * to * 4 times, 1CH 4N sl.1 K1 in N PSSO K2 tog in N, 4N 1CH 1N 1MB 1N rep from * to * to last 10 sts, 1CH 7N 1CH 1N. **Round 4** 2N * 1CH 2N 3PT 2N 1CH 3N * rep from * to * 4 times, 1CH 2N 1PT sl.1 K1 in PT PSSO K2tog in PT 1PT 2N 1CH 3N rep from * to * to last 10 sts, 1CH 2N 3PT 2N 1CH 1N. **Round 5** 3CH * 2N 1PT 1CH 1PT 2N 5CH * rep from * to * 4 times, 2N sl.1 K1 in PT PSSO K2tog in PT 2N 5CH rep from * to * to last 9 sts, 2N 1PT 1CH 1PT 2N 2CH. **Round 6** 2N * 1CH 2N 3PT 2N 1CH 3N * rep 4 times, 1CH 1N sl.1 K1 in N PSSO K2tog in N, 1N 1CH 3N rep from * to * to last 10 sts, 1CH 2N 3PT 2N 1CH 1N. **Round 7** 1RB 1N * 1CH 7N 1CH 1N 1RB 1N * rep from * to * 4 times, 1CH sl.1 K1 in N PSSO K2tog in N, 1CH

1N 1RB 1N rep from * to * to last 10 sts, 1CH 7N 1CH 1N. **Round 8** K2N * 1CH 7N 1CH 3N * rep from * to * 4 times, sl.1 K1 in CH PSSO K2tog in CH, 3N repeat from * to * to last 10 sts, 1CH 7N 1CH 1N. **Round 9** K61CH sl.1 K1 in CH PSSO K2tog in CH, K in CH to end. **Round 10** change to No. 11 needles and K60 sts in CY sl.1 K1 in CY PSSO K2tog in CY, K to end. **Round 11** purl in CY. **Round 12** K59 in CY, inc in next st K1 inc in next st, K to end in CY. **Round 13** complete neck in CH, K60 inc in next st, K1 inc in next st, K to end. **Round 14** K61 inc in next st, K1 inc in next st, K to end. **Round 15** K62 inc in next st, K1 inc in next st, K to end. **Round 16** K63 inc in next st, K1 inc in next st, K to end. **Round 17** K64 inc in next st, K1 inc in next st, K to end. **Round 18** K65 inc in next st, K1 inc in next st, K to end. **Round 19** K66 inc in next st, K1 inc in next st, K to end. **Round 20** K67 inc in next st, K1 inc in next st, K to end. **Round 21** K68 inc in next st, K1 inc in next st, K to end. **Round 22** K69 inc in next st, K1 inc in next st, K to end. **Round 23** K70 inc in next st, K1 inc in next st, K to end. **Round 24** K71 inc in next st, K1 inc in next st, K to end. **Round 25** K72 inc in next st, K1 inc in next st, K to end. Cast off.

ARMBANDS

Using size 11 needles and CH yarn, with right side facing pick up 155 sts. **Row 1** P in CH. **Row 2** K in CY. **Row 3** K in CY. **Row 4** K in CY. **Row 5** P in CH. **Row 6** K in CH. **Row 7** P in CH. **Row 8** K 10 inc in next st, K4 inc in next st, K4 inc in next st till last 21 sts, inc in next st, K4 inc in next st, K4 inc in next st K10. **Row 9** purl in CH. Cast off. Make up.

Turkish Carnation Cardigan

DETAILS

Dark Turkish Carnation Cardigan shown with antique amethyst necklace from Asprey, silk harem trousers from Arabesque, shoes by Bellesco from Bally and cloisonné handmirror from People's Park, Singapore. Photographed in the Roof Bar of the Hilton Hotel, London. Light Turkish Carnation Cardigan shown with Turkish trousers from Arabesque, *bustier* top from Friends and antique beaded Chinese slippers. Photographed in the Royal Horseguards Hotel, London. Dark cardigan knitted to order by Beatrice Bellini Handknits, London.

MATERIALS: Dark Cardigan

A ANI Real Shetland Wool (2-ply knits to 3-ply pattern), 10 oz in Oxford Grey, **knitted double**

B ANI Real Shetland Wool (2-ply knits to 3-ply pattern), 2 oz each of Purple and Navy, **knitted together**

C ANI Mohair, 1 25 gm ball in Dark Brown and ANI Fine Shetland 2-ply Lace Weight, 1 oz in Navy, **knitted together**

D ANI Mohair, 1 25 gm ball in Mauve, and ANI Fine Shetland 2-ply Lace Weight, 1 oz in Navy, **knitted together**

E ANI Mohair, 2 25 gm balls in Navy, and ANI Fine Shetland 2-ply Lace Weight, 2 oz in Black, **knitted together**

F ANI Real Shetland Wool (2-ply knits to 3-ply pattern), 2 oz in Green Lovat, **knitted double**

G ANI Real Shetland Wool (2-ply knits to 3-ply pattern), 3 oz each of Tweed Brown and Forest Green, **knitted together**

H Sunbeam Aran Knit Wool, 7 50 gm balls in Black

J ANI Real Shetland Wool (2-ply knits to 3-ply pattern), 1 oz in Periwinkle, **knitted double**

K ANI Real Shetland Wool (2-ply knits to 3-ply pattern), 2 oz in Aqua Green, **knitted double**

L ANI Real Shetland Wool (2-ply knits to 3-ply pattern), 1 oz in Royal Blue, **knitted double**

MATERIALS: Light Cardigan

A Sirdar Sherpa (knits as Aran), 6 50 gm balls in Safari

B Sirdar Sherpa (knits as Aran), 3 50 gm balls in Shingle

C Sirdar Caprine, 2 25 gm balls in Birch Mist

D Sirdar Caprine, 2 25 gm balls in Heather Mist

E Sirdar Superwash 4-ply, 5 25 gm balls in Sandalwood **knitted double**

F Sirdar Superwash 4-ply, 2 25 gm balls in Wine, **knitted double**

G Sirdar Fontein Crepe, 6 25 gm balls in Chinese Fig, **knitted double**

H Sirdar Fontein Crepe, 7 25 gm balls in Hawthorn, **knitted double**

J Sirdar Fontein Crepe 4 25 gm balls in Azure, **knitted double**

K Sirdar Fontein Crepe, 4 25 gm balls in Baby Blue, **knitted double**

L Sirdar Fontein Crepe, 4 25 gm balls in Saxe, **knitted double**

Needles: One pair British No. 7 (American No. 8, 4½ mm). One circular needle 1 metre long, British No. 5 (American No. 8, 5½ mm). One circular needle 1 metre long, British No. 7 (American No. 8, 4½ mm).

MEASUREMENTS One size, to fit chest measurements 32–42 in (81–107 cm). **Length** at centre back: 30 in (76 cm). **Body width:** 28 in (71 cm). **Sleeve length** 16½ in (42 cm).

TENSION 5 sts and 5 rows to 1 in (2.5 cm) on size 5 needles.

DESCRIPTION The cardigan is knitted in one piece, starting at the back lower hem, casting on on either side for the sleeves and continuing down the fronts. It has separately made ribbed cuffs and welt. The front bands and neckband are made in one separate piece. The cardigan is knitted from the chart, which has been marked off to assist the knitter to keep her place. Each yarn or combination of yarns has a special mark on the chart. The cardigan is knitted on No. 5 circular needles. The front bands and neckband are knitted on No. 7 circular needles and the welt and cuffs on No. 7 ordinary needles.

ABBREVIATIONS **See** page 176.

SPECIAL NOTE When working the design use separate small balls for each colour, twisting the yarns when changing colour in order to avoid a hole. Colour 'A' yarn is carried right across the pattern, and should be knitted in every fourth stitch. Where necessary knit in other colours likewise.

BACK AND FRONT

Using No. 5 circular needles cast on 128 sts. (Rope edge.) Work from chart in stocking stitch 90 rows ending with a purl row. **Next row**: cast on 64 sts in colour A (Rope edge) for a total of 192 sts. Go to the other end of the circular needle and with a fresh ball of colour A yarn cast on 64 sts (Rope edge) for total of 256 sts. Return to beginning of first cast-on and knit across according to chart. Continue according to chart for a total of 54 rows ending with a purl row. **Next row**: pattern 113 sts. Cast off *loosely* in colour A. Pattern 113 sts. Continue according to chart down fronts, joining new wools for right front, for a further 7 rows. **Next row**: increase one stitch on each side of neck edge, continuing according to chart; inc 1 stitch on each side of neck every following 6 rows 7 times until there are 121 sts on each front. Pattern two more rows. **Next row**: Cast off 64 sts loosely on outsides of both fronts, continuing in chart across the remainder of fronts. Continue following chart for two rows. **Next row**: Increase 1 st at both neck edges on this and every 6th row a total of 7 times (128 sts). Continue according to chart to bottom of fronts. Cast off loosely.

CUFFS

With No. 7 needles and right side of work facing, pick up and K 53 sts in H. **Row 1** K1 * P3 K1 * rep from * to * to end. **Row 2** P1 * K3 P1 * rep from * to * to end. Rep last 2 rows 5 times more. **Row 13** K1 * P2 K2tog * rep from * to * to end. **Row 14** P1 * K2 P1 * rep from * to * to end. **Row 15** K1 * P2 K1 * rep from * to * to end. **Row 16** as row 14. **Rows 17–19** work in rib in G. **Row 20** rib in F. Cast off in rib in F.

LOWER WELT

With No. 7 needles and right side of work facing, pick up and K 205 sts in H. **Row 1** K1 * P3 K1 * rep from * to * to end. **Row 2** P1 * K3 P1 * rep from * to * to end. Rep last 2 rows 7 times more. Work 3 rows G. Work 1 row F. Cast off in F.

FRONT BAND AND NECKBAND

With No. 7 needles and right side of work facing, pick up and K 305 sts in H. **Row 1** K1 * P3 K1 * rep from * to * to end. **Row 2** P1 * K3 P1 * rep from * to * to end. Rep last 2 rows 5 times more. Work 3 rows G. Work 1 row F. Cast off in F. Darn in all ends, press and make up.

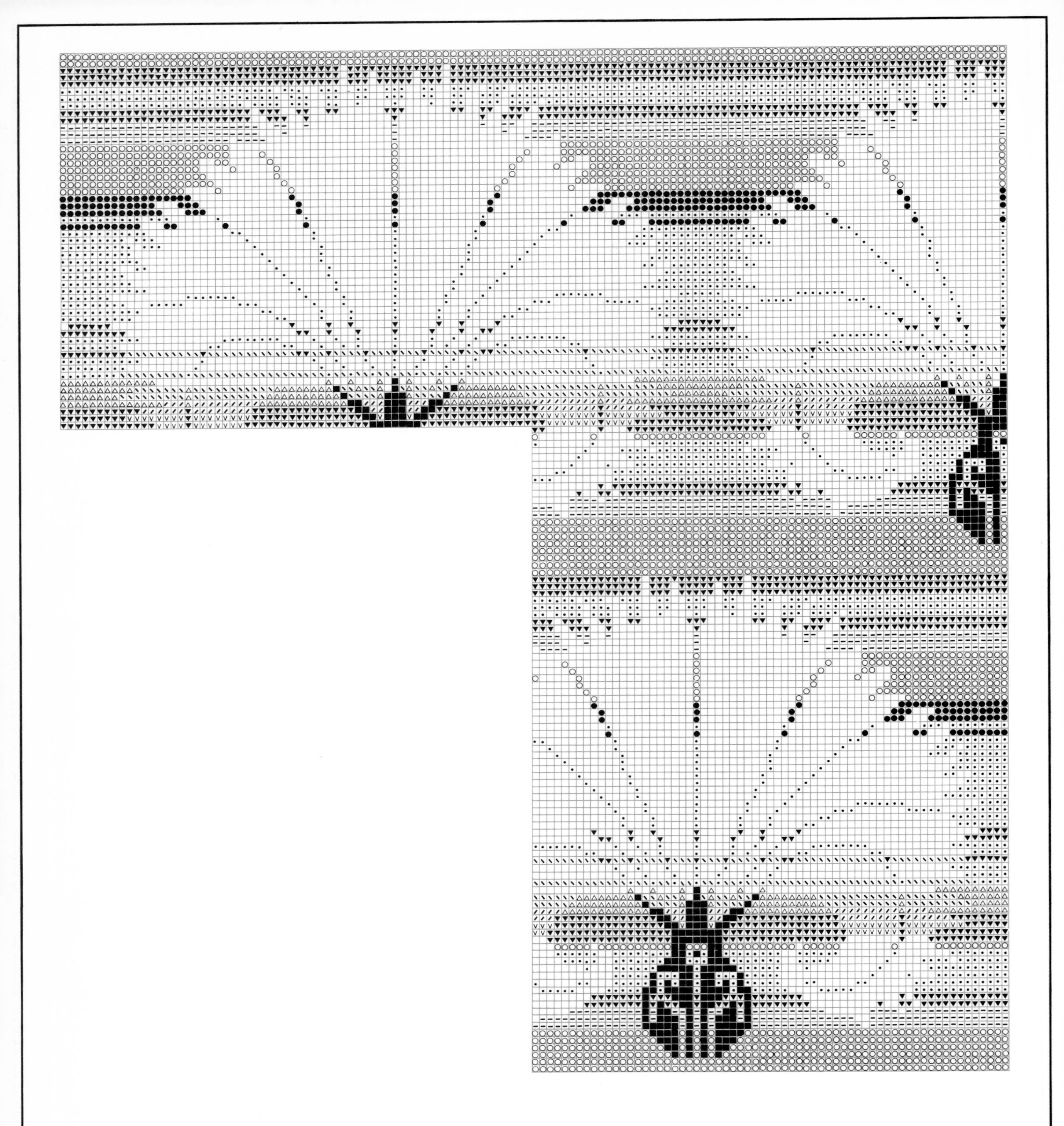

Continued on page 139

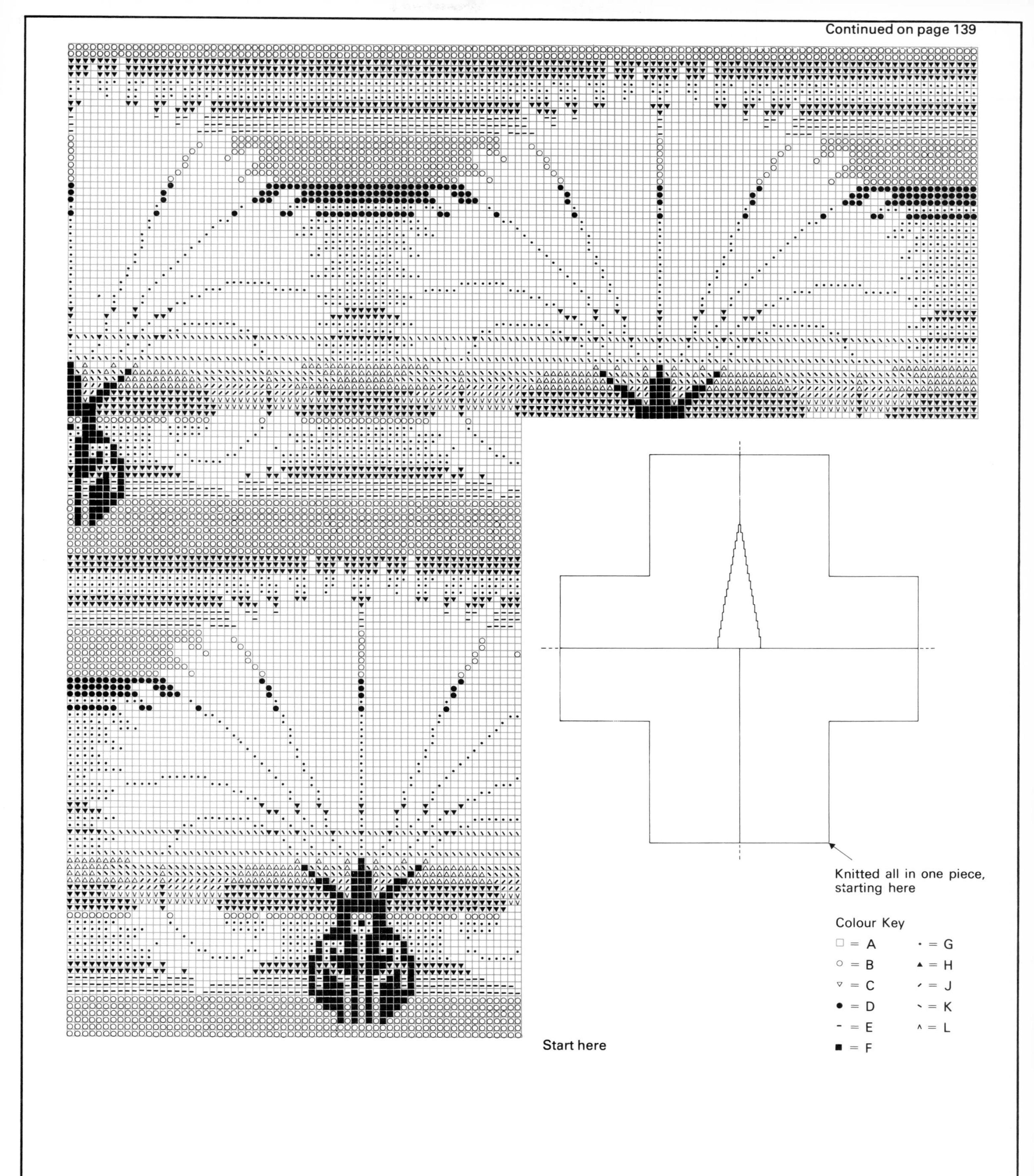

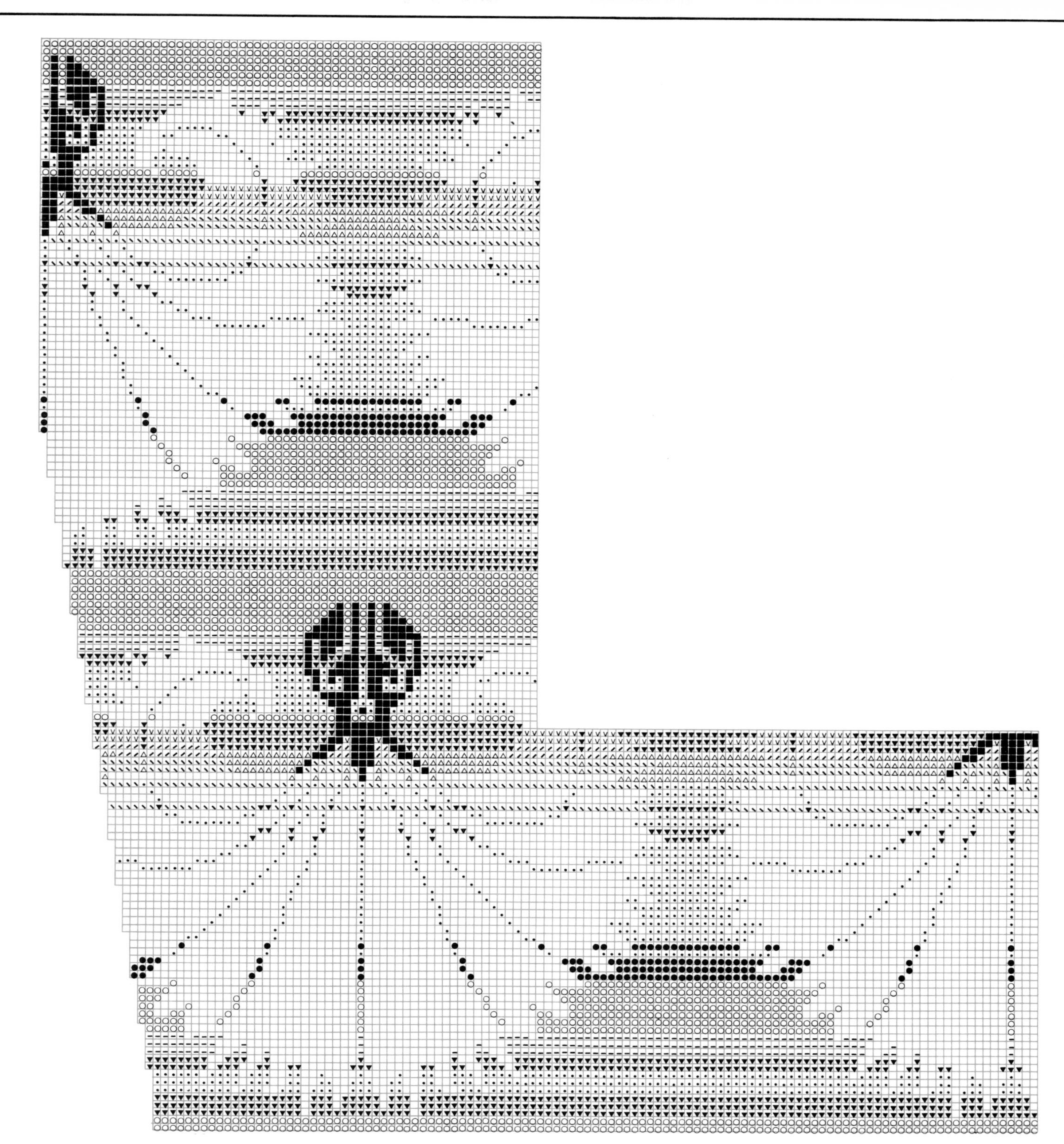

Continued

Jennifer Kiernan

Jennifer Kiernan studied fashion design at Ravensbourne College of Art and fabric design at the Royal College of Art in London before going on to specialize in the bold, bright knits that are her trademark. The shapes are exciting but never excessive, the colours so exuberant that they make you feel as good as you look.

'I've always liked bold shapes and patterns, because I find them easier to 'read' and to relate to. My fabric designs are very big and clear, and my fashion designs have always been based on very strong shapes. You can see elements of both in my handknits, but the main thing I've carried over from my fabric training is a love of decorating surfaces. That's really the way I design my handknits – *I decorate the surface with colours and patterns that accentuate the shape.*

Each of my handknit collections is loosely based on a theme, a pattern or combination of patterns, like random dashes or checks and stripes. The patterns are always abstract or geometric, partly because I don't like figurative designs, and partly because of the sheer logic of what you can get out of the yarn and needles. Every collection also has a colour theme. The check and stripe collection used the dark end of the spectrum, shading from neutral to dark maroon, and the triangle collection that these two sweaters were taken from was based on the warm natural shades. Each collection features about twenty basic designs which are presented in a number of colourways. I showed the *Ribs and Triangles* sweater in deep peony red and light tan as well as rust, so you can see how much variety there is in a collection.

Of course, my ideas about colour and everything else change over the six months it takes to do a collection. My original design ideas develop into others, but there are commercial considerations as well. For example, the *Triangle* sweater was one of the first I did for the triangle collection, and it shows the bold design style I'm happiest with. The shape is quite extreme – a chunky triangle accentuated with pleats, biggish shoulders and strong colour patterns. Artistically it's very satisfying, but a lot of people won't buy wide sweaters, and the pleats are a bit bulky for wearing underneath a tailored jacket, so my shapes got narrower as the collection went along. *Ribs and Triangles* was one of the last designs in the series, and here the shape is classic although the colours and patterns aren't. The body and sleeves are slimmer, the shoulders aren't so large, and it has a round neck instead of a yoke. It's a much simpler, tighter, more commercial design.

I tend to use patterns and colours to balance the overall design. For instance, after I'd done the body of the *Ribs and Triangles* sweater, I felt it needed the contrast of a second stitch pattern. I did the yoke first, then carried the stitch on to the sleeves, in order to relate them to the main body of the sweater. It's worked out very well because the strong line across the shoulders emphasizes the difference in the patterns, and brings out the colours more. The *Triangle* sweater needed something very definite to balance the triangular shape and patterns, so I picked out the colours on the body and did a wide yoke in mixed stripes, finished with a coloured yarn tie at the neck.

Sympathetic is a word I use a lot when I talk about colour, because although my colourings are very vivid they're not shocking. I like to work with sympathetic shades of the same basic colours, just occasionally adding a contrast to make the eye realize what's happening. You have to introduce enough variety to keep things interesting, but if you use too many colours the design gets too hectic, so everything has got to be carefully worked out.

The colours also have to be sympathetic to the design. The softer the background colour, the more the stitch pattern will stand out. Black deadens stitch patterns, but on the other hand it's a very good background for colours because it really throws them out. Grey and navy aren't as hard as black, so they're very useful if you want a dark background that shows off the colours *and* the stitches.

If you make up the sweaters in the colours I've given, you'll find that *Triangles* is very adaptable – it looks good with trousers or easy skirts in black, soft green and all the other contrast colours on the body. Because it's more classic, *Ribs and Triangles* looks smartest worn with a skirt. And I would choose a skirt in a darker tone of the rust since the background is such a strong colour that it needs to be carried through.

When I'm working on a design I lay all the yarns out and play around with them until I get them to do what I want. I carry quite a large stock of odd colours and odd packets of yarn, so I can be sure of getting just the right *shade* of a colour. You should always choose your colours in broad daylight. Hold the balls of wool against each other to see what they do – some jump against each other, some blend beautifully. Look at the secondary colours against the background, and lay strands across the balls of wool so you can see how they'll look when they're knitted up. And obviously – hold the colours against your face, to make sure they're sympathetic to *you*!'

Triangles

DETAILS Shown with earrings and bracelet from Detail.

MATERIALS **Yarn:** Sirdar Superwash Double Knitting; 15 25 gm balls in colour A, 1 25 gm ball each of colours B, C and D used on sleeves, 1 25 gm ball each of colours E, F, G, H, J, K and L. **Needles:** No. 7 British (No. 6 American, 4½ mm), No. 5 British (No. 8 American, 5½ mm).

MEASUREMENTS **To fit** 32–38 in bust (81–97 cm) bust. **Length** 22 in (56 cm). **Sleeve seam** 16 in (42 cm).

TENSION 17 sts and 22 rows to 4 in (10 cm) over stocking stitch on No. 5 needles.

ABBREVIATIONS **See** page 176. **Colours:** A=Pebble Beige, B=Medoc, C=Chinese Blue, D=Honey Beige, E=Festive Scarlet, F=Gorse, G=Saxe, H=Black, J=Peppermint, K=Nut Brown, L=Horse Chestnut.

BACK AND FRONT (Alike)
With No. 7 needles and A cast on 71 sts. **1st row** K1 * P1 K1 rep from * to end. **2nd row** P1 * K1 P1 rep from * to end. Rep these 2 rows for 8 cm ending with a 2nd row. **Next row** K twice into first st, * K1 K twice into next st, rep from * to end (107 sts). Change to No. 5 needles and cont in patt as follows. *** **1st row** (wrong side) P1A K20A P1AH K20A P1AL K20A P1AF K20A P1AC K20A P1AB P1A. **2nd row** K1A K2AB P19A K2AC P19A K2AF P19A K2AL P19A K2AH P19A K1A. **3rd row** P1A K18A P3AH K18A P3AL K18A P3AF K18A P3AC K18A P3AB P1A. Cont in this way, working one more st into the two colours tog on each row until the 21st row. **21st row** P1A P21AH P21AL P21AF P21AC P21AB P1A. **22nd row** K to end in A only. *** Rep from *** to *** once, using D for H, G for L, E for F, J for C and K for B. Rep these 44 rows once more. Beg with a K row, work 3 rows in reverse st.st, ending with a wrong side row. Cont in g.st, K 2 rows each in B, C, F, L, H, D, G.

SHAPE NECK AND SHOULDERS
Next row with E, cast off 5, K until 34 sts on needle, K2tog, turn and leave rem sts on spare needle. **Next row** with E, K2tog, K to end. **Next row** with J, cast off 5, K to last 2 sts, K2tog. **Next row** with J, K2tog, K to end. Rep the last 2 rows 3 times more, but using K, B and C in turn instead of J. Cast off rem 6 sts. Return to the sts on spare needle. Join in E, cast off 25, K to end. **Next row** with E, cast off 5, K to last 2 sts, K2tog. Cont to match first side.

RIGHT SLEEVE
With No. 7 needles and A cast on 35 sts and work in rib as on BACK for 8 cm, ending with a 2nd row. **Next row** K2 K twice into next st, rep from * to last st, K1 (67 sts). Change to No. 5 needles and cont in patt as follows. ** **1st row** (wrong side) K43A P1AB K23A. **2nd row** P23A K2AB P42A. **3rd row** K41A P3AB K23A. Cont in this way, working one more st into 2 colours on every row until 21st row. **21st row** K23A P21AB K23A. **22nd row** with A, P23 K21 P23. ** Rep from ** to ** once using D instead of B, then once more but using C instead of B. Work 3 rows in reverse st.st in A, then cast off loosely.

LEFT SLEEVE
Work as for right sleeve until there are 67 sts. **1st row** K23A, P1AB, K43A. **2nd row** P42A K2AB P23A. Cont to match right sleeve, reversing patt as shown but using the same colours as right sleeve.

NECKBAND
Join right shoulder seams. With No. 7 needles and A yarn, with right side facing pick up and K 95 sts round neck. Beg first row with K1, work 4 rows in rib as on Back. **Next row** rib 4 (cast off 2, rib 6) twice, cast off 2, rib 4 (cast off 2, rib 6) to last 5 sts, cast off 2 rib 3. **Next row** rib to end, casting on 2 sts over each 2 cast off. Rib 2 more rows. Cast off in rib.

TO MAKE UP
Press lightly with a warm iron over a damp cloth. Join left shoulder seam and neckband. Sew in sleeves with centre of sleeve to shoulder seam. Join side and sleeve seams. Using one strand each of any nine of the colours used, make a plait to tie round neck. Thread plait through holes at neck to tie at centre front.

Ribs and Triangles

DETAILS Photographed at the Park Lane Hotel, London.

MATERIALS **Yarn:** Sirdar Superwash Double Knitting, 15 25 gm balls in main colour A, 1 25 gm ball each in colours B, C, D, E, F, G. **Needles:** No. 7 British (No. 6 American, 4½ mm), No. 5 British (No. 8 American, 5½ mm). **Also** two stitch holders.

MEASUREMENTS **To fit** 32–36 in (81–91 cm) bust. **Length** 25 in, 63 cm. **Sleeve seam:** 18½ in, 47 cm.

TENSION Approximately 16 sts and 22 rows to 4 in (10 cm) over main patt on No. 5 needles. On sleeve patt, approximately 20 sts and 32 rows to 4 in (10 cm).

ABBREVIATIONS **See** page 176. **Colours:** A=Horse Chestnut, B=Camel, C=Festive Scarlet, D=Gorse, E=Saxe, F=Cyclamen, G=Peppermint.

BACK
With No. 7 needles cast on 73 sts. **1st row** K1 * P1 K1 rep from * to end. **2nd row** P1 * K1 P1 rep from * to end. Rep these 2 rows for 7 cm ending with a 1st row. **Next row** work twice into first st, * P2 work twice into next st, rep from * to end (98 sts). Change to No. 5 needles and cont in patt as follows. **1st row** K1 * K3 P1 K4 P3 K4 P1 rep from * to last st, K1. **2nd row** P1 * K2 P3 K3 P3 K2 P3 rep from * to last st, P1. **3rd row** K1 * K3 P3 K2 P3 K2 P3 rep from * to last st, K1. **4th row** P1 * K4 P1 K3 P1 K4 P3 rep from * to last st, P1. **5th row** K1 * K3 P13 rep from * to last st, K1. **6th row** as 4th row. **7th row** as 3rd row. **8th row** as 2nd row. **9th row** as 1st row. **10th row** P1 * P5 K3 P8 rep from * to last st, P1. These 10 rows form the patt. Cont in patt with coloured triangles as follows. Wind off just over a yard of each of B, C and D. This will make it easier to work than having complete balls of each colour. **11th row** working as 1st row, patt 16A K1B patt 31A K1C patt 31A K1D, patt 17A. **12th row** working as 2nd row, patt 17A K1D P1D patt 30A K1C P1C patt 30A, K1B, P1B, patt 15A. **13th row** patt 14A, K1B P2B, patt 29A K1C P2C patt 29A, K1D P2D patt 17A. **14th row** patt 17A K3D P1D patt 28A K3C P1C patt 28A K3B P1B patt 13A. **15th row** patt 12A K1B P4B patt 27A K1D P4D patt 17A. **16th row** patt 17A K4D patt 28A K4C patt 28A K4B patt 13A. **17th row** patt 14A P3B patt 29A P3C patt 29A P3D, patt 17A. **18th row** patt 17A K2D patt 30A K2C patt 30A K2B, patt 15A. **19th row** patt 16A P1B patt 31A P1C patt 31A P1D patt 17A. **20th row** as 10th row. **21st–30th rows** work in patt but work the coloured triangles on 2nd, 4th and 6th patts instead of 1st, 3rd and 5th, using E, F and G. **31st–40th rows** as 11th–20th rows but using C, D and B in that order. **41st–50th rows** as 21st–30th rows but

using F, G and E in that order. **51st–60th rows** as 11th–20th rows, but using D, B and C. **61st–70th rows** as 21st–30th rows but using G, E and F. **71st–90th rows** as 11th–30th rows. **91st–110th rows** as 31st–50th rows, inc one st at end of last row (99 sts). Cont with yoke in patt as follows. **1st and 2nd rows** with A, K to end. **3rd row** with B, K1 * sl.1 K1 rep from * to end. **4th row** with B, K1 * yfwd sl.1 ybk K1 rep from * to end. Rep these 4 rows 3 times more, but using C, D and E instead of B. **Shape Shoulders** still working in patt, using F, G and B cast off 6 sts at beg of next 12 rows. Leave rem 27 sts on stitch holder.

FRONT

As given for Back until 12 rows of yoke have been worked. **Shape neck, next row** working in patt to match back, patt 44 turn and cont on these sts. Cast off 2 sts at beg of next and foll 3 alt rows but at the same time when work measures the same as back to shoulder (i.e. 16 rows of yoke patt) shape shoulders by casting off 6 sts at beg of next and foll 5 alt rows. Slip centre 11 sts onto stitch holder, rejoin yarn and patt to end. Cont to match first side.

SLEEVES

With No. 7 needles and A yarn cast on 35 sts and work in rib as on Back for 7 cm, ending with a 1st row. **Next row** P twice into each st to last st, P1 (69 sts). Change to No. 5 needles and cont in patt as on yoke, working in the same colour sequence, inc one st at each end of 5th and every foll 10th row until there are 81 sts, then cont without shaping until sleeve measures 47 cm from beg, ending with a wrong-side row. Cast off *loosely*.

NECKBAND

Join left shoulder seam. With No. 7 needles and A yarn and right side of work facing, K back neck sts, pick up and K 16 sts down left front neck, K front neck sts, pick up and K 17 sts up right front neck (71 sts). Beg 1st row with K1, work 3 rows in K1 P1 rib. **4th row** [(P1 K1) twice, yfwd, K2tog] 8 times, K1 [sl.1 K1 PSSO yfwd, (K1 P1) twice] 3 times, sl.1 K1 PSSO yfwd K1 P1. Rib 3 more rows. Cast off in rib.

TO MAKE UP

Press lightly with a warm iron over a damp cloth. Join right shoulder seam and neckband. Sew in sleeves with centre of sleeve to shoulder seam. Join side and sleeve seams. With nine strands of colours as required make a plait long enough to tie at neck. Thread through holes in neckband to tie at centre front.

Jaime Ortega

Jaime Ortega's designs reflect a dancer's instinct for line, flow, shape and suppleness. To him, movement is a vital part of design – the movement of colours and textures on a shape, and the way the shape as a whole moves with the body. His designs have an air of relaxed refinement, and a sporty elegance reminiscent of Chanel.

'I don't like to think of myself as a 'knitwear designer'. When you say 'knitwear', people think of jumpers – they don't think of style. In my work I concentrate on shape and texture and I use them to design stylish clothes – the fact that they're knitted is incidental.

There's no reason why a knitted design can't be as stylish, or do as much for you, as a design in fabric. It should do even more, since it's much more comfortable and has many more texture possibilities. But you have to remember that fabrics and knits are not the same. They have different properties, and you can't expect a fabric pattern to work for a knit. A knit can't carry a lot of seams, tucks, zips and gathers. So if you want a knitted design to have a shape, you've got to literally build it in as you knit.

But that's only half the story, because shape alone is not enough. I don't think you can draw a line between knits and textiles – you need to add the texture dimension, use your yarn to help the shape along. That's why I like *bouclé* – it's springy and it has a lot of body, so it doesn't grip or clutch. I suppose you could say it *floats* next to the skin, which is why you don't need a perfect figure to wear it. It has a good colour density – it catches the light, and throws off different *shades* of a colour, not just one flat tone. It's light and comfortable to wear, and it has a beautiful surface texture – soft and subtle.

I don't like flat surfaces, and I like to use lots of detail – bright touches of contrast colour, or different stitches – to add movement to a design. Traditional ribbing and cabling are much more interesting in chunky yarns than in fine ones. If the basic design is good, you can add lots of detail without making it look fussy.

Personally, I never feel completely relaxed in a jacket. But at the same time, however casual the occasion, I always like to look well put together. This top is the ideal solution – wear it with a shirt and slacks, or rollneck and jeans, and you have an effortless total look. Practicality is important, too, so there is a pouch for holding things in. And it's warm enough so you don't have to wear a coat over it. After all, no man who has a really nice sweater likes to cover it up!'

Jogger

DETAILS Shown with sweatshirt, matching drawstring trousers and Nike running shoes from Gymnasium, London, YMCA.

MATERIALS **Yarn:** Sirdar Sheba, 7 50 gm balls in colour A, 5 balls in colour B, 4 balls in colour C, 1 ball in colour D. **Needles:** No. 3 British (No. 10 American, 6½ mm). **Also**, 150 cm of narrow elastic.

MEASUREMENTS **To fit** chest sizes 38–40 in (97–102 cm). **All round** 45 in (114 cm). **Length** 23 in (58.5 cm). **Sleeve seam** 19 in (48 cm).

TENSION 12 sts and 14 rows to 4 in (10 cm) over reverse st.st on No. 3 needles.

ABBREVIATIONS See page 176. **Colours:** A=Silver Cloud, B=Sable, C=Bordeaux, D=Flame.

BACK
With No. 3 needles and C yarn cast on 70 sts and working in reverse st.st throughout, cont until work measures 2 in (5 cm) from beg ending with a K row. Join in A yarn, K 1 row then beg with a K row cont until work measures 13 in (34 cm) from beg, ending with a K row. Join in C, K 1 row then beg with a K row cont until work measures 16 in (40 cm) from beg, ending with a K row. Join in D and K 2 rows. Join in B, K 1 row, ** beg with a K row cont until work measures 24 in (62 cm) from beg, ending with a K row. **Shoulder shaping:** cast off 8 sts at beg of next 6 rows. With D, K 1 row then cast off rem 22 sts knitwise.

FRONT
Work as given for Back to **. **Divide for neck:** next row K35, turn and leave rem sts on spare needle. **Next row:** P1 P2tog, P to end. Work 1 row straight. Cont as set, dec at neck edge on next and every alt row until 24 sts rem, cont to same length as Back to shoulders, ending at armhole edge. **Shoulder shaping:** cast off 8 sts at beg of next and foll 2 alt rows. Return to the sts which were left. Rejoin yarn and K to end. **Next row:** P to last 3 sts, P2tog P1. Cont to match first side.

SLEEVES

With No. 3 needles and C yarn cast on 52 sts and work 2 in (5 cm) in rev st.st ending with a K row. Join in B, K 1 row then beg with a K row cont until work measures 8 in (20 cm) from beg, ending with a K row. Change to C, K 1 row then beg with a K row cont until work measures 10 in (25 cm) from beg, ending with a K row. Join in D and K 2 rows. Join in A, K 1 row then beg with a K row cont until work measures 18 in (46 cm) from beg, ending with a K row. Change to C, K 1 row then beg with a K row cont until work measures 20 in (51 cm) from beg, ending with a K row. Change to D, K 2 rows and cast off *loosely*.

POUCH

With No. 3 needles and B yarn cast on 28 sts and beg with a P row cont in rev st.st until work measures 6 in (16 cm) ending with a K row. Cast off loosely purlwise. With right side of work facing and C, pick up and K20 sts along side of pouch. Beg with a K row work in rev st.st for 2 in (5 cm). Cast off. Work along other side in the same way.

NECKBAND

With No. 3 needles and C yarn pick up and K27 sts down left front neck, 1 st from centre front, 27 sts up right front neck, 22 sts across back neck (77 sts). Cast off knitwise.

TO MAKE UP

Join shoulder seams. Sew cast off edges of sleeves to armholes, with centre of sleeve to shoulder seam. Join side and sleeve seams. Sew pouch on front. Turn hem on sleeves and bottom edge in half to inside and slip stitch. Thread elastic through cuffs and bottom edge to gather as required.

Note

For your nearest stockist of Sirdar yarns, write to Sirdar Ltd, PO Box 31, Alverthorpe, Wakefield, Yorks.

Creative Dressing Collection

Presenting our capsule collection of handknits, a clutch of good looks with plenty of potential, specially designed for Creative Dressing. Where two alternative figures are given, the first figure refers to the smaller size, and the figures in brackets to the larger size.

Cunard

How's this for syncopated chic – a French jazz sweater from the Twenties, sleek and sparkly in silver and Art Deco greens. The fun of period designs like this is the way they can take you back in time. Slip it on and you'll look and feel like a flapper. The sweater's straight lines skim the body, making you look taller and slimmer, and the simple shrug top is as nonchalantly elegant as a cardigan by Chanel. If you make them up to wear together, sweater and cardigan should be the same length, so choose the two-button top. If you want to make the cardigan to wear on its own, there's a slightly longer three-button version as well. **Yes sir, that's my baby!**

DETAILS Sweater shown with earrings from Butler & Wilson. Sweater and cardigan shown with earrings, necklace and opera glasses from Butler & Wilson, shoes by Manolo Blahnik from Zapata and skirt from Simpsons.

MATERIALS **Yarn for sweater:** Sirdar Super Prelude (4-ply equivalent) 10(10) 20 gm balls in main colour, Crystal. Sirdar Wash'N'Wear 4-ply, 4(5) 20 gm balls in light contrast, Peppermint, and 4(5) 20 gm balls in dark contrast, Bottle. **Yarn for cardigans:** Sirdar Wash'N'Wear 4-ply, 9(10) 20 gm balls in Peppermint. Sirdar Super Prelude, 2(3) 20 gm balls in Crystal. **Needles:** No. 10 British (No. 3 American, 3¼ mm), No. 12 British (No. 1 American, 2¾ mm). **Also** one stitch holder, 2 or 3 buttons for cardigan.

MEASUREMENTS Two sizes, to fit bust 32–34 (36–38) in (81–86(91–97) cm). **Sweater length**: 22½(23½) in (57(59) cm). **Sleeve length:** with cuff extended, 19 in (48 cm). **Two-button cardigan length:** 22(23) in (56(58.5) cm). **Three-button cardigan length:** 24½(25½) in (63(65) cm).

TENSION 14 sts and 17 rows to 2 in (5 cm) over pattern, 15 sts and 19 rows to 2 in (5 cm) over st.st.

ABBREVIATIONS **See** page 176. **Colours**: M=main colour, L=Light contrast, D=dark contrast.

SWEATER BACK
Cast on 121(133) sts with No. 12 needles and M. **1st row** K2 * P1 K1, rep from * to last st, K1. **2nd row** K1 * P1 K1, rep from * to end. Rep these 2 rows for 3 in (7.5 cm). Change to No. 10 needles. Working in st.st, proceed in Fair Isle pattern as follows. **Row 1** * 1M 1L 4M 1L 4M 1L, rep from * to last st, 1M. **Row 2** 1M * 2L 3M 1L 3M 2L 1M, rep from * to end. **Row 3** * 1M 3L 2M 1L 2M 3L, rep from * to last st, 1M. **Row 4** 1M * 4L 1M 1L 1M 4L 1M rep from * to end. **Row 5** * 1L 1M 4L 1M 4L 1M, rep from * to last st, 1L. **Row 6** 1L * 2M 3L 1M 3L 2M 1L, rep from * to

end. **Row 7** * 1L 3M 2L 1M 2L 3M rep from * to last st, 1L. **Row 8** 1L * 4M 1L 1M 1L 4M 1L rep from * to end. **Rows 9–16** as for rows 1–8, but substituting D for L. These 16 rows form the pattern. Rep them 5 times more, then work rows 1–8 again. **Armhole shaping:** cast off 4 sts at beg of next 2 rows, then dec at both ends of every row until 97(103) sts remain. Continue until work is 7(8) in or 18(20) cm from beg of armholes, measured on the straight, finishing after a wrong side row. **Shoulder shaping:** cast off 9(10) sts at beg of next 4 rows and 10(11) sts at beg of following 2 rows. 41 sts. Change to No. 12 needles and M, K 1 row. Beg with 2nd row, work 9 rows K1 P1 rib as given for lower edge. Cast off in rib.

SWEATER FRONT

Work as Back until armhole shaping is complete, 97(103) sts. Work 6(11) rows thus, finishing after an 8th (16th) row of pattern. **Neck shaping:** work 28(31), turn, leaving remaining sts unworked. Proceed on 1st set of sts for left side. Continue until work matches Back to outer shoulder. **Shoulder shaping:** cast off 9(10) sts at beg of next 2 side-edge rows. Work to side edge. Cast off remaining 10(11) sts. Place central 41 sts on a stitch holder. Join yarn to inner edge of remaining sts and complete right side of neck to correspond with left, reversing shoulder shaping.

FRONT NECKBAND

With right sides of work facing, using No. 12 needles and M, K up 49 sts along left front neck edge. With same needle work sts of centre front thus: K5 ML K5 ML K2 ML K5 ML K7 ML K5 ML K2 ML K5 ML K5. K up 49 sts along right front neck edge, 147 sts. **1st row** (wrong side) K1 P1 alternately 23 times, K2tog, P1, K2tog, now P1 K1 alternately 22 times, P1 K2tog, P1 K2tog, then P1 K1 alternately 23 times. **2nd row** as 1st row of rib. **3rd row** rib 45 sts as set, K2tog, P1, K2tog, rib 43, K2tog, P1, K2tog, rib 45. **4th row** rib 44, P2 K1 P2 rib 41, P2 K1 P2 rib 44. Work 5 more rows, decreasing in this manner each side of st at each corner of neck on every wrong side row. Cast off in rib.

SLEEVES (Alike)

Cast on 59(71) sts with No. 12 needles and M. Work 5 in (13 cm) K1 P1 rib, finishing after a 1st row of rib and increasing at both ends of last row. Change to No. 10 needles and patt. Inc at both ends of the 5th and every 4th row following until there are 69(81) sts, then of every 8th(10th) row until there are 91(99) sts. Proceed until the 8th row of the 8th patt has been worked. **Shape top:** cast off 4sts at beg of next 2 rows. Dec at both ends of next and every alternate row until 53(61) sts remain. Work 1 row. Dec at both ends of every row until 25 sts remain. Cast off.

TO MAKE UP

Press on the wrong side under a dry cloth, omitting ribbing. Join side, shoulder and sleeve seams. Set sleeves into armholes. Press seams.

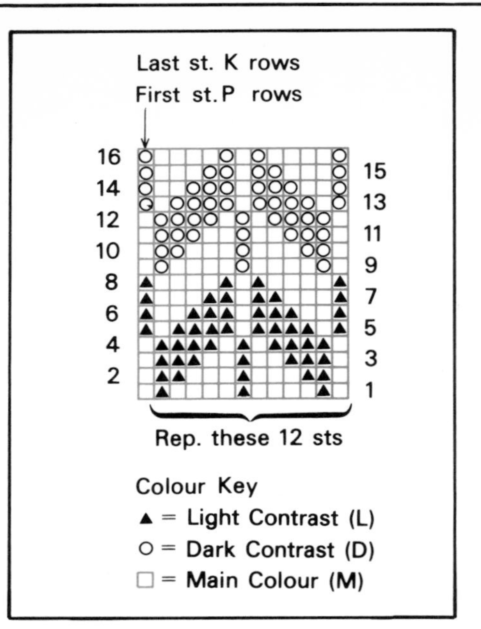

THREE-BUTTON CARDIGAN, BACK

Cast on 127(141) sts with No. 12 needles and M. Work 2¾ in (7 cm) K1 P1 rib as given for lower edge of Sweater. Change to No. 10 needles and L and st.st. Proceed until work measures 16½ in (42 cm), finishing after a wrong side row. **Armhole shaping:** cast off 8 sts at beg of next 2 rows (111 (125) sts. Continue until work measures 8(8½) in or 20(21) cm from beg of armholes, finishing after a P row. **Shoulder**

shaping: cast off 9(11) sts at beg of next 6 rows and 8(9) at beg of following 2 rows, 41 sts. Change to No. 12 needles and M. K 1 row. Beg with 2nd row, work 9 rows K1 P1 rib. Cast off in rib.

LEFT FRONT

Cast on 61(69) sts with No. 12 needles and M. Work 2¾ in (7 cm) in K1 P1 rib. Change to No. 10 needles and L, and st.st. Proceed until work measures 6 in (15 cm) **Neck shaping:** dec at front edge on next right-side row and every 8th row following. Proceed thus until work matches Back to armholes. **Armhole shaping:** cast off 8 sts at beg of next side-edge row. Continue to dec at front edge on every 8th row until 35(42) sts remain. Proceed until work matches Back to outer shoulder. **Shoulder shaping:** cast off 9(11) sts at beg of next 3 side-edge rows. Work to side edge. Cast off remaining 8(9) sts. **Right Front:** work as for Left Front, reversing shapings.

FRONT BORDERS

With right side of work facing, using No. 12 needles and M, K up 23 sts along edge of ribbing on right front. Break M. With L, K up 28 sts along remainder of straight edge and 145(151) along shaped edge as far as shoulder. Change to M. Purl 1 row. Work 3 rows K1 P1 rib. In next row make buttonholes thus: rib to last 39 sts, cast off 3, rib 13 (including st already on needle after casting off), cast off 3, rib 13, cast off 3, rib 4. In next row cast on 3 sts over each buttonhole. Rib 4 rows, cast off in rib. Work left front border to match, omitting buttonholes.

ARMHOLE BORDERS

Join shoulder seams and seams of borders. With right side of work facing, using No. 12 needles and L, K up 131(147) sts along armhole, leaving cast-off sts under arms free. Change to M. Purl 1 row. Work 9 rows K1 P1 rib. Cast off in rib. Complete second armhole in same way.

TO MAKE UP

Press on the wrong side under a dry cloth. Join side seams. Sew side edges of armhole borders to cast-off sts under arms. Press seams.

TWO-BUTTON CARDIGAN

As for three-button cardigan, with following exceptions. **Back**: proceed until work measures 13¾ in (35 cm), finishing after a wrong-side row. **Left and Right Fronts**: proceed until work measures 3½ in (9 cm). **Front Borders**: with right side of work facing, using No. 12 needles and M, K up 23 sts along edge of ribbing on right front. Break M. With L, K up 14 sts along remainder of straight edge and 145(151) along shaped edge as far as shoulder. Change to M. Purl 1 row. Work 3 rows K1 P1 rib. In next row make buttonholes thus: rib to last 23 sts, cast off 3, rib 13 (including st already on needle after casting off), cast off 3, rib 4. In next row cast on 3 sts over each buttonhole. Rib 4 rows. Cast off in rib. Work left front borders to match, omitting buttonholes.

Henley

Splash out with style in this sporty regatta blazer, an adaptation of those brightly coloured boating coats that gave the *blaze*-er its name. *Henley* shows how successfully the softly tailored look translates into knitting. While ladies' flannel blazers are often too boxy and boyish, this knitted version is a shameless flatterer. The fabric follows the natural lines of the body so the waist and shoulders preserve their feminine proportions, and you don't have any difficulty achieving a proper fit. Traditional tailoring requires a great deal of skill and time, but this design couldn't be easier. The fabric is all in garter stitch, and the smart braid trim that gives the edges a pleasingly rounded finish removes the need for knitted borders. You might like to knit Henley in red or dark blue trimmed with white, two popular combinations on the river during the regatta season. If you prefer the cruise boat look, knit the blazer in cream or navy, trim with matching braid and finish with shiny brass buttons. Whatever you choose, it'll be a nice way of showing that you're not just one of the fellows.

DETAILS Shown with glasses from Butler & Wilson, deck chair from Supotco and traditional Welsh tapestry lap rug.

MATERIALS **Yarn:** Sirdar Pullman Chunky Knitting, 20(21:22:23:24) 50 gm balls. **Needles:** No. 4 British (No. 9 American, 6 mm). **Trimming:** 3 buttons, 4 metres of wool braid.

MEASUREMENTS **To fit:** 32(34:36:38:40) in (81(86:91:97:102) cm) bust. **Actual size:** 36(38:40:42:44) in (91(97:102:107:112) cm). **Full length:** 26(26:26½:27:27) in (66(66:67:69:69) cm). **Sleeve seam:** 17 in (44 cm) all sizes.

TENSION 7 sts and 14 rows to 2 in (5 cm).

ABBREVIATIONS **See** page 176.

BACK

Using No. 4 needles, cast on 64(68:70:74:78) sts. (Note that first row is right side.) Proceed in g.st until work measures 18½(18½:18½:19:19) in or 47(47:47:48:48) cm, ending with a wrong side. **Shape armholes:** cast off 3 sts at beg of next 2 rows. Work 3 rows dec 1 st each end of every row. Work 1 row, then dec 1 st each end of next and every alt row to 46(48:50:52:54) sts. Cont straight on these sts until work measures 26(26:26½:27:27) in or 66(66:67:69:69) cm ending with a wrong side. **Shape shoulders:** cast off 5 sts at beg of next 4 rows. Cast off 4(4:5:5:6) sts at beg of next 2 rows. Cast off rem 18(20:20:22:22) sts.

LEFT FRONT

Using No. 4 needles, cast on 32(34:35:37:39) sts. **1st row** (right side) sl.1 K to end. **2nd row** sl.1 ML K to end. Rep 1st and 2nd rows 3 times more. 36(38:39:41:43) sts. Cont in g.st until work measures 13 in or 33 cm, ending with a wrong-side row. **Shape front lapel: next row** sl.1 K to last st, ML K1. Work 9 rows straight. Rep the last 10 rows until work measures same as the Back to armhole, ending with a wrong-side row. **Shape armhole:** cast off 3 sts at beg of next row. Work 1 row then dec 1 st at same edge on the foll 3 rows, then dec 1 st at same edge on the foll 3(4:4:5:6) alt rows **at the same time** continuing to inc at front edge for lapel until 7 sts have been inc in all. 34(35:36:37:38) sts. Cont on these sts until 13 rows less than Back have been worked to shoulder, ending with a right-side row. **Shape neck:** cast off 10(11:11:12:12) sts, K to end. Work 10 rows dec 1 st at neck edge on every row. 14(14:15:15:16) sts. Work 2 rows straight. **Shape shoulder:** cast off 5 sts at beg of next and foll alt row. Work 1 row. Cast off rem 4(4:5:5:6) sts.

RIGHT FRONT

Using No. 4 needles, cast on 32(34:35:37:39) sts. **1st row** (right side) sl.1 K to end. **2nd row** sl.1 K to last st, ML K1. Rep 1st and 2nd rows 3 times more. 36(38:39:41:43) sts. Cont in g.st until work measures 4 in or 10 cm, ending with a wrong-side row. **Make buttonholes** as follows. ** **Next row** sl.1 K2, cast off 2 sts, K to end. **Next row** sl.1 K to last 3 sts, cast on 2 sts, K3. Work 24 rows. ** Rep from ** to ** once more, then make a 3rd buttonhole. Complete to correspond with Left Front, reversing all shapings.

SLEEVES (Alike)

Using No. 4 needles, cast on 30(30:30:32:32) sts. Proceed in g.st inc 1 st each end of 11th(9th:7th:7th:5th) and every following 12th(12th:10th:10th:10th) row to 46(48:50:52:54) sts. Cont straight until work measures 17½ in (44 cm), ending with a wrong-side row. **Shape top:** cast off 3 sts at beg of next 2 rows. Work 4(4:6:6:6) dec 1 st each end of next and every alt row. 36(38:38:40:42) sts. Work 24(24:24:24:20) rows dec 1 st each end of next and every foll 4th row. 24(26:26:28:32) sts. Work 8(8:8:10:14) rows dec 1 st each end of next and every alt row. 16(18:18:18:18) sts. Cast off.

COLLAR

Using No. 4 needles, cast on 70(72:72:74:74) sts. Work 2 rows in g.st then proceed as follows. **1st row** sl.1 K to last 6 sts, turn. **2nd row** sl.1 K to last 6 sts, turn. **3rd row** sl.1 K to last 10 sts, turn. **4th row** sl.1 K to last 10 sts, turn. **5th row** sl.1 K to last 14 sts, turn. **6th row** sl.1 K to last 14 sts, turn. **7th row** sl.1 K to end. Work 18 rows in g.st across all sts. Cast off.

POCKETS (2 Alike)

Using No. 4 needles, cast on 24 sts. Work 34 rows in g.st. Cast off.

TO MAKE UP

Sew side, shoulder and sleeve seams. Sew in sleeves. Placing ends of shaped edge of collar 9(10:10:11:11) sts in from edge of lapel, sew collar evenly in position. Bind top edge of pockets with wool braid. Sew pockets in position placing them 1 in or 2.5 cm from lower edge and 2 in or 5 cm from side seams. Bind all round outer edge of blazer with wool braid. Sew on buttons. Fold collar and lapels to right side and press.

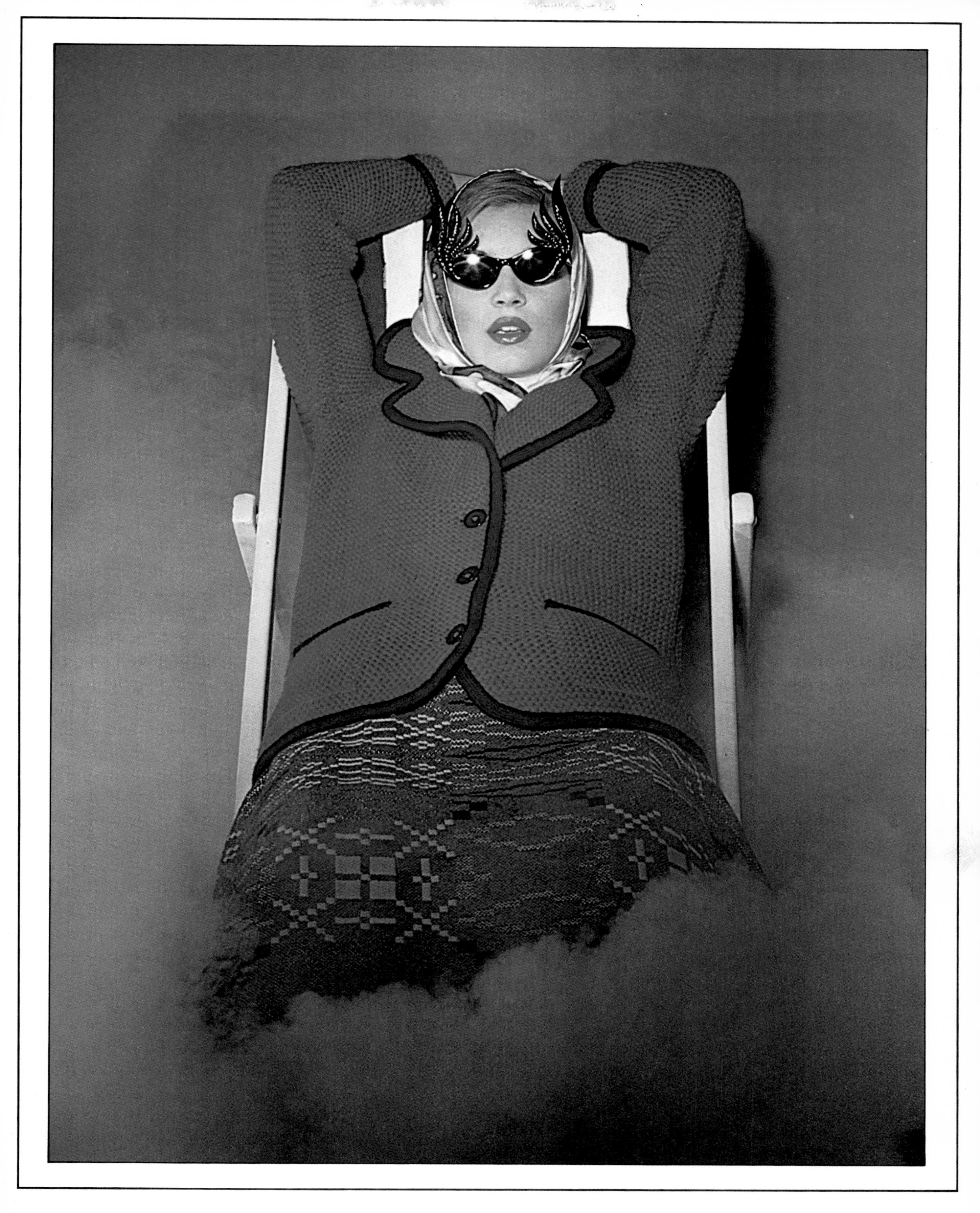

Shady Lady

A clever use of colour enhances the beauty of traditional stitches in this Aran-inspired design by Suzanne Barnacle. The shades and stitches used on the dress relate to the lines of the body. The sequence of shades draws the eye to the top panel where the cream and sandalwood bands balance the shoulders and make them look broader. The zigzag centre panel makes the waist look longer and slimmer, and the sweeping diagonals at the bottom give the hem an illusion of movement. With its straight lines, easy shoulders and V- or slash- neckline, the dress makes a perfect cool-weather casual to wear loose or with a slender belt. Turn the shady dress into a lady-like ensemble by adding the flattering long tunic that softens the colour contrast on the dress and links up to the colour on the sleeves, adding the subtle interest of two stitch patterns in the same shade. The weight of the yarn is quite light, so although the stitches make the fabric look thick, it's not too hot or heavy. Wrap yourself up in shades of your favourite colour, or knit the pieces to wear separately. The tunic makes a wonderful fashion basic, and you can knit the dress in neon colours if you want to be a bright young thing.

Shady Lady Dress

DETAILS Dress shown here with earrings from Butler & Wilson and boots by Bally. Photographed at Dukes Hotel, London. Dress and tunic shown with H.E. Kiewe's Royal Fair Isle.

MATERIALS **Yarn:** Sirdar Wash'N'Wear Double Crepe; 12 40 gm balls in main colour A, 8 40 gm balls in colour B, 4 40 gm balls in colour C, 6 40 gm balls in colour D, 3 40 gm balls in colour E.. **Needles:** one pair each No. 10 British (No. 3 American, 3¼ mm), No. 9 British (No. 4 American, 3¾ mm), No. 8 British (No. 5 American, 4 mm).

MEASUREMENTS **To fit** 34–37 in bust (86–94 cm bust). **Length** approximately 43 in (109 cm). **Sleeve length** approximately 17½ in (44.5 cm).

TENSION 12 sts to 2 in (5 cm) on No. 8 needles over st. st.

ABBREVIATIONS See page 176. **Colours:** A=Dark Chocolate, B=Sandalwood, C=Nut Brown, D=Chinese Fig, E=Driftwood.

BACK, V-NECK VERSION

Using No. 9 needles and A yarn, cast on 120 sts. **1st row sl.1 K to end. Rep this row twice. Change to No. 8 needles and proceed in first patt as follows. **1st row** (right side) * P3 K3 rep from * to end. **2nd and every foll alt row** K all the knit sts and P all the purl sts. **3rd row** P2 * K3 P3 rep from * to last 4 sts, K3 P1. **5th row** P1 * K3 P3 rep from * to last 5 sts, K3 P2. **7th row** * K3 P3 rep from * to end. **9th row** K2 * P3 K3 rep from * to last 4 sts, P3 K1. **11th row** K1 * P3 K3 rep from * to last 5 sts, P3 K2. **12th row** as 2nd row. These 12 rows form patt. Cont in first patt until work measures 11 in or 28 cm, ending on wrong side. Break off A yarn. Join B yarn and proceed as follows. **1st row** K60 ML K60, 121 sts. **2nd row** sl.1 * P1 K1 rep from * to end. **3rd row** sl.1 K1 * P1 K1 rep from * to last st, K1. **4th row** as 2nd row. Break off B yarn. Join in C yarn and proceed as follows. **Next row** (right side) K15 K2tog (K28 K2tog) 3 times, K14, 117 sts. Cont in C yarn and proceed in 2nd patt as follows. **1st row** (wrong side) K5 * P3 K5 rep from * to end. **2nd row** P5 * K3 P5, rep to end. **3rd row** as 1st row. **4th row** K. **5th row** K1 * P3 K5 rep from * to last 4 sts, P3 K1. **6th row** P1 * K3 P5 rep from * to last 4 sts, K3 P1. **7th row** as 5th row. **8th row** K. These 8 rows form 2nd patt. Cont until 87 rows of 2nd patt have been worked in all. Break off C yarn. Join in B yarn and work as follows. **1st row** K. **2nd row** sl.1 * P1 K1 rep from * to end. **3rd row** sl.1 K1 * P1 K1 rep from * to last st, K1. **4th row** as 2nd row. Break off B yarn. Join in D yarn and work as follows. **Next row** K57 K2tog K58, 116 sts. Cont in D yarn and proceed in 3rd patt as follows. **1st row** (wrong side) * P2 K2 rep from * to end. **2nd row** P1 * K2 P2 rep from * to last 3 sts, K2 P1. **3rd row** K1 * P2 K2 rep from * to last 3 sts, P2 K1. **4th row** * K2 P2 rep from * to end. **5th row** as 4th row. **6th row** as 3rd row. **7th row** as 2nd row. **8th and 9th rows** as 1st row. **10th row** as 3rd row. **11th row** as 2nd row. **12th and 13th rows** as 4th row. **14th row** as 2nd row. **15th row** as 3rd row. **16th row** as 1st row. These 16 rows form 3rd patt. Cont in patt until 3rd patt measures 9 in or 23 cm, ending on wrong side. Break off D yarn. Join in B yarn and work as follows. **1st row** K57 K2tog K57, 115 sts. **2nd row** sl.1 * P1 K1 rep from * to end. **3rd row** sl.1 K1 * P1 K1 rep from * to last st, K1. **4th row** sl.1 rib 6, work 2tog, rib 57, work 2tog, rib 27. 113 sts. Break off B yarn. Join in E yarn and proceed in 4th patt as follows. **1st row** (right side) P1 * K7 P1 rep from * to end. **2nd row** K1 * P7 K1 rep from * to end. **3rd row** P2 K5 * P3 K5 rep from * to last 2 sts, P2. **4th row** K2 P5 * K3 P5 rep from * to last 2 sts, K2. **5th row** P3 K3 * P5 K3 rep from * to last 3 sts, P3. **6th row** K3 P3 * K5 P3 rep from * to last 3 sts, K3. **7th row** P4 K1 * P7 K1 rep from * to last 4 sts, P4. **8th row** K4 P1 * K7 P1 rep from * to last 4 sts, K4. **9th row** as 2nd row. **10th row** as 1st row. **11th row** as 4th row. **12th row** as 3rd row. **13th row** as 6th row. **14th row** as 5th row. **15th row** as 8th row. **16th row** as 7th row. These 16 rows form 4th patt. ** Cont in patt until work measures approximately 41½ in or 105 cm, ending on wrong side. *** Break off E yarn. Join in D yarn and work as follows. **1st row** K. **2nd row** sl.1 * P1 K1 rep from * to end. **3rd row** sl.1 K1 * P1 K1 rep from * to last st, K1. Rep 2nd and 3rd rows until work measures 43 in or 109 cm, ending with a 2nd row. **Shape shoulders:** keeping continuity of rib patt cast off 6 sts at beg of next 8 rows, then 7 sts at beg of next 4 rows. Cast off rem 57 sts.

FRONT, V-NECK VERSION

Work exactly as given for Back from ** to **. **Shape neck:** keeping continuity of patt divide for neck as follows. **Next row** patt 53, turn and leave rem 60 sts on a spare needle. Dec once at neck edge on 2nd row from division and every foll alt row until 38 sts rem. ** Cont without shaping until work measures approximately 41½ in or 105 cm, ending on wrong side. Break off E yarn. Join in B yarn and work as follows. **1st row** K. **2nd row** sl.1 * P1 K1 rep from * to last st, K1. **3rd row** sl.1 * P1 K1 rep from * to last st, K1. Rep 2nd and 3rd rows until work matches the Back at side edge, ending on wrong side. **Shape shoulders:** keeping continuity of the rib patt cast off 6 sts at beg of next and foll 3 alt rows. Work 1 row. Cast off 7 sts at beg of next and foll alt row. With right side facing, rejoin E yarn to rem 60 sts. Cast off centre 7 sts, patt to end. Complete 2nd side to correspond with first, reversing shapings.

BACK AND FRONT, SLASH-NECK VERSION

Work exactly as given for the Back of V-neck version from ** to ***. **Shape shoulders:** cont in E yarn and work as follows. **1st row** patt to last 6 sts, turn. **2nd row** as 1st row. **3rd and 4th rows** patt to last 12 sts, turn. **5th and 6th rows** patt to last 18 sts, turn. **7th and 8th rows** patt to last 24 sts, turn. **9th and 10th rows** patt to last 31 sts, turn. **11th and 12th rows** patt to last 38 sts, turn. **13th row** work to end. **14th row** work across all sts. Break off E yarn. Join in B yarn. Change to No. 10 needles and work as follows. **1st row** K. **2nd row** sl.1 * P1 K1 rep from * to end. **3rd row** sl.1 K1 * P1 K1 rep from * to last st, K1. Rep 2nd and 3rd rows until work measures 43 in or 109 cm at side edge, ending with a 2nd row. Cast off evenly in rib. Place a marker 5 in or 13 cm from side edge along cast-off edge.

SLEEVES (Alike) FOR BOTH VERSIONS

Using No. 10 needles and D yarn cast on 49 sts. **1st row** (right side) sl.1 K1 * P1 K1 rep from * to last st, K1. **2nd row** sl.1 * P1 K1 rep from * to end. Rep 1st and 2nd rows 8 times, then 1st row again. **Next row** sl.1 rib 2 ML (rib 2 ML) 22 times, rib 2 (72 sts). Change to No. 8 needles and proceed in patt as follows. **1st row** sl.1 P1 K2 * P2 K2 rep from * to end. **2nd row** as 1st row. **3rd row** sl.1 * K2 P2 rep from * to last 3 sts, K3. **4th row** sl.1 * P2 K2 rep from * to last 3 sts, P2 K1. **5th row** sl.1 K1 P2 * K2 P2 rep from * to end. **6th row** as 5th row. **7th row** as 4th row. **8th row** as 3rd row. **9th and 10th rows** as 1st row. **11th row** as 4th row. **12th row** as 3rd row. **13th and 14th rows** as 5th row. **15th row** as 3rd row. **16th row** as 4th row. These 16 rows form patt. Keeping continuity of patt inc once at each end of next and every foll 8th row until there are 90 sts. Cont without shaping until work measures 16 in or 41 cm, ending on wrong side. Break off D yarn. Join in B yarn and work as follows. **Next row** sl.1 K44 ML K45. 91 sts. Now proceed as follows. **1st row** (wrong side) sl.1 * P1 K1 rep from * to end. **2nd row** sl.1 K1 * P1 K1 rep from * to last st, K1. Rep 1st and 2nd rows 4 times, then 1st row again. Cast off evenly in rib.

TO MAKE UP V-NECK VERSION

Read pressing instructions on yarn label. Join shoulder seams. **Neck border:** using No. 10 needles and B yarn cast on 11 sts. **1st row** sl.1 K1 * P1 K1 rep from * to last st, K1. **2nd row** sl.1 * P1 K1 rep from * to end. Rep 1st and 2nd rows until when slightly stretched strip fits all around neck from cast-off row, up right side, along back and down left side, ending with a 2nd row. Sew in position as you go along. Cast off evenly in rib. Sew in position right neck border to cast-off row of the Front. Catch down left neck border to base of right border.

TO COMPLETE BOTH VERSIONS

For slash-neck version, press following instructions on yarn label and join shoulder seams to markers. To complete both versions, fold sleeves in half lengthways and place a marker at centre. Sew sleeves in position along armhole edge placing marker at shoulder seam. Join side and sleeve seams.

Shady Lady Tunic

MATERIALS **Yarn:** Sirdar Wash'N'Wear Double Crepe; 17(18) 40 gm balls. **Needles:** one pair each No. 10 British (No. 3 American, 3¼ mm), No. 9 British (No. 4 American, 3¾ mm), No. 8 British (No. 5 American, 4 mm).

MEASUREMENTS **To fit** 32–34 (36–38) in bust (81–86 (91–97) cm) bust. **Length:** approximately 39 (40) in (99 (102) cm).

TENSION 12 sts to 2 in (5 cm) on No. 8 needles.

BACK
Using No. 9 needles cast on 117(133) sts. **1st row** sl.1 K to end. Rep 1st row twice. Change to No. 8 needles and proceed in patt as follows. **1st row** (right side) sl.1 K to end. **2nd row** sl.1 K4 * P3 K5 rep from * to end. **3rd row** sl.1 P4 K3 * P5 K3 rep from * to last 5 sts, P4 K1. **4th row** as 2nd row. **5th row** as 1st row. **6th row** sl.1 * P3 K5 rep from * ending last rep K1. **7th row** sl.1 K3 * P5 K3 rep from * to last st, K1. **8th row** as 6th row. These 8 rows form patt. Cont in patt until work measures 31 in or 79 cm ending on wrong side. **Shape armholes:** cast off 12 sts at beg of next 2 rows (93(109) sts). Place markers at inner edge of cast-off rows. Cont on these sts until armhole measures 8(9) in or 20(23) cm, ending on wrong side. **Shape shoulders:** cast off 5(6) sts at beg of next 6 rows, then 4(5) sts at beg of next 6 rows. Cast off rem 39(43) sts.

POCKET LININGS (Both Alike)
Using No. 8 needles cast on 37 sts and proceed in st.st for 44 rows. Leave sts on a spare needle.

LEFT FRONT
Using No. 9 needles cast on 53(61) sts. **1st row** sl.1 K to end. Rep 1st row twice. Change to No. 8 needles and proceed in 8 row patt as for the Back until work measures 18 in or 46 cm, ending on wrong side. **Place pocket: next row** sl.1 patt 10(18), slip next 37 sts on a length of yarn, and in place of these patt across sts from one pocket lining, patt 5. Cont in patt until work measures same as the Back to armhole, ending on wrong side. **Shape armholes and front edge: next row** cast off 12 sts, patt to last 2 sts, K2tog. Place a marker. Keeping armhole edge straight cont in patt dec once at front edge every foll 4th row from previous dec until 27(33) sts rem. Cont without shaping until work matches the Back at armhole edge, ending on wrong side. **Shape shoulder:** cast off 5(6) sts at beg of next and foll 2 alt rows. Work 1 row. Cast off 4(5) sts at beg of next and foll 2 alt rows.

RIGHT FRONT
Work exactly as given for Left Front noting that 'place pocket' will read as follows. **Next row** sl.1 patt 4, slip next 37 sts on a length of yarn and in place of these patt across sts of 2nd pocket lining, patt. 11(19) sts. **To make up** follow pressing instructions on yarn label, then join shoulder seams.

FRONT BORDERS (Alike)
Using No. 10 needles cast on 24 sts. **1st row** sl.1 K to end. Rep 1st row until strip when slightly stretched fits along front edge to centre back of neck. Sew in position as you go along. Join at centre back.

ARMHOLE BORDERS (Alike)
With right side facing and No. 10 needles pick up and knit 96(108) sts along armhole edge between markers. **1st row** sl.1 K to end. Rep 1st row until work fits to end of cast-off row. Place a marker at both ends of last row. Cont as before until work matches first part. Cast off evenly.

POCKET BORDERS (Alike)
With right side facing and No. 10 needles, rejoin yarn to 37 sts of pocket and work as follows. **1st row** sl.1 K to end. Rep 1st row 16 times. Cast off loosely.

TO COMPLETE
Fold the front borders in half to wrong side and loosely slip-hem in position. Catch together at lower edge. Fold the armhole borders in half at markers to wrong side and loosely slip-hem in position. Catch down at cast-off edges. Catch down the pocket borders neatly to right side and the pocket linings to wrong side. Join side seams to armhole borders.

Curvey

Very much in the Schiaparelli tradition, this provocative design by Jane Sidey uses strong contrasts of shape and colour to turn a simple cardigan into something striking, sophisticated – and suggestive. The stark combination of black and white is softened here through the use of fluffy *bouclé* yarn that replaces hard-edged chic with a tantilizing suppleness. Snaking sinuously up the front over concealed pockets, and around the back of the neck, the white band hints at frankly feminine curves disguised by the jacket's straight lines. Good looks, an air of mystery and a sense of humour are three things men like most in a woman. If you aim to look good, Curvey will help you to dress the part.

DETAILS Shown with earrings from Detail. Photographed at the Park Lane Hotel, London.

MATERIALS **Yarn:** 17(19) 50 gm balls of Sirdar Sheba in black, 3(3) 50 gm balls of Sirdar Sheba in cream. **Needles:** No. 4 British (No. 9 American, 6 mm), No. 6 British (No. 7 American, 5 mm). **Also** hook fastenings for front.

MEASUREMENTS **To fit:** 32–34 (36–38) in bust (81–86 (92–97) cm) bust. **Actual measurement** all around under arms, 37(41½) in (94(105) cm). **Length:** 24(24¾) in (61(63) cm). **Sleeve seam:** for both sizes, 17 in (43 cm).

TENSION: 7 stitches to 2 in (5 cm).

ABBREVIATIONS **See** page 176. **Colours:** B=black, C=cream.

BACK
Cast on 66(74) sts with B and No. 6 needles. Work 5 rows st.st. K 1 row on the wrong side to mark hemline. Work 6 rows st.st. Change to No. 4 needles. Proceeding in st.st, continue until work measures 15¾ in (40 cm) from hemline, after a P row. **Shape armholes:** cast off 2 sts at beg of next 2 rows, then dec at both ends of every row until 48(52) sts remain. Proceed until work is 19(21) cm from beg of armholes, measured on the straight after a P row. **Shape shoulders** cast off 5 sts at beg of next 4 rows and 4(6) at beg of following 2 rows. Break yarn and leave remaining 20 sts on a spare needle.

LEFT FRONT
Begin by working pocket back. Cast on 15 sts with No. 4 needles and B. Work 14 rows st.st. Break yarn and leave sts on a spare needle. For main part, cast on 33(37) sts with B and No. 6 needles. Work 5 rows st. st. Break yarn. Cast on 3 sts, then K sts of main part as follows. **1st row** all K. **2nd row** sl.1, K2, P to end. Rep these 2 rows twice more. Change to No. 4 needles. Proceed in st.st with border as set until work measures 6¼ in (16 cm) from hemline, finishing after a wrong side row. **Join pocket:** next row K 9, place remaining sts on a spare needle and K sts of pocket back. Beg with a P row, work 28 more rows on these 24 sts. Break yarn. Join yarn to inner edge of remaining 27(31) sts. Work 29 rows. **Next row** work 12(16), place needle holding sts of side edge in front of work and P 1 st from each needle together until 15 are joined. P rem 9 sts. Proceed until work matches Back to armholes, finishing at side edge. **Shape armhole:** cast off 2 sts at beg of next row, work back then dec at side edge on every row until 27(29) sts remain. Proceed until work is 12(14) cm from beg of armhole, measured on the straight,

inc at both ends of next and every 12th row following until there are 48(52) sts. Proceed until work measures 17 in (43 cm) from hemline, finishing after a P row. **Shape top:** cast off 2 sts at beg of next 2 rows. Dec at both ends of next and every 4th row following until 32(36) sts remain. Work 3 rows. Dec at both ends of next and every alternate row until 24 sts remain for both sides. Dec at both ends of next 6 rows. Cast off remaining 12 sts.

NECKBAND

Join shoulder seams. With right side of work facing, using B and No. 6 needles, K the 5 sts from safety-pin at right front neck, K up 16 along remainder of right front neck, K sts of back, K up 16 along left front neck and K remaining 5 (62 sts). **Next row** * K3 K2tog, rep from * 3 times more, K22 ** K2tog, K3, rep from ** 3 times more (54 sts). K 1 row. Cast off knitwise on the wrong side.

FRONT TRIMMING

Work 2 pieces alike, 2nd side being sewn on in reverse. **Note:** the end of yarn at cast-on edge represents the outer edge. Cast on 7 sts with C and No. 4 needles. Slipping 1st st on every row, * work 6 rows g.st on all sts. Slipping next st at turn to avoid leaving a hole in the work, work 4 sts, turn, work back. Rep from * 3 times more. Proceed without shaping until work measures 6¼ in (16 cm) along outer edge, finishing at outer edge. **Next row** sl 1 K1, inc in next st, K to end. **Next row** sl.1, K to end. Rep these 2 rows until there are 12 sts, finishing at outer edge. **Now work 6 sts, turn, work back. Work 2 rows on all sts. ***Rep from ** 3 times more. Work 10 rows without shaping. Rep from ** to *** 4 times. **Next row** sl.1 K1, K2tog, K to end. Work 3 rows. Rep the last 4 rows. 10 sts. Work 11 rows, thus finishing at inner edge. **** Work 5 sts, turn, work back. Work 2 rows on all sts. Rep from **** 3 times more, finishing at inner edge. **Next row** sl.1 K1, K2tog, K to end. Work 3 rows. Rep the last 4 rows. 8 sts. Proceed without shaping until work measures 20 in (51 cm) unstretched when laid flat, finishing at outer edge. **Shape for collar: 1st row** sl.1 K1, inc in next st, K to end. **2nd row** sl.1, K to end. Rep these 2 rows until there are 18 sts, finishing at outer edge. **Shape fold of collar:** ***** sl.1 K10, turn, work back. Work 4 rows on all sts. Rep from ***** 6 times more. Cast off.

finishing at front edge. **Shape neck:** cast off 3 sts (1 on needle after casting off), work 4 more, and place these 5 sts on a safety-pin. Work to end. Now dec at neck edge on every row until 14(16) sts remain. Proceed until work matches Back to outer shoulder. **Shape shoulder:** cast off 5 sts at beg of next 2 side-edge rows. Work to side edge. Cast off remaining 4(6) sts.

RIGHT FRONT

Cast on 33(37) sts with B and No. 6 needles. Work 5 rows st.st, K 1 row. Cast on 3 sts. Proceed as for Left Front, reversing position of border, until work measures 6¼ in (16 cm) from hemline, after a wrong side row. Now K27(31) and place remaining 9 sts on a safety-pin. Turn. Now work 28 more rows on these sts. Break yarn. Work pocket back as for Left Front, but do not break yarn. Now K sts of pocket back, then proceed across the 9 sts at side edge. Work 28 more rows on these sts. Now work 9, place sts of front section in front of work and P 1 st from each needle at the same time until 15 are joined. P remaining 12(16). Complete to match Left Front, reversing shapings.

SLEEVES (Alike)

Cast on 36(40) sts with B and No. 6 needles. Work 5 rows st.st. K 1 row. Work 6 rows st.st. Change to No. 4 needles. Proceeding in st.st,

TO MAKE UP

Join side, shoulder and sleeve seams. Set sleeves into armholes, gathering sleeve head 3¼ in (8 cm) each side of shoulder seams. Turn up hem on lower and sleeve edges. Turn in and hem the garter stitch borders on front edges. Sew pocket backs in position. Press seams and hems. Sew on trimming as in photograph, cast-on edges to lower front edges of garment, and allowing wide edges of outer sections of trimming to overlap pocket openings by 1½ in (4 cm). Outer edges of trimming should be left free to 7 in (18 cm) from shoulders. Join cast-off edges of collar sections and sew collar to neck edge. Sew hooks and eyes at neck and lower edge.

Winning Streaks

Random stripes and stipples give a rich tapestry look to a design that's as easy to knit as it is to wear. Just sit back and let the colours in the yarn do the work for you. The pattern couldn't be simpler – plain stocking stitch rising to a yoke in twisted ribs that keep the coat neat and firm across the shoulders. The two yarns give a better coat weight than a single double knitting yarn and the fabric is firm enough to take patch pockets if you want to add them. The instructions give you a choice of coat and jacket lengths – both go well with casual or tailored separates, and you can wear the longer version as a coat-dress. Two 4-ply yarns are knitted together to create the fabric – one colour carries all the way through, the other changes as you go. Random fabrics are like roulette – the more you plan, the more you lose. You needn't limit yourself to the five colours we've used here, and you don't have to balance the stripes. Let yourself go, be adventurous – and you're bound to come up with a coat full of winning streaks.

DETAILS Shown with earrings from Butler & Wilson, belt and hadbag by Mulberry, shoes by Maud Frizon from Browns and tights by Aristoc.

MATERIALS **Yarn:** for coat length, Sirdar Wash'N'Wear 4-ply, 23(25:27) 20 gm balls in colour A, 7(8:9) 20 gm balls in colours B, D and E, 10(11:12) 20 gm balls in colour C. For jacket length: 18(20:22) 20 gm balls in colour A, 6(7:8) 20 gm balls in colours B, D and E, 9(10:11) 20 gm balls in colour C. **Needles:** No. 9 British (No. 4 American, 3¾ mm), No. 7 British (No. 6 American, 4½ mm). **Also:** cable pin. **Trimming:** 6 buttons for the coat, 5 buttons for the jacket.

MEASUREMENTS **To fit**: 34(36:38) in bust (86(91:97) cm) bust. **Length of coat** from top of shoulder when completed, 42(42:43) in, (107(107:109) cm). **Length of jacket** from top of shoulder when completed, 26(26:27) in (66(66:69) cm). **Sleeve seam**: 17½(18:18½) in (44(46:47) cm).

TENSION 5 sts and 7 rows to 1 in (2.5 cm) over stocking stitch using 2 strands of yarn.

ABBREVIATIONS **See** page 176. **Colours:** A=Black, B=Sweet Apricot, C=Wedgewood, D=Coral Reef, E=Celadon.

COAT BACK
With No. 7 needles and using one strand A and one strand B, cast on 102(106:110) sts. Work in st.st for 13 rows. **Next row** K for hemline. Continue in st.st, beg with a K row. Now work in a stripe sequence as follows: 10 rows A and B, 6 rows A and C, 2 rows A and B, 6 rows A and C, 22 rows A and D, 10 rows A and E. These 56 rows form the patt and are repeated. Work 4 patts in all. **Shape armholes:** continuing in A and E, cast off 3(5:4) sts at beg of next 2 rows, then dec 1 st at each end of the foll 2 rows (92(92:98) sts remain). Break E. Join in B. Finish in A and B in cable patt as follows. **1st row** P2 K4 to last 2 sts, P2. **2nd row** K2 P4 to last 2 sts, K2. **3rd row** P2, cable 4 as follows, sl. next 2 sts on

cable pin to back, K2 then K2 sts on cable pin, to last 2 sts, P2. **4th row** as 2nd row. These 4 rows form cable patt and are repeated. Work straight until armholes measure 9(9:10) in or 23(23:25) cm from beg. **Shape shoulders:** cast off 7 sts at beg of next 6 rows and 7(7:9) sts at beg of foll 2 rows. Cast off rem sts.

LEFT FRONT

With No. 7 needles and A and D, cast on 60(63:66) sts. Work in st.st for 13 rows. **Next row** K for hemline. Continue in st.st beg with a K row. Work in the stripe patt as follows: 10 rows A and D, 6 rows A and B, 2 rows A and C, 6 rows A and B, 22 rows A and E, 10 rows A and C. These 56 rows form the patt. Work 4 patts in all. **Shape armhole:** continuing in A and C, cast off 8(11:8) sts at beg of next row, then dec 1 st at this same edge on next 2 rows. 50(50:56) sts rem. Work 1 row. Break C. Join in D. Finish in A and D in cable patt as for Back. Work straight until measurement is 38(38:38½) in or 97(97:98) cm from hemline, finishing at centre edge. **Shape neck:** cast off 10 sts at beg of next row then dec 1 st at this edge on every row 12(12:16) times. When armhole measures the same as that of Back to shoulder, shape shoulder. Cast off at armhole edge 7 sts 3 times and 7(7:9) sts once.

RIGHT FRONT

With No. 7 needles and A and C, cast on 60(63:66) sts. Work in st.st for 13 rows. **Next row** K for hemline. Cont in st.st, beg with a K row. Work in the stripe patt as follows: 10 rows A and C, 6 rows A and E, 2 rows A and D, 6 rows A and E, 22 rows A and B, 10 rows A and D. Now continue as for Left Front reversing the shapings and working the cable yoke patt in A and E.

RIGHT SLEEVE

With No. 9 needles and A and C, cast on 56(56:62) sts. ** Work in cable patt as for yoke. Work 3 in (7.5 cm), finishing with a right-side row. Inc to 60(60:66) sts as follows. **Next row** P1(1:4) * P twice into next st, P17. Rep from * twice more, P twice into next st, P0(0:3). Change to No. 7 needles. Work st.st. ** Work in the striped patt as for Left Front. Inc 1 st at each end of every 14(10:12) rows until there are 72(78:82) sts. When sleeve measures 17½(18:18½) in (44(46:47) cm) from beg finishing with a P row, shape top of sleeve. **Next row** cast off 8(11:8), work to end. **Next row** cast off 3(5:4), work to end. Now dec 1 st at beg only of next 10(10:12) rows, then dec 1 st at each end of every row until 15(16:20) sts remain. Cast off.

LEFT SLEEVE

With No. 9 needles and A and C, cast on 56(56:62) sts. Work from ** to ** as for Right Sleeve, then work in the stripe patt as for Right Front. Inc 1 st at each end of every 14(10:12) rows until there are 72(78:82) sts. When sleeve measures 17½(18:18½) in (44(46:47) cm) from beg finishing with a P row, shape top of sleeve. **Next row** cast off 3(5:4), work to end. **Next row** cast off 8(11:8), work to end. Now finish the shaping as for top of Right Sleeve.

BORDERS

With No. 9 needles and A and C, cast on 11 sts. Work in K1 P1 rib as follows. **1st row** K1 P1 to last st, K1. **2nd row** P1 K1 to last st, P1. These 2 rows are repeated. Work straight for 12(12:12½) in (30.5(30.5:32) cm). **Next row** rib 4, cast off 3 sts for buttonhole, rib 4. **Next row** rib 4, cast off 3 sts, rib 4. Form 5 more buttonholes at intervals of 5 in, 12.5 cm, measured from centre of buttonhole. When border measures 38(38:38½) in or 97(97:98) cm from beg, cast off in rib. Rep for other border, omitting buttonholes.

TO MAKE UP

Backstitch shoulder, side and sleeve seams, oversewing cuffs. Set right sleeve into right armhole and left sleeve into left armhole. Fold along hemline at bottom of coat and slip stitch down. Oversew borders to centre front edges. **Knit collar:** with right side of work facing, No. 9 needles and A and C, K up 115(117:127) sts round neck, beg and ending halfway across front borders. Work in K1 P1 rib as for borders, beg with a 1st row. Work straight for 5½(5½:6) in (14(14:15) cm). Cast off in rib. Press seams under a dry cloth with a warm iron. Sew on buttons.

JACKET

Back and front: work as for back and front of coat but only working 2 patts in all (instead of 4 patts) before shaping armholes. **Sleeves:** work as for sleeves of coat. **Borders:** work in K1 P1 rib as for borders of coat but work 1(1:1½) in or 2(2:4) cm only before first buttonhole. Form 4 more buttonholes at intervals of 5 in (12·5 cm) measured from centre of buttonhole. When border measures 22(22:22½) in (56(56:57) cm) from beg, cast off in rib. Rep for other border, omitting buttonholes. Make up and knit collar as for coat.

Fruity

Hit the jackpot in Suzanne Barnacle's fruity sweater, a cheery batwing top with cherries, lemons and plums tumbling down the sleeves. *Fruity* shows how effective an imaginative use of surface decoration can be. Large and vivid motifs like these often overwhelm a design when given a central placement, but by working the fruits on to the sleeves you can use colour and pattern lavishly without detracting from the simplicity of the shape. The second yarn texture adds another dimension of interest, and the contrast yarns are not carried across the back so the fabric is supple and easy to work. You can knit the top plain, or vary the design by changing the fruit colours – use green to turn the lemons into limes, and pink to turn the plums into guavas. The black ground acts as a perfect foil to the colours, so you can pack them in as brightly as they do on market fruit stalls.

MATERIALS **Yarn**: Sirdar Wash'N'Wear Double Crepe; 10 40 gm balls of colour A, 1 40 gm ball of colour B. Sirdar Sheba; 1 50 gm ball each of colours C, D, E. **Needles:** one pair each No. 8 British (No. 5 American, 4 mm), No. 10 British (No. 3 American, 3¼ mm).

MEASUREMENTS One size, to fit 32–38 in bust (81–97.5 cm) bust. **Length:** approximately 24 in (61 cm). **Sleeve** at centre, approximately 18 in (46 cm).

TENSION 12 sts to 2 in (5 cm) on No. 8 needles.

ABBREVIATIONS **See** page 176. **Colours:** A=Black, B=Sea Grass, C=Flame, D=Narcissus, E=Bordeaux.

NOTE When working motifs **do not** carry colour across wrong side of work, but wind off small quantities, twisting yarn when changing colour to avoid a hole.

BACK

** Using No. 10 needles and A yarn, cast on 101 sts. **1st row** (right side) sl.1 K1 * P1 K1 rep from * to last st, K1. **2nd row** sl.1 * P1 K1 rep from * to end. Rep 1st and 2nd rows 14 times. Change to No. 8 needles and proceed in st.st until work measures 12 in or 30 cm, ending on wrong side. **Shape armholes:** cast off 6 sts at beg of next 2 rows ** (89 sts). Cont on these sts until work measures 24 in or 61 cm, ending on wrong side. **Shape shoulders:** cast off 9 sts at beg of next 6 rows (35 sts). Cast off rem sts.

FRONT

Work exactly as given for Back from ** to **. Cont on these sts until work is 14 rows less than the Back to armhole edge. **Shape neck: next row** K38, turn and leave rem 51 sts on a spare needle. Work 11 rows dec once at neck edge on

every row (27 sts). Work 2 rows. **Shape shoulder:** cast off 9 sts at beg of next and foll 2 alt rows. With right side facing, rejoin yarn to rem 51 sts and cast off centre 13 sts (38 sts). Complete to correspond with first side reversing shapings.

SLEEVES (Alike)

Using No. 10 needles and A yarn cast on 47 sts. **1st row** (right side) sl.1 K1 * P1 K1 rep from * to last st, K1. **2nd row** sl.1 * P1 K1 rep from * to end. Rep 1st and 2nd rows 8 times, then 1st row again. **Next row** sl.1, rib 4, ML (rib 2, ML) 19 times, rib 4 (67 sts). Change to No. 8 needles and work 2 rows st.st. Proceed in patt and at the same time shape sides as follows. **1st row** (right side) K1A ML 7A 3C (21A 3C) twice, 7A ML 1A. 69 sts. **2nd row** P8A 5C (19A 5C) twice, 8A. **3rd row** K1A ML 7A 5C (19A 5C) twice, 7A ML 1A. 71 sts. **4th row** P9A 5C (19A 5C) twice, 9A. **5th row** K1A ML 8A 5C (19A 5C) twice, 8A ML 1A. 73 sts. **6th row** P11A 3C (21A 3C) twice, 11A. **7th row** K1A ML 71A ML 1A. 75 sts. **8th row** P75A. **9th row** K1A ML 7A 3C 5A 3C (13A 3C 5A 3C) twice, 7A ML 1A (77 sts). **10th row** P8A 5C 3A 5C (11A 5C 3A 5C) twice, P8A. **11th row** K1A ML 7A 5C 3A 5C (11A 5C 3A 5C) twice, 7A ML 1A (79 sts). **12th row** P9A 5C 3A 5C (11A 5C 3A 5C) twice, 9A. **13th row** K1A ML 8A 5C 3A 5C (11A 5C 3A 5C) twice, 8A ML 1A (81 sts). **14th row** P11A 3C 5A 3C (13A 3C 5A 3C) twice, 11A. Work 18 rows in A yarn only inc once at both ends of next and every foll alt row as before (99 sts). **33rd row** K1A ML 14A 6D (16A 6D) 3 times, 11A ML 1A (101 sts). **34th row** P12A 9D (13A 9D) 3 times, 14D. **35th row** K1A ML 12A 12D (10A 12D) 3 times, 9A ML 1A (103 sts). **36th row** P12A 12D (10A 12D) 3 times, 13A. **37th row** K1A ML 11A 13D (9A 13D) 3 times, 11A ML 1A (105 sts). **38th row** P14A 12D (10A 12D) 3 times, 13A. **39th row** K1A ML 12A 12D (10A 12D) 3 times, 13A ML 1A (107 sts). **40th row** P15A 12D (10A 12D) 3 times, 14A. **41st row** K1A ML 12A 12D (10A 12D) 3 times, 15A ML 1A (109 sts). **42nd row** P18A 11D (11A 11D) 3 times, 14A. **43rd row** K1A ML 16A 7D (15A 7D) 3 times, 18A ML 1A (111 sts). Work 10 rows in A yarn only, inc once at both ends of 2nd and every foll alt row as before (121 sts). Work 3 rows straight. **57th row** K1A ML 12A 7E (15A 7E) 4 times, 12A ML 1A (123 sts). **58th row** P13A 9E (13A 9E) 4 times, 13A. **59th row** K12A 11E (11A 11E) 4 times, 12A. **60th row** P12A 11E (11A 11E) 4 times, 12A. **61st row** K1A ML 11A 11E (11A 11E) 4 times, 11A ML 1A (125 sts). **62nd row** P13A 11E (11A 11E) 4 times, 13A. **63rd row** K13A 11E (11A 11E) 4 times, 13A. **64th row** P13A 11E (11A 11E) 4 times, 13A. **65th row** K1A ML 12A 11E (11A 11E) 4 times, 12A ML 1A (127 sts). **66th row** P14A 11E (11A 11E) 4 times, 14A. **67th row** K14A 11E (11A 11E) 4 times, 14A. **68th row** P15A 4E 1A 4E (13A 4E 1A 4E) 4 times, 15A. **69th row** K1A ML 15A 3E 1A 3E (15A 3E 1A 3E) 4 times, 15A ML 1A (129 sts). Work 15 rows in A yarn only, inc once at both ends of 4th and every foll 4th row as before (135 sts). **85th row** K1A ML 17A 3C (21A 3C) 4 times, 17A ML 1A (137 sts). **86th row** P18A 5C (19A 5C) 4 times, 18A. **87th row** K18A 5C (19A 5C) 4 times, 18A. **88th row** P18A 5C (19A 5C) 4 times, 18A. **89th row** K1A ML 17A 5C (19A 5C) 4 times, 17A ML 1A (139 sts). **90th row** P20A 3C (21A 3C) 4 times, 20A. **91st row** K139A. **92nd row** P139A. **93rd row** K1A ML 15A 3C 5A 3C (13A 3C 5A 3C) 4 times, 15A ML 1A (141 sts). **94th row** P16A 5C 3A 5C (11A 5C 3A 5C) 4 times, 16A. **95th row** K16A 5C 3A 5C (11A 5C 3A 5C) 4 times, 16A. **96th row** P16A 5C 3A 5C (11A 5C 3A 5C) 4 times, 16A. **97th row** K1A ML 15A 5C 3A 5C (11A 5C 3A 5C) 4 times, 15A ML 1A (143 sts). **98th row** P18A 3C 5A 3C (13A 3C 5A 3C) 4 times, 18A. Continuing in A yarn only work straight until sleeve measures 18 in (46 cm) at centre, ending on wrong side. Cast off evenly. Place a marker at centre of cast-off edge.

COLLAR

Using No. 10 needles and A yarn, cast on 33 sts. **1st row** (right side) K2 * P1 K1 rep from * to last st, K1. **2nd row** K1 * P1 K1 rep from * to end. Keeping rib patt correct cast on 8 sts at beg of next 8 rows (97 sts). Work straight until the collar measures 4 in or 10 cm at side edge, ending on wrong side. Break off A yarn. Change to No. 8 needles, join in B yarn and work as follows. **1st row** K. **2nd row** K1 * P1 K1 rep from * to end. Cast off evenly in rib.

TO MAKE UP

Read pressing instructions on yarn label. Using B yarn, embroider leaves and stalks of fruit motifs on sleeves. Join shoulder seams. Sew sleeves in position placing marker at shoulder seam. Join side sleeve seams. Sew collar in position around neck edge. Using 4 strands of B yarn make a twisted cord 36 in or 91 cm in length. Place a marker at centre. Using C yarn make 2 small pompons. Attach one to each end of cord. Sew cord in position at centre of back neck under collar.

Inshallah

Berber stripes and Moorish colours give *Inshallah* the richness of an oriental carpet, and the mixture of crepe, mohair and *bouclé* yarns make for a fabric that's magically warm and light. The shawl's triangular shape is much more interesting than the usual rectangle, giving you lots of opportunities for imaginative dressing. Drape it over the head as they do in the Casbah, throw it over one shoulder gypsy fashion, or wear it low across the back, with the ends twisted round your arms. The stripes curve around the body adding variety to the pattern, and it looks just as nice worn with either side out. Like Scheherazade's stories, this shawl will see you through 1,001 nights in style.

MATERIALS **Yarn**: Sirdar Caprine 2; 6 25 gm balls in colour A, 4 25 gm balls in colour B. Sirdar Sheba; 4 50 gm balls in colour C. Sirdar Wash'N'Wear Double Crepe; 2 40 gm balls in colour D, 1 40 gm ball in colour E. **Needles**: No. 3 British (No. 10 American, 6½ mm).

MEASUREMENTS **Width across**: 72 in (182 cm). **Point to cast-on edge**: 36 in (91 cm).

TENSION over pattern, 24 sts and 48 rows to 8 in (20 cm).

ABBREVIATIONS **See** page 176. **Colours:** A=Heather Mist, B=Silver Mist, C=Bordeaux, D=Medoc, E=Wedgwood.

Using A, cast on 216 sts. **1st row** using A, K2tog, K until 2 sts rem, K2tog. **2nd row** all K. Rep these 2 rows once in C, once in A, twice in D, once in B, once in C, twice in E, once in B, then twice in D. (192 sts). Cont in this 24 row stripe pattern, dec 1 st at both ends of every right side row until 2 sts rem, K2tog and fasten off. Using A, B and C only, make a 6 in (15 cm) fringe for each of the two shorter sides.

Dutch Treat

A trim of lace and tulips turns this tidy two-piece by Suzanne Barnacle into a real Dutch treat. The simple lines of the top and skirt make for well-groomed good looks that are practical as well as pretty. Both pieces are quite easy to knit, and the peplum in raised chevron stitch requires no shaping. The versatile top can be worn loose or lightly gathered at the waist, as a button-front blouse over a camisole or as a cardigan when teamed with a toning tie-neck blouse. The hearts and flowers add a final feminine touch – be sure to embroider the back as well. Knit *Dutch Treat* in the colours given here, or in your favourite pastels – pale peach, soft jonquil yellow, or perhaps misty forget-me-not blue.

MATERIALS **Yarn:** for cardigan, 18(19:19:20) 25 gm balls of Sirdar Superwash 4-ply in main colour; for embroidery, 1 25 gm ball each of Sirdar Superwash 4-ply in Peppermint (095), Coral Reef (078) and Powder Blue (010). For skirt, 16(17:17:18) 25 gm balls of Sirdar Superwash 4-ply in main colour. **Needles:** one pair each No. 10 British (No. 3 American, 3¼ mm), No. 11 British (no. 2 American, 3 mm), No. 12 British (No. 1 American, 2¾ mm). **Trimming:** 7 buttons, elastic for waist of skirt.

MEASUREMENTS **To fit:** 32(34:36:38) in bust (81(86:91:97) cm) bust. **Actual size:** 34(36:38:40) in (86(91:97:102) cm). **Skirt to fit** 34(36:38:40) in (86(91:97:102) cm) hips. **Actual size:** 36(38:40:42) in (91(97:102:107) cm). **Length of cardigan:** 20½(21:21½:22) in (52(53:54:56) cm). **Sleeve length:** 17½ in (44 cm), all sizes. **Length of skirt:** 28(28½:28½:29) in (71(72:72:74) cm).

TENSION 15 sts to 2 in (5 cm).

ABBREVIATIONS See page 176. **Colours:** A=main colour, B=Peppermint, C=Coral Reef, D=Powder Blue.

CARDIGAN BACK
Using No. 10 needles and yarn A, cast on 141(151:161:171) sts. **1st row** (wrong side) sl.1 P to last st, K1. **2nd row** sl.1 * yfwd K3 sl.1 K2tog PSSO K3 yfwd K1 rep from * to end. **3rd row** as 1st row. **4th row** sl.1 K1 yfwd K2 sl.1 K2tog PSSO K2 yfwd K1 P1, rep from * to end. **5th row** sl.1 * P9 K1 rep from * to end. **6th row** sl.1 * K2 yfwd K1 sl.1 K2tog PSSO K1 yfwd K2 P1 rep from * to end. **7th row** sl.1 * P9 K1 rep from * to end. **8th row** sl.1 * K3 yfwd sl.1 K2tog PSSO yfwd, K3 P1, rep from * to end. Rep 1st to 8th rows 4 times more. **Next row** sl.1 P9(4:7:3) P2tog (P8(8:7:7) P2tog), 12(14:16:18) times, P9(4:7:3) (128(136:144:152) sts). Proceed in st. st until work measures 13 in (33 cm), ending on wrong side. **Shape armholes:** cast off 4 sts at beg of next 2 rows. Dec once at both ends of next 7 rows. Work 1 row. Dec once at both ends of next and foll 7(9:11:13) alt rows. Cont until work measures 20½(21:21½:22) in (52(53:54:56) cm), ending on wrong side. **Shape shoulders:** cast off 8(8:9:9) sts at beg of next 4 rows, then 8(9:8:9) sts at beg of next 2 rows. Cast off rem 42(44:46:48) sts.

CARDIGAN LEFT FRONT

Using No. 10 needles and A yarn, cast on 73(75:83:85) sts. **1st row (wrong side) sl.1 K1(3:1:3), P to last st, K1. **2nd row** sl.1 * yfwd K3 sl.1 K2tog PSSO K3 yfwd K1, rep from * to last 2(4:2:4) sts, K2(4:2:4). **3rd row** as 1st row. **4th row** sl.1 * K1 yfwd K2 sl.1 K2tog PSSO K2 yfwd K1 P1, rep from * to last 2(4:2:4) sts, K2(4:2:4). **5th row** sl.1 K2(4:2:4) * P9 K1, rep from * to end. **6th row** sl.1 * K2 yfwd K1 sl.1 K2tog PSSO K1 yfwd K2 P1, rep from * to last 2(4:2:4) sts, K2(4:2:4) sts. **7th row** as 5th row. **8th row** sl.1 * K3 yfwd sl.1K2tog PSSO yfwd K3 P1, rep from * to last 2(4:2:4) sts, K2(4:2:4). Rep 1st to 8th rows 4 times more. ** **Next row** sl.1 K4, slip these 5 sts on to a thread, P7(16:5:9), P2tog P15(33:11:18) 3(1:5:3) times, P2tog P7(16:5:8), K1. 64(68:72:76) sts. Proceed on these 64(68:72:76) sts in st.st until work matches the Back at side edge, ending on wrong side. **Shape armhole:** cast off 4 sts at beg of next row. Work 1 row. Dec once at armhole edge in next 7 rows. Work 1 row. Then dec once at armhole edge in next and foll 7(9:11:13) alt rows. Cont until work measures 18(18½:19:19½) in (46(47:48:50) cm), ending on right side. **Shape neck:** cast off 8(8:9:9) sts, work to end. Work 13(14:14:15) rows dec once at neck edge on every row (24(25:26:27) sts). Cont until work matches the Back at armhole edge, ending on wrong side. **Shape shoulder:** cast off 8(8:9:9) sts at beg of next and foll alt row. Work 1 row. Cast off rem 8(9:8:9) sts.

CARDIGAN RIGHT FRONT

Work exactly as given for the left front from ** to **. **Next row** sl.1 P7(16:5:8) P2tog (P15(33:11:18) P2tog) 3(1:5:3) times, P7 (16:5:9), slip last 5 sts on to a thread (64(68:72:76) sts).

SLEEVES (Alike)

Using No. 12 needles and A yarn, cast on 57(61:61:65) sts. **1st row** (right side) sl.1 K1 * P1 K1, rep from * to last st, K1. **2nd row** sl.1 * P1 K1 rep from * to end. Rep 1st and 2nd rows 14 times, then 1st row again. **Next row** sl.1 rib 4(6:2:4) ML (rib 4, ML) 12(12:14:14) times, rib 4(6:2:4) (70(74:76:80) sts). Change to No. 10 needles and proceed in st.st, inc once at each end of 5th(5th:7th:7th) and every foll 12th(12th:10th:10th) row until there are 90(94:98:102) sts. Cont without shaping until work measures 17½ in (44 cm), ending on wrong side. **Shape top:** cast off 4 sts at beg of next 2 rows. Work 52(56:60:64) rows dec once at each end of next and every alt row (30 sts). Cast off evenly.

TO MAKE UP

Follow pressing instructions on label, then join shoulder and side seams. Join sleeve seams. Insert sleeves.

BORDERS

Button border: with right side facing and No. 12 needles, rejoin A yarn to 5 sts on a thread at left Front. **1st row** sl.1 K1 ML K1 ML K2. 7 sts. **2nd row** sl.1 K1 ML K3 ML K2. 9 sts. Now proceed in K1 P1 rib as for sleeve cuff until strip when slightly stretched fits along front edge to start of neck shaping, ending on wrong side. Cast off in rib. **Buttonhole border:** with right side facing and No. 12 needles, rejoin A yarn to 5 sts on a thread at right Front. **1st row** sl.1 K1 ML K1 ML K2. 7 sts. **2nd row** sl.1 K1 ML K3 ML K2. 9 sts. Now proceed as for button border with the addition of 7 buttonholes, first to come on 3rd rib row, seventh to come 2 rows below neck shaping and remainder spaced evenly between. **To make buttonhole: 1st row** (right side) sl.1 rib 2, cast off 3, rib to end. **2nd row** sl.1 rib 2, cast on 3, rib 3. First mark position of buttons on Left Front with pins to ensure even spacing, then work buttonholes to correspond.

COLLAR

Using No. 12 needles and A yarn, cast on 57(59:59:61) sts. **1st row** (right side) sl.1 K1 * P1 K1 rep from * to last st, K1. Cast on 8 sts at beg of next 8 rows taking inc sts into rib patt (121(123:123:125) sts). Work 2 rows. Change to No. 10 needles and proceed in st.st as follows. **1st row** sl.1 K to end. **2nd row** sl.1 K1, P to last 2 sts, K2. Rep 1st and 2nd rows once. **Shape collar: 1st inc row** sl.1 K7(8:8:9) ML (K15 ML) 7 times, K8(9:9:10) (129(131:131:133) sts). Work 1 row. **2nd inc row** sl.1 K7(8:8:9) ML (K16 ML) 3 times, K17 ML (K16, ML) 3 times, K8(9:9:10) (137(139:139:141) sts). Work 3 rows. **3rd inc row** sl.1 K7(8:8:9) ML (K17 ML) 3 times, K19 ML (K17 ML) 3 times, K8(9:9:10) (145(147:147:149) sts). Work 3 rows. **4th inc row** sl.1 K7(8:8:9) ML (K18 ML) 3 times, K21 ML (K18 ML) 3 times, K8(9:9:10) (153(155:155:157) sts). Work 3 rows. **5th inc row** sl.1 K7(8:8:9) ML (K19 ML) 3 times, K23 ML (K19 ML) 3 times, K8(9:9:10) (161(163:163:165) sts). Work 1 row. **Next row** sl.1 K2tog, K to last 3 sts, sl.1 K1 PSSO K1. **Next row** sl.1 K2tog, K to last 3 sts, sl.1 K1 PSSO K1. **Next row** sl.1 K2tog, K to last 3 sts, sl.1 K1 PSSO K1. Cast off loosely.

TO COMPLETE CARDIGAN

Using B, C and D yarns, embroider flowers on the front and back. Using B and C yarn, embroider flower heads on the collar. Sew collar in position all round neck, starting and ending at centre of front borders. Using 4 strands of A yarn, make a twisted cord 50(52:54:56) in or 127(132:137:142) cm long. Knot ends. Thread through holes at waist.

SKIRT BACK AND FRONT (Alike)

Using No. 11 needles, cast on 200(208:216:224) sts. **1st row** all K. Rep 1st row 20 times. Change to No. 10 needles and proceed in st.st until work measures 4 in (10 cm), ending on wrong side. Proceeding in st.st dec on next row and foll 12th row as follows. **1st dec row** K2tog tbl, K58 K2tog tbl, K1 K2tog K70(78:86:94) K2tog tbl, K1 K2tog K58 K2tog. Work 11 rows. **2nd dec row** K2tog tbl, K56 K2tog tbl, K1 K2tog K68(76:84:92) K2tog tbl, K1 K2 tog, K56 K2tog. Work 11 rows. **3rd dec row** K2tog tbl, K54 K2tog tbl, K1 K2tog K66(74:82:90), K2tog tbl, K1 K2tog K54 K2tog. Cont dec 6 sts thus on every following 12th row until 134(142:150:158) sts remain. Proceed without shaping until work measures approximately 20 in (51 cm), ending on wrong side and lifting work and measuring while hanging to allow dropping. **Next row** K2tog tbl, K36 K2tog tbl, K1 K2tog K48(56:64:72) K2tog tbl, K1 K2tog K36 K2tog. Cont dec as before on every following 12th row until 104(112:120:128) sts remain. Proceed until work is 1 in (2·5 cm) shorter than required length, ending on wrong side. Change to No. 11 needles and work 1 in (2·5 cm) in K1 P1 rib. Cast off in rib. **To make up:** join side seams using a fine back stitch. Cut elastic to fit waist and join in a ring. Sew ring inside waist ribbing using a herringbone stitch over elastic to form a casing. Press seams.

Cabaret

If you want to be a star, just slip into this sparkly glitter vest and twinkle the night away. A knitted version of the brocade waistcoat, *Cabaret* can turn a workday dress into a special-date-after-work outfit, and you can use it to add a bit of dazzle to the evening clothes you already have. Wear it with a velvet theatre suit and spangly tights, dress it up with a long skirt and lots of jewellery, or team it with its matching bow tie, evening trousers and a

tailored silky shirt. The clever fabric is much easier to knit than it looks, and the band of elastic hugs the waistcoat in to the small of the back, giving a smart svelte finish. Try substituting plain white and silver-threaded yarns for the black, or mix buff, cream and gold-threaded yarns for a top that sparkles like champagne. *Cabaret* looks just as good on men, but if you can't coax him into trying on your evening waistcoat, knit him one of his own in day-time shades.

DETAILS Shown with suede evening mules by Charles Jourdan, Balinese trousers in Tootal's Robia cotton stripe voile D3070 and wing collar shirt from Barney's Underground, New York. Photographed at the Royal Horse-guards Hotel, London.

MATERIALS **Yarn:** Sirdar Super Prelude, 3(4:4) 20 gm balls in colour A. Sirdar Fontein Crepe 3(3:3) 25 gm balls in colours B and C. Sirdar Fontein Crepe, 1 20 gm ball in colour A for bow tie. **Needles:** No. 12 British (No. 1 American, 2¾ mm), No. 10 British (No. 3 American, 3¼ mm). **Trimming:** 4 buttons and shirring elastic for waistcoat, elastic for bow tie.

MEASUREMENTS **To fit:** 34(36:38) in (86(91:97) cm) bust/chest. **Actual size:** 36(38:40) in (91(97:102) cm). **Length** at back, excluding border, 19(19½:20) in (48(49:51) cm).

TENSION 15 sts to 2 in (5 cm) on No. 10 needles.

ABBREVIATIONS **See** page 176. **Colours:** A=Black (Super Prelude), B=Black (Fontein Crepe), C=Hawthorn (Fontein Crepe).

NOTE: When changing colours, carry yarn not in use up side of work.

BACK

Using No. 10 needles and A yarn, cast on 135(143:151) sts and work 2 rows in st.st. Proceed in patt as follows. Join in B yarn. **1st row** (right side) using B yarn, sl.1 K2 * sl.1 K3, rep from * to end. **2nd row** sl.1 P2 * sl.1 P3, rep from * to last 4 sts, sl.1 P2 K1. **3rd row** as 1st row. **4th row** sl.1 P to last st, K1. **5th row** using A yarn, sl.1 K2 * sl.1 K3, rep from * to end. **6th row** sl.1 P2 * sl.1 P3, rep from * to last 4 sts, sl.1 P2 K1. **7th row** as 5th row. **8th row** sl.1 P to last st, K1. Join in C yarn. **9th row** sl.1 K2 * sl.1 K3, rep from * to end. **10th row** sl.1 P2 * sl.1 P3, rep from * to last 4 sts, sl.1 P2 K1. **11th row** as 9th row. **12th row** sl.1 P to last st, K1. These 12 rows form patt. Cont in patt until work measures 11 in (28 cm), ending on wrong side. **Shape armholes:** cast off 6 sts at beg of next 2 rows. Keeping continuity of patt, work 6 rows dec 1 st each end of every row (111(119:127) sts). Work 16 rows dec 1 st each end of next and every alt row (95(103:111) sts). Cont in patt on these sts until work measures 19(19½:20) in (48(49:51) cm), ending on wrong side. **Shape shoulders:** cast off 10(11:11) sts at beg of next 4 rows. Cast off 10(10:12) sts at beg of next 2 rows. Cast off rem 35(39:43) sts.

LEFT FRONT

Using No. 10 needles and A yarn, cast on 3 sts and work 2 rows in st.st. Proceed in patt, shaping point as follows. Join in B yarn. **1st row** (right side) using B yarn inc in first st, sl.1 inc in last st. **2nd row** inc in first st, P1 sl.1 P1, inc in last st. **3rd row** inc in first st, K2 sl.1 K2, inc in last st. **4th row** inc in first st, P to last st, inc in last st. **5th row** using A yarn, inc in first st, * sl.1 K3, rep from * to last 2 sts, sl.1 inc in last st. **6th row** inc in first st, P1 * sl.1 P3, rep from * to last 3 sts, sl.1 P1 inc in last st. **7th row** inc in first st, K2 * sl.1 K3, rep from * to last 4 sts, sl.1 K2 inc in last st. **8th row** inc in first st, P to last st, inc in last st. Join in C yarn. **9th row** using C yarn, inc in first st, * sl.1 K3, rep from * to last 2 sts, sl.1 inc in last st. **10th row** inc in first st, P1 * sl.1 P3, rep from * to last 3 sts, sl.1 P1 inc in last st. **11th row** inc in first st, K2 * sl.1 K3, rep from * to last 4 sts, sl.1 K2 inc in last st. **12th row** inc in first st, P to last st, inc in last st (27 sts). These 12 rows form the patt rep. Keeping continuity of patt cont to inc 1 st each end of every row to 55 sts, thus ending on wrong side. ** **Next row** cast on 4(6:8) sts, patt to end. **Next row** patt to end. **Next row** cast on 4(6:8) sts, patt to end (63(67:71) sts). Cont in patt on these sts until work measures 11 in or 28 cm at side edge, ending with same patt row as on the Back. **Shape armhole and front slope: next row** cast off 6 sts, patt to last 2 sts, K2tog. **Next row** patt to end. Work 7 rows dec 1 st at armhole edge on every row. Work 1 row. Work 14 rows dec 1 st at armhole edge on next and every alt row **at the same time** dec 1 st at front edge on every foll 6th(6th:4th) row from last dec until 30(32:34) sts rem. Cont straight until work measures same as the Back to shoulder, ending on wrong side. **Shape shoulder:** cast off 10(11:11) sts at beg of next and foll alt row. Work 1 row. Cast off rem 10(10:12) sts.

RIGHT FRONT

Work exactly as given for Left Front to **. Work 1 row thus ending on right side. **Next row** cast on 4(6:8) sts, patt to end. **Next row** patt to end. **Next row** cast on 4(6:8) sts, patt to end. 63(67:71) sts. Cont in patt on these sts and complete to correspond with the Left Front, reversing all shapings.

ARMBANDS (Alike)

Sew shoulder seams. Using No. 12 needles and A yarn, with right side facing, pick up and knit 131(139:147) sts evenly along armhole edge. **1st row** sl.1 * P1 K1 rep from * to end. **2nd row** sl.1 K1 * P1 K1, rep from * to last st, K1. Rep 1st and 2nd rows twice more, then 1st row again. Cast off in rib. **To make up:** read pressing instructions on yarn label, sew side seams.

LEFT FRONT BORDER (Lady's)

Using No. 12 needles and A yarn, cast on 11 sts. **1st row** sl.1 K1 * P1 K1, rep from * to last st, K1. **2nd row** sl.1 * P1 K1, rep from * to end. Rep 1st and 2nd rows until border when slightly stretched fits along lower edge from left side seam to base of point, ending with a 1st row. ** **Shape point as follows: 1st row** rib 2. **2nd row** sl.1 K1. **3rd row** rib 4, turn. **4th, 6th, 8th and 10th rows** sl.1 rib to end. **5th row** rib 6, turn. **7th row** rib 8, turn. **9th row** rib 10, turn. Cont in rib on all sts until border when slightly stretched fits along front lower point, up front edge and round to centre back of neck. Cast off in rib. Sew border in position. Mark positions for 4 buttons, top button 2 in (5 cm) below start of front slope, first button just above last inc at lower front, spacing remaining two buttons evenly between. ** **Left Front Border and Lower Back:** using No. 12 needles and A yarn, cast on 11 sts. **1st row** sl.1 K1 * P1 K1, rep from * to last st, K1. **2nd row** sl.1 * P1 K1, rep from * to end. Rep 1st and 2nd rows until border when slightly stretched fits along lower back edge from left side seam to base of point, ending with a 2nd row. Shape point as given for the Left Front and work buttonhole to correspond with markers as follows. **1st row** rib 4, cast off 3 sts, rib to end. **2nd row** rib to end casting on 3 sts over those cast off. Complete as for Left Front border. Sew border in position, join borders at back of neck. Sew on buttons. Thread 4 rows of shirring elastic on wrong side along lower back edge.

RIGHT FRONT BORDER (Men's)

Using No. 12 needles and A yarn, cast on 11 sts. **1st row** sl.1 K1 * P1 K1, rep from * to last st, K1. **2nd row** sl.1 * P1 K1, rep from * to end. Rep 1st and 2nd rows until border when slightly stretched fits along lower edge from right side seam to base of point, ending with a 2nd row. Work as given for the Left Front border of lady's version from ** to **. **Left Front Border (Men's):** using No. 12 needles and A yarn, cast on 11 sts. **1st row** sl.1 K1 * P1 K1, rep from * to last st, K1. **2nd row** sl.1 * P1 K1 rep from * to end. Rep 1st and 2nd rows until border when slightly stretched fits along lower edge from right side seam to base of point, ending with a 1st row. Work as given for the Left Front of the lady's version and work buttonholes to correspond with markers as follows. **1st row** rib 4, cast off 3 sts, rib to end. **2nd row** rib to end casting on 3 sts over those cast off. Complete as for right front border. Sew border in position. Join borders at back of neck. Sew on buttons. Thread 4 rows of shirring elastic on wrong edge.

BOW TIE

Using No. 10 needles and A yarn (used double throughout), cast on 33 sts and work 33 rows in garter stitch (every row knit). Cast off. **To make up:** using double yarn gather tie through centre, draw up tightly, wind yarn 5 times round centre and fasten off securely. Attach a length of elastic to back of tie and join at back.

Allsorts

Most men are as shy of adding colour to their clothes as they are of adding water to their whisky, but this sweater scattered with cheery hints of colour will appeal to the most conservative tastes. The easy-to-knit stitch gives you as much colour and pattern as you'd get on a standard Fair Isle without the complications of Fair Isle knitting, and the possibilities are endless. *Allsorts* has been knitted with five contrast colours, but it looks just as good with four or three. Experiment with soft colours against a light background, or bright colours against dark yarns. Try an all-over two-tone pattern, or reverse the pattern midway – sleeves in green on blue, body in blue on green. Show him how much fun colourful clothes can be, and he might even throw out his red socks!

DETAILS Shown here with shirt and trousers by Paul Smith.

MATERIALS **Yarn:** Sirdar Wash'N'Wear Double Crepe, 9(9:10:10) 40 gm balls in main colour A, 1 40 gm ball in each of colours B, C, D, E, F. **Needles:** No. 10 British (No. 3 American, 3¼ mm), No. 6 British (No. 7 American, 5 mm).

MEASUREMENTS **To fit:** 36(38:40:42) in (91(97:102:107) cm) chest. **Actual size:** 39(41:43:45) in (99(104:109:114) cm). **Length:** 26(26½:26:27) in (66(67:67:69) cm). **Sleeve length:** all sizes, 19 in (48 cm).

TENSION 12 sts to 2 in (5 cm) on No. 6 needles.

ABBREVIATIONS **See** page 176. **Colours:** A=Light Navy, B=Mellow Gold, C=Medoc, D=Harlequin, E=Turquoise, F=Nut Brown.

BACK

Using No. 10 needles and A yarn, cast on 109(115:121:127) sts. **1st row (right side) sl.1 K1 * P1 K1, rep from * to last st, K1. **2nd row** sl.1 * P1 K1, rep from * to end. Rep 1st and 2nd rows 14 times, then 1st row again. **Next row** sl.1 rib 8(8:7:10) ML (rib 13(14:15:15) ML 7 times, rib 9(8:8:11). 117(123:129:135) sts. Change to No. 6 needles and proceed in patt as follows. **1st row** (right side) in A, K. **2nd row** in A, K. **3rd row** in B, * K1 sl.1p, rep from * to last st, K1. **4th row** in B, * K1 yfwd sl.1p ybk, rep from * to last st, K1. **5th row** in A, K. **6th row** in A, K. **7th row** in C, * sl.1p K1, rep from * to last st, sl.1p. **8th row** in C, * sl.1p ybk K1 yfwd, rep from * to last st, sl.1p. **9th–16th row** as 1st–8th row, but substituting D yarn for B and E yarn for C. **17th–24th rows** as 1st–8th rows, but substituting F yarn for B and B yarn for C. **25th–32nd rows** as 1st–8th rows, but substituting C yarn for B and D yarn for C. **33rd–40th rows** as 1st–8th rows, but substituting E yarn for B and F yarn for C. These 40 rows form patt. ** Cont in patt until work measures 26(26½:26½:27) in (66(67:67:69) cm), ending on wrong side. **Shape shoulders:** cast off 5(5:5:6) sts at beg of next 10 rows, then 5(5:6:5) sts at beg of foll 2 rows, then 4(5:6:5) sts at beg of foll 4 rows. Cast off rem 41(43:43:54) sts.

FRONT

Work exactly as given for the Back from ** to **. Cont in patt until work is 16 rows less than the Back at side edge. **Next row** patt 51(53:56:58) turn and work to end. Leave rem 66(70:73:77) sts on a thread. Work 13 rows dec once at neck edge on every row (38(40:43:45) sts). Work 1 row. **Shape shoulders:** keeping neck edge straight, cast off 5(5:5:5) sts at beg of next and foll alt rows. Work 1 row. Cast off 5(5:6:6) sts at beg of next row. Work 1 row. Cast off 4(5:6:5) sts at beg of next and foll alt row. With right side facing, rejoin appropriate colour to rem sts, cast off centre 15(17:17:19) sts, work to end. Complete to correspond with first side, reversing shapings.

SLEEVES (Alike)

Using No. 10 needles and A yarn, cast on 55(55:57:57) sts. **1st row** (right side) sl.1 K1 * P1 K1, rep from * to last st, K1. **2nd row** sl.1 * P1 K1, rep from * to end. Rep 1st and 2nd rows 16 times, then 1st row again. **Next row** sl.1 rib 5(4:2:2) ML (rib 4(3:4:3) ML) 11(15:13:17) times, rib 5(5:2:3) (67(71:71:75) sts). Change to No. 6 needles and proceed in 40 row patt as for the Back. **Shape sides:** keeping continuity of patt, inc once at each end of 13th (9th:13th:9th) and every foll 8th(8th:6th:6th) row until there are 89(95:101:107) sts. Cont in patt without shaping until sleeve measures 19 in or 48 cm, ending on wrong side. **Shape top:** cast off 4 sts at beg of next 12 rows. Cast off rem 41(47:53:59) sts. Follow pressing instructions on yarn label, then join right shoulder seam.

NECK BORDER

With right side facing, No. 10 needles and A yarn, pick up and knit 22(22:23:23) sts down left side of neck, pick up and knit 13(15:15:17) sts across centre, pick up and knit 22(22:23:23) sts up right side of neck and 38(40:40:42) sts across back neck. 95(99:101:105) sts. **1st row** (wrong side) sl.1 * P1 K1, rep from * to end. **2nd row** sl.1 K1 * P1 K1, rep from * to last st, K1. Rep 1st and 2nd rows 14 times. Cast off loosely in rib. **To complete:** join left shoulder seam. Fold the neck border in half to wrong side and slip hem in position. Fold sleeves in half lengthways and mark centre. Sew sleeve in position along armhole edge with marked point at shoulder seam. Join side and sleeve seams.

Burlington

Come the winter, a cardigan is man's best friend – but who says you can't teach an old dog new tricks? A textured stitch and extra length are all you need to give the classic cardigan a bit of dash. Besides adding surface interest, the stitch gives the fabric extra body so the pockets can stand up to lots of wear, and the cardigan doesn't cling round the middle. Add a bit of length to the body if necessary – a hem that hits just past the seat is flattering as well as functional. Knit *Burlington* in smart black or camel, or in cream with a sporty navy stripe through the ribbing at cuffs and edge. Whatever you choose, this cosy cardigan by Joan Proudfoot is sure to warm his heart.

MATERIALS **Yarn:** Sirdar Pullman Chunky Knitting, 20(21:22) 50 gm balls. **Needles:** one pair each No. 5 British (No. 8 American, 5½ mm), No. 3 British (No. 10 American, 6½ mm). **Trimming:** 6 buttons.

MEASUREMENTS **To fit:** 38(40:42) in chest, (97(102:107) cm) chest. **Actual size:** 40(42:44) in (102(107:112) cm). **Full length:** 27(27:27½) in (69(69:70) cm). **Sleeve length:** 19 in (48 cm) for all sizes.

TENSION 7 sts to 2 in (5 cm) on No. 3 needles.

ABBREVIATIONS **See** page 176.

BACK

Using No. 5 needles, cast on 81(85:89) sts. **1st row** sl.1 K1 * P1 K1, rep from * to last st, K1. **2nd row** sl.1 * P1 K1, rep from * to end. Rep 1st and 2nd rows 4 times more, inc 1 st at centre of last row (82(86:90) sts). Change to No. 3 needles and proceed in patt as follows. **1st row** (right side) sl.1 K1 * P2 K2, rep from * to end. **2nd row** sl.1 P1 * K2 P2, rep from * to last 4 sts, K2 P1 K1. **3rd row** as 2nd row. **4th row** as 1st row. These 4 rows form patt. Cont in patt until work measures 18 in (46 cm), ending with a wrong-side row. **Shape armholes:** keeping continuity of patt, cast off 4 sts at beg of next 2 rows. Work 4 rows, dec 1 st each end of every row (66(70:74) sts). Work 6(8:10) rows, dec 1 st each end of next and every alt row (60(62:64) sts). Cont in patt on these sts until work measures 27(27:27½) in (69(69:70) cm), ending on a wrong-side row. **Shape shoulders:** cast off 7 sts at beg of next 4 rows. Cast off 6 sts at beg of next 2 rows. Cast off rem 20(22:24) sts.

POCKET LININGS

(Make 2.) Using No. 3 needles cast on 20 sts and work 18 rows in st.st. Break yarn, leave these sts on a thread.

LEFT FRONT

Using No. 5 needles, cast on 39(41:43) sts. **1st row** sl.1 K1 * P1 K1, rep from * to last st, K1. **2nd row** sl.1 * P1 K1, rep from * to end. Rep 1st and 2nd rows 4 times more. Change to No. 3 needles and proceed in patt as follows. **For 38 and 42 in or 97 and 107 cm sizes only: 1st row** (right side) sl.1 K1 * P2 K2, rep from * to last st, K1. **2nd row** sl.1 * P2 K2, rep from * to last 2 sts, P1 K1. **3rd row** sl.1 P1 * K2 P2, rep from * to last st, K1. **4th row** sl.1 * K2 P2, rep from * to last 2 sts, K2. These 4 rows form patt. **For 40 in and 102 cm size only: 1st row** (right side) sl.1 K1 * P2 K2, rep from * to last 3 sts, P2 K1. **2nd row** sl.1 * K2 P2, rep from * to last 4 sts, K2 P1 K1. **3rd row** sl.1 P1 * K2 P2, rep from * to last 3 sts, K3. **4th row** sl.1 * P2 K2, rep from * to end. These 4 rows form patt. **For all 3 sizes:** cont in patt until work measures 6 in or 15 cm, ending on a wrong-side row. **Place pocket** as follows. **Next row** patt 9, slip next 20 sts on to a thread, patt across 20 sts of first pocket lining, patt to end. Cont in patt until work measures 18 in or 46 cm, ending on a wrong-side row. **Shape armhole and front slope: next row** cast off 4 sts, patt to last 2 sts, K2tog. **Next row** patt to end. Dec 1 st at armhole edge on the foll 5 rows, then the foll 2(3:4) alt rows, **at the same time** dec 1 st at front edge on every foll 5th(4th:4th) row from last dec until 20 sts rem. Cont straight until work measures same as the Back to shoulder, ending on a wrong-side row. **Shape shoulder:** cast off 7 sts at beg of next and foll alt row. Work 1 row. Cast off rem 6 sts.

RIGHT FRONT

Using No. 5 needles, cast on 39(41:43) sts. **1st row** sl.1 K1 * P1 K1, rep from * to last st, K1. **2nd row** sl.1 * P1 K1, rep from * to end. Rep 1st and 2nd rows 4 times more. Change to No. 3 needles and proceed in patt as follows. **38 and 42 in or 97 and 107 cm sizes only: 1st row** (right side) sl.1 * K2 P2, rep from * to last 2 sts, K2. **2nd row** sl.1 P1 * K2 P2, rep from * to last st, K1. **3rd row** sl.1 * P2 K2, rep from * to last 2 sts, P1 K1. **4th row** sl.1 K1 * P2 K2, rep from * to last st, K1. These 4 rows form patt. **40 in or 102 cm size only: 1st row** (right side) sl.1 * P2 K2, rep from * to end. **2nd row** sl.1 P1 * K2 P2, rep from * to last 3 sts, K3. **3rd row** sl.1 * K2 P2, rep from * to last 4 sts, K2 P1 K1. **4th row** sl.1 K1 * P2 K2, rep from * to last 3 sts, P2 K1. These 4 rows form patt. **For all 3 sizes:** cont in patt until work measures 6 in or 15 cm, ending on a wrong-side row. **Place pocket** as follows. **Next row** patt 10(12:14), slip next 20 sts on to a thread, patt across 20 sts of second pocket lining, patt to end. Complete to correspond with Left Front, reversing all shapings.

SLEEVES (Alike)

Using No. 5 needles, cast on 41(45:45) sts. **1st row** sl.1 K1 * P1 K1, rep from * to last st, K1. **2nd row** sl.1 * P1 K1, rep from * to end. Rep 1st and 2nd rows 6 times more, inc 1 st at centre of last row (42(46:46) sts). Change to No. 3 needles and proceed in patt as follows. **1st row** (right side) sl.1 K1 * P2 K2, rep from * to end. **2nd row** sl.1 P1 * K2 P2, rep from * to last 4 sts, K2 P1 K1. **3rd row** as 2nd row. **4th row** as 1st row. These 4 rows form patt. Cont in patt inc 1 st each end of 5th and every foll 6th row to 60(62:64) sts working inc sts into patt. Cont straight until work measures 19 in (48 cm), ending with a wrong-side row. **Shape top:** keeping continuity of patt, cast off 4 sts at beg of next 2 rows. Work 8 rows dec 1 st each end of next and foll 4th row. 48(50:52) sts. Work 18(16:18) rows dec 1 st each end of next and every alt row (30(34:34) sts). Work 8(10:10) rows dec 1 st each end of every row (14 sts). Cast off.

TO MAKE UP
Read pressing instructions on yarn label. Sew shoulder, side and sleeve seams. Sew in sleeves. **Front border:** using No. 5 needles, cast on 7 sts. **1st row** sl.1 (K1 P1) twice, K2. **2nd row** sl.1 (P1 K1) 3 times. ** **3rd row** sl.1 K1 P1, cast off 2 sts, K1. **4th row** sl.1 P1, cast on 2 sts, K1 P1 K1. Rep 1st and 2nd rows 7 times. ** Rep from ** to ** 4 times more, then 3rd and 4th rows once. Now rep 1st and 2nd rows until border is of sufficient length to fit up front to shoulder, across back neck, and down right front to lower edge. Cast off in rib. **Pocket tops (alike):** using No. 5 needles join yarn to 20 sts left on a thread and work as follows. **Next row** sl.1 K9 ML K10. 21 sts. **1st row** sl.1 * P1 K1, rep from * to end. **2nd row** sl.1 K1 * P1 K1, rep from * to last st, K1. Rep 1st and 2nd rows once more, then 1st row again. Cast off in rib. **To complete:** sew front border in position placing buttonhole end to left front. Sew side edges of pocket tops in position, slip stitch pocket linings in position. Sew on buttons.

Getaway

Looking for something different? This button-front sweater with optional shirt collar will help you to escape from the usual V-, roll- and crew-neck classics. Sleek through the body and ribbed across the shoulders for a flattering fit, *Getaway* gives you lots of stylish options. Knit it collarless in tweedy yarns for a casual weekend look, strike a sporty note by using contrast colours for the body and sleeves, or opt for smart simplicity with body, sleeves and collar in black or navy.

MATERIALS **Yarn:** for collarless version, Sirdar Pullman Chunky Knitting, 15(16:17) 50 gm balls. For version with collar, 16(17:18) 50 gm balls. **Needles:** one pair each No. 5 British (No. 8 American, 5½ mm), No. 3 British (no. 10 American, 6½ mm). **Trimming**: 3 buttons.

MEASUREMENTS **To fit**: 38(40:42) in chest (97(102:107) cm) chest. **Actual size:** 40(42:44) in (102 (107:112) cm). **Full length:** 26(26½:27) in (66(67:69) cm). **Sleeve length:** 19 in (48 cm) for all sizes.

TENSION 7 sts to 2 in (5 cm) on No. 3 needles.

ABBREVIATIONS **See** page 176.

COLLARLESS SWEATER BACK
Using No. 5 needles cast on 69(73:77) sts. **1st row** P3(2:1) * K3 P3, rep from * to last 6(5:4) sts, K3 P3 (2:1). **2nd row** K3(2:1) * P3 K3, rep from * to last 6(5:4) sts, P3 K3(2:1). Rep 1st and 2nd rows 7 times more. ** Change to No. 3 needles and proceed in st.st until work measures 18 in (46 cm), ending with a purl row. Place a marker each end of last row, then rep 1st and 2nd rows of rib until work measures 26(26½:27) in (66(67:69) cm), ending with a 2nd row. **Shape shoulders:** cast off 8(8:9) sts at beg of next 4 rows. Cast off 8(9:8) sts at beg of next 2 rows. Cast off rem 21(23:25) sts.

FRONT
Work exactly as given for Back until ** is reached. Change to No. 3 needles and proceed in st.st until work measures 18 in (46 cm), ending with a knit row. **Next row** P31(33:35) sts, cast off 7 sts, P to end. Place a marker each end of last row. Proceed on first set of 31(33:35) sts for left side, leave rem sts on a spare needle. **1st row** P3(2:1) * K3 P3, rep from * to last 4(1:4) sts, K4(1:4). **2nd row** P4(1:4) * K3 P3, rep from * to last 3(2:1) sts, K3(2:1). Rep 1st and 2nd rows until work measures 23½(24:24½) in or 60(61:62) cm, ending with a 1st row. **Shape neck: next row** cast off 3 sts, patt to end. Work 4(5:6) rows dec 1 st at neck edge on every row. 24(25:26) sts. Cont in patt on these sts until work measures 26(26½:27) in (66(67:69) cm) ending on wrong side. **Shape shoulder:** cast off 8(8:9) sts at beg of next and foll alt row. Work 1 row. Cast off rem 8(9:8) sts. With right side facing rejoin yarn to rem 31(33:35) sts from spare needle and proceed as follows. **1st row** K4(1:4) * P3 K3, rep from * to last 3(2:1) sts, P3(2:1). **2nd row** K3(2:1) * P3 K3 rep from * to last 4(1:4) sts, P4(1:4). Rep 1st and 2nd rows and complete to correspond with left side, reversing shapings.

SLEEVES (Alike)
Using No. 5 needles, cast on 37(37:43) sts. **1st row** P2 * K3 P3, rep from * to last 5 sts, K3 P2. **2nd row** K2 * P3 K3, rep from * to last 5 sts, P3 K2. Rep 1st and 2nd rows 7 times more. Change to No. 3 needles and proceed in st.st inc 1 st

each end of 5th and every foll 10th(8th:8th) row to 49(53:57) sts. Cont without shaping until work measures 19 in (48 cm), ending with a purl row. **Shape top: 1st row** K to last 6 sts, turn. **2nd row** sl.1 P to last 6 sts, turn. **3rd row** sl.1 K to last 12 sts, turn. **4th row** sl.1 P to last 12 sts, turn. **5th row** sl.1 K to last 18 sts, turn. **6th row** sl.1 P to last 18 sts, turn. **7th row** sl.1 K to end. **8th row** P across all sts. Cast off. **To make up:** read pressing instructions on yarn label. Sew shoulder seams. Sew in sleeves between markers. Sew side and sleeve seams.

NECK BORDER

With right side facing, using No. 5 needles, join yarn to right front neck edge, pick up and knit 18 sts evenly along right side of neck, 21(23:25) sts across back neck, and 18 sts along left side of neck (57(59:61) sts). **1st row** sl.1 * P1 K1, rep from * to end. **2nd row** sl.1 K1 * P1 K1, rep from * to last st, K1. Rep 1st and 2nd rows twice more, then 1st row again. Cast off in rib.

BUTTONHOLE BORDER

With right side facing, using No. 5 needles, join yarn to neck edge of left side of front opening, pick up and K 31 (35:39) sts evenly to base of opening. **1st row** sl.1 * P1 K1, rep from * to end. **2nd row** sl.1 K1 * P1 K1, rep from * to last st, K1. Rep 1st and 2nd rows once more. **5th row** sl.1 (P1 K1) twice, * P2tog yrn (P1 K1) 4(5:6) times, rep from * once more, P2tog yrn (P1 K1) twice. **6th row** as 2nd row. Rep 1st and 2nd rows once more, then 1st row again. Cast off in rib.

BUTTON BORDER

With right side facing, using No. 5 needles, join yarn to base of right side of front opening, pick up and K31(35:39) sts evenly to neck edge. **1st row** sl.1 * P1 K1, rep from * to end. **2nd row** sl.1 K1 * P1 K1, rep from * to last st, K1. Rep 1st and 2nd rows 3 times more, then 1st row again. Cast off in rib. **To complete:** sew side of button border to base of opening. Sew buttonhole border over top of button border. Sew on buttons.

SWEATER WITH COLLAR

Work exactly as given for the Back, Front and Sleeves of collarless version. **To make up:** read pressing instructions on yarn label. Sew shoulder seams. Sew in sleeves between markers. Sew side and sleeve seams.

BUTTONHOLE BORDER

With right side facing, using No. 5 needles, join yarn to neck edge of left side of front opening, pick up and K23(27:31) sts evenly to base of opening. **1st row** sl.1 * P1 K1, rep from * to end. **2nd row** sl.1 K1 * P1 K1, rep from * to last st, K1. Rep 1st and 2nd rows once more. **5th row** sl.1 P1 K1 * P2tog yrn (P1 K1) 3(4:5) times, rep from * once more, P2tog yrn, P1 K1. **6th row** as 2nd row. Rep 1st and 2nd rows once more, then 1st row again. Cast off in rib.

BUTTON BORDER

With right side facing, using No. 5 needles, join yarn to base of right side of front opening, pick up and knit 23(27:31) sts evenly to neck edge. **1st row** sl.1 * P1 K1, rep from * to end. **2nd row** sl.1 K1 * P1 K1, rep from * to last st, K1. Rep 1st and 2nd rows 3 times more, then 1st row again. Cast off in rib.

COLLAR

With right side facing, using No. 5 needles, join yarn to centre of button border, pick up and knit 23 sts evenly along right side of neck, 21(23:25) sts across back neck and 23 sts along left side of neck to centre of buttonhole border. 67(69:71) sts. **1st row** sl.1 * P1 K1, rep from * to end. **2nd row** sl.1 K1 * P1 K1, rep from * to last st, K1. Rep 1st and 2nd rows until collar measures 5 in (13 cm), ending with a 1st row. Cast off loosely in rib. **To complete:** sew side of button border to base of opening. Sew buttonhole border over top of button border. Sew on buttons.

Blaze

Here's a brilliant blend of colour and pattern – Mary Wear's slipover slashed front and back with a blaze of fourteen colours, shaded in the Florentine manner to look like watered silk. The plain background shade highlights the beauty of the stitches, a fine chevron and rib pattern that cuts across the colour band, giving a wonderful *trompe l'oeil* effect. The stitch has a natural tendency to meander, so you'll have to count the stripe stitches carefully as you go. Less experienced knitters can do the slipover in stocking stitch or reverse stocking stitch, moving the stripe along one stitch each time without counting, although in this case the lines in the colour band will be straight rather than wavy. You can knit the band in a single contrast colour, just two or three colours, or in yarns with different textures. Whatever you choose, you'll be blazing a very stylish trail.

MATERIALS **Yarn:** Sirdar Wash'N'Wear 4-ply, 6 20 gm balls in main colour, plus oddments of different-coloured yarns equivalent to 3 20 gm balls for contrast stripe. **Needles:** No. 10 British (No. 3 American, 3¼ mm), No. 12 British (No. 1 American, 2¾ mm).

MEASUREMENTS one size, to fit chest measurements 36–38 in (91–97 cm). **Length:** 23 in (57.5 cm).

TENSION 7½ sts and 10½ rows to 1 in (2·5 cm) (cast on 20 sts and work in stocking stitch for 25 rows, and measure over 2 in (5 cm)).

ABBREVIATIONS See page 176. **Colours:** MC=main colour. Other colours used in stripe: A=Honeydew, B=Celadon, C=Peppermint, D=Flamenco, E=Crystal Rose, F=Chinese Fig, G=Coral Reef, H=Pebble Beige, J=Wedgwood, K=Begonia, L=Sugar Plum, M=Bottle, N=Opaline, P=Black. Used here in pattern of 3-row stripes separated by random arrangement of 3 × 1 row and 2 × 1 row stripes. Uniform stripes of 2, 3 or 4 rows can also be used.

NOTE This is a 2-row pattern worked over 16 sts with an edge stitch at each end as follows. K1 * K2tog K3 KC K3 KC K3 sl.1 K1 PSSO K3 * rep from * K1. The second pattern row is purl with K1 at each end. The colour stripe is 30 sts wide and moves 1 stitch at each end. Counting **must** be carried out accurately as the stitches are knitted on to the right-hand needle. Stripe colours should be tied to main colour yarn at beginning and end of stripe. When more than one row is worked in one colour, twist stripe colour around main colour to avoid making a hole. Ribbing can be plain or striped, as shown here. Colour sequence in ribbing for hem: cast on and work 1 row M, 2 rows E, 3 rows MC, 1 row each L and J, 5 rows MC, 1 row each N, A and B, 8 rows MC. For neck: 1 row MC, 1 row each K and G, 3 rows MC, 1 row C, 1 row and cast off in M. For armholes: 1 row MC, 1 row each in D and P, 2 rows MC, 1 row C, 1 row and cast off in M.

BACK

Using No. 12 needles and plain or contrast yarn, cast on 136 sts. **1st row** – K3 P2 K2 to last st K1. **2nd row** K1 P2 * K2 P2 rep from * to last st K1. Rep these 2 rows 10 times, then 1st row once. **24th row** – rib 4, work into front and back of next st * rib 13 work into front and back of next st. Rep from * 8 times, work twice into next st rib 5 (146 sts). Change to No. 10 needles and pattern. **1st row** K1 in MC, join contrast for stripe, K2tog K3 KC K3 KC K3 sl.1 K1 PSSO K3 K2tog K3 KC K3 KC K3 sl.1 K1 PSSO K1 (31 sts on right hand needle). Join on second ball of MC, K2, K2 tog K3 KC K3 KC K3 sl.1 K1 PSSO K3. **2nd row** K1 in MC, P to stripe, P30 stripe colour, K1 MC. **3rd row** MC K1 K2tog (2 sts on needle), contrast K3 KC K3 RC K3 sl.1 K1 PSSO K3 K2tog K3 KC K3 RC K3 sl.1 K1 PSSO K2 (32 sts on RH needle), MC K1 rep patt to end, K1. **4th row** MC K1 P to stripe, P 30 stripe colour, MC, P1 K1. **5th row** MC K1 K2tog K1 (3 sts) contrast K2 KC K3 KC K3 sl.1 K1 PSSO K3, K2tog K3 KC K3 KC K3 sl.1 K1 PSSO K3 (33 sts on RH needle), MC, rep pattern to end K1. **6th**

row as 4th, ending MC P2 K1. Continue thus, with 1 MC stitch more on RH needle each pattern row before stripe, 30 sts stripe, 1 st less MC after stripe, until there are 57 MC sts before stripe and 59 sts after, finishing with a P row. (14 in (35.5 cm) approx from edge). **Shape Armholes** Cast off 9 sts on next row, work pattern until there are 48 sts on RH needle including 1 st left after casting off, work 30 sts stripe and 58 sts after stripe. **2nd row** cast off 10 sts, P to stripe, P30, P to end (126 sts). Now decrease at each end of alternate rows thus: **Row 1** K2 sl.1 K1 PSSO (this is the decrease as st will not be replaced owing to armhole cast off) K3 K2tog K3 etc until there are 48 sts on RH needle, patt stripe 30 sts, patt MC to last 7 sts K2tog K5 (124 sts). **Row 2** P MC to stripe, P30 stripe, P MC to end (124 sts). **Row 3** K1 sl.1 K1 PSSO (the dec as above) K3 K2tog etc to stripe (48 sts), patt 30 sts stripe, patt MC to last 6 sts K2tog K4 (122 sts). **Row 4** as row 2 (122 sts). **Row 5** sl.1 K1 PSSO K3, patt to stripe (48 sts), work stripe (30 sts), patt to last 5 sts K2tog K3 (120 sts). **Row 6** as row 2 (120 sts). **Row 7** sl.1 K1 PSSO K2 patt to stripe (48 sts), work stripe (30 sts), patt to last 4 sts K2tog K2 (118 sts). **Row 8** as row 2 (118 sts). **Row 9** sl.1 K1 PSSO K1 K2tog K3 etc to stripe (48 sts), work stripe (30 sts), patt to last 3 sts K2tog K1 (116 sts). **Row 10** as row 2 (116 sts). **Row 11** sl.1 K1 PSSO K2tog patt to stripe (48 sts), work stripe (30 sts), patt to last 2 sts K2 tog (114 sts). **Row 12** as row 2 (114 sts). Continue on these 114 sts – MC 49 sts, stripe 30 sts, MC 35 sts – moving on 1 st each pattern row as before, until there are 84 MC sts and 30 sts stripe finishing with a P row. **Shape Shoulders**Pattern to last 9 sts, turn and P to last 9 sts and turn. Pattern to last 18 sts, turn and P to last 18 sts and turn. Pattern to last 27 sts, turn and P to last 36 sts and turn. Place centre 42 sts on threads pr a stitch holder for shoulders.

FRONT

Work ribbing as for back, including increase row to 146 sts. Change to No. 10 needles and pattern. **1st row** MC pattern 115 sts, join contrast and patt 30 stripe sts, join second ball of MC, K1. **2nd row** K1 MC P30 stripe P114 MC K1. **3rd row** MC pattern 114 sts, stripe 30 sts K2 MC. **4th row** MC K1 P1 P30 sts stripe P113 K1 MC. **5th row** MC pattern 113 sts stripe pattern 30 sts MC K3. **6th row** MC K1 P2 P30 sts stripe P 112 sts K1. Continue thus until there are 59 sts before stripe and 57 after, but check against back that it is the same length to armhole. **Shape Armholes** Cast off 9 sts, patt until there are 48 sts on RH needle, 58 sts after. Cast off 10 sts, P to stripe, P30, P to end (126 sts). Work 6 decreases each end of pattern row 6 times as back: **1st row** 46 sts RH needle to stripe, 48 sts after, P back (124 sts). **3rd row** 44 sts RH needle to stripe, 48 sts after, P back (122 sts.) **5th row** 42 sts RH needle to stripe, 48 sts after, P back (120 sts). **7th row** 40 sts RH needle to stripe, 48 sts after, P back (118 sts). **9th row** 38 sts RH needle to stripe, 48 sts after, P back (116 sts). **11th row** 36 sts RH needle to stripe, 48 sts after, P back (114 sts). Continue straight as for back until there are 17 sts before stripe and 67 sts after. **Shape Neck** MC pattern 16, stripe 30, pattern 68. Leave these 68 sts on a thread for right shoulder and bottom neck. Return to neck edge of first 46 sts for left shoulder. **Row 1** P30 stripe, MC P15 K1. **Row 2** patt 15 MC 31 stripe (to allow for decrease at side of neck) ending K2tog K3 KC K3 K5 (46 sts). **Row 3** P2tog P29 stripe MC to last st K1 (45 sts). **Row 4** patt 14 MC 31 stripe ending K3 KC K3 K4. **Row 5** as row 3 (44 sts). **Row 6** patt 13 MC 31 stripe ending K3 KC K3 K3. **Row 7** as row 3 (43 sts). **Row 8** patt 12 MC 31 stripe ending K3 KC K5. **Row 9** as row 3 (42 sts) **Row 10** patt 11 MC 31 stripe ending K3 KC K4. **Row 11** as row 3 (41 sts). **Row 12** patt 10 MC 31 stripe ending K3 KC K3. **Row 13** as row 3 (40 sts). **Row 14** patt 9 MC 31 stripe ending K3 KC K2. **Row 15** as row 3 (39 sts). **Row 16** patt 8 MC 31 stripe ending K3 KC K1. **Row 17** as row 3 (38 sts). **Row 18** patt 7 MC 31 stripe ending sl.1 K1 PSSO K8. **Row 19** as row 3 (37 sts). **Row 20** patt 6 MC 31 stripe ending sl.1 K1 PSSO K7. **Row 21** as row 3 (36 sts). **Row 22** patt 5 MC 30 stripe ending sl.1 K1 PSSO K5 stripe colour, 1 MC. **Row 23** P1 MC, 30 stripe, 5 MC. **Row 24** Patt 4 MC 30 stripe ending sl.1 K1 PSSO K4 (stripe colour), 2 MC. **Row 25** P2 MC 30 stripe 4 MC. **Row 26** patt 3 MC 30 stripe ending sl.1 K1 PSSO K3 (stripe colour), 3 MC. **Row 27** P3 MC 30 stripe 3 MC. **Row 28** patt 2 MC 30 stripe ending sl.1 K1 PSSO K2 (stripe colour), 4 MC. **Row 29** P4 MC 30 stripe 2 MC. **Row 30** patt 1 MC 30 stripe ending sl.1 K1 PSSO K1 (stripe colour) 5 MC. **Row 31** P5 MC 30 stripe 1 MC. **Row 32** Patt 30 stripe ending sl.1 K1 PSSO (stripe colour) K6 MC. **Row 33** P6 MC 30 stripe colour. **Row 34** Patt 30 stripe ending sl.1 K1 PSSO (stripe colour) K6 MC, finish at neck edge when armhole should be the same length as back armhole. **Shape Shoulder.** P6 MC, 21 stripe, turn, pattern back. P6 MC, 12 stripe, turn, pattern back. P6 MC, 3 stripe, turn, pattern back (neck edge). Transfer 36 sts left on thread for left shoulder on back (with stripe) to spare needle. With point at neck edge, place left shoulders together, right sides inside, and cast off with a No. 10 needle, knitting shoulders together. Slip 36 stitches for right shoulder from thread to a No. 10 needle with point at shoulder end, leaving 22 stitches on thread for neck. **1st row** K1 P to end. **2nd row** pattern starting sl.1 K1 PSSO K3 KC K3 sl.1 K1 PSSO K3 etc (45 sts). **3rd and alternate rows** K1 P to end. **4th row** pattern starting sl.1 K1 PSSO K2 KC K3 (44 sts). **6th row** pattern starting sl.1 K1 PSSO K1 KC K3 (43 sts). **8th row** sl.1 K1 PSSO KC K3 (42 sts). **10th row** pattern starting K4 sl.1 K1 PSSO K3 (41 sts). **12th row** pattern starting K3 sl.1 K1 PSSO K3 (40 sts). **14th row** pattern starting K2 sl.1 K1 PSSO K3 (39 sts). **16th row** pattern starting K1 sl.1 K1 PSSO K3 (38 sts). **18th row** pattern starting sl.1 K1 PSSO K3 (37 sts). **20th row** pattern starting sl.1 K1 PSSO K2 K2tog K3 (36 sts). Continue in pattern without further shaping starting K3 K2tog K3 KC K3 KC K3 until work is same length as back armhole, finishing at neck edge. **Shape Shoulder** pattern to last 9 sts, turn and P back. Pattern to last 18 sts, turn and P back. Pattern to last 27 sts, turn and P back. Leave stitches on a thread or stitch holder.

NECK EDGING

Using a No. 12 needle and MC, with right side of work facing, commence by slipping the 42 sts left for back neck on to needle, pick up and K 42 sts down left side of neck, 22 sts from thread for centre neck, pick up and K 42 sts along right side of neck (148 sts). Work in K2 P2 rib as for bottom edge of sweater for 8 rows. Cast off in rib.

ARMHOLE EDGINGS

Slip stitches for right shoulder on to 2 No. 10 needles with points at neck edge, and with right sides together, using a No. 10 needle, knit together as for left shoulder. Using a No. 12 needle, with right side of work facing you, pick up and K 180 sts. Work in K2 P2 rib as for bottom edge of sweater for 8 rows. Cast off in rib. Work other armhole to match.

TO MAKE UP

Darn in ends of yarn where stripe colours were joined on. Pin out to shape and press lightly. Join edges of neck ribbing on right shoulder. Join underarm seams including armhole ribbing and darn in any ends. Press seams.

Greensleeves

Here's Nature's own recipe for relaxation – six soothing shades of green, knitted together in a serene variation of a classic multi-coloured Fair Isle design. Tonal colourings like this replace the liveliness of contrast colours with the calmer play of shade on shade, giving a softer, more subtle look to traditional Fair Isles. The use of one primary colour in depth adds extra richness to the fabric, and the delicacy of the shading lightens the intricacy of the motifs. To all men who don't like strongly patterned sweaters and don't like plain ones either, *Greensleeves* will be a joy and delight.

DETAILS Shown here with shirt, tie and trousers by Paul Smith.

MATERIALS **Yarn:** Sirdar Wash'N'Wear 4-ply crepe, 10(11) 20 gm balls in main colour A, 2(2) 20 gm balls each in colours B and C, 3(3) 20 gm balls each in colours D and E, 1(1) 20 gm ball in colour F. **Needles:** No. 12 British (No. 1 American, 2¾ mm), No. 9 British (No. 4 American, 3¾ mm).

MEASUREMENTS **To fit** chest 34–36(38–40) in (86–91(97–102) cm). **Actual size:** 36(40) in (91(102) cm). **Length to shoulder:** 25½(26½) in (65(67) cm). **Sleeve length** for both sizes, 19 in (48 cm).

TENSION 15 sts to 2 in (5 cm) on No. 9 needles.

ABBREVIATIONS **See** page 176. **Colours:** A=Bottle Green, B=Woodland Green, C=Celadon, D=Montego Lime, E=French Fern, F=Peppermint.

BACK

** Using No. 12 needles and A yarn, cast on 134(150) sts. **1st row** (right side) sl.1 K1 * P2 K2, rep from * to end. **2nd row** sl.1 P1 * K2 P2, rep from * to end. Rep 1st and 2nd rows 10 times, inc once in centre of last row, 135(151) sts.

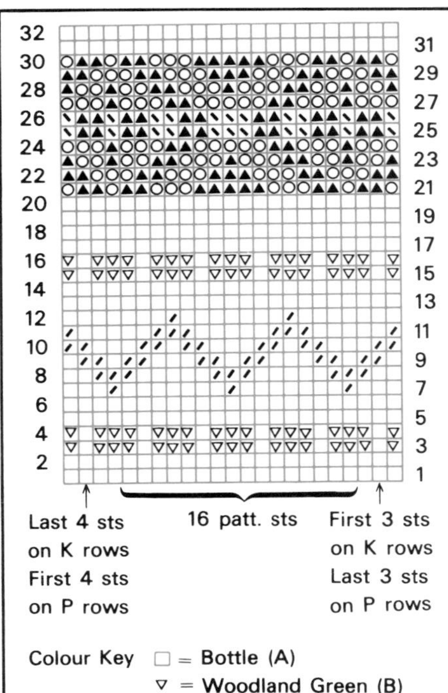

Change to No. 9 needles and work rows 1 to 32 from chart, rep the 16 patt sts 8(9) times and first 3 sts and last 4 sts on K rows, and first 4 sts and last 3 sts on P rows as indicated. Read odd rows K from right to left and even rows P from left to right.** Cont in patt until work measures 25½(26½) in (65(67) cm), ending on wrong side. **Shoulder shaping:** keeping continuity of patt, cast off 6(7) sts at beg of next 8 rows. Then 7 sts at beg of next 4 rows, then 7(8) sts at beg of next 2 rows. Leave rem 45(51) sts on a thread.

FRONT

Work exactly as given for Back from ** to **. Cont in patt until Front is 48(56) rows less than Back at side edge, thus ending on wrong side. **Shape neck: next row** patt 67(75), turn, and leave rem sts on a thread, work to end. Dec once at neck edge on next and every foll alt row until 45(50) sts rem. Cont straight on these sts until Front matches Back at side edge, ending on wrong side. **Shoulder shaping:** cast off 6(7) sts at beg of next and foll 3 alt rows. Work 1 row. Cast off 7 sts at beg of next and foll alt row. Work 1 row. Cast off rem 7(8) sts. With right sides facing, rejoin appropriate colour to rem sts, cast off centre st, patt to end. Complete to correspond with first side, reversing shapings.

SLEEVES (Alike)

Using No. 12 needles and A yarn cast on 62(66) sts. **1st row** – (right side) sl.1 K1 * P2 K2, rep from * to end. **2nd row** sl.1 P1 * K2 P2, rep from * to end. Rep 1st and 2nd rows 8 times, then 1st row again. **Next row** sl.1 (rib 6(4) ML) 9(15) times, rib 7(5) (71(81) sts). Change to No. 9 needles and work rows 1 to 32 from chart, rep the 16 patt sts 4(5) times and first 3(0) sts and last 4(1) sts on K rows and first 4(1) sts and last 3(0) sts on P rows **at the same time. Shape sides:** keeping continuity of patt, inc once at each end of 11th(5th) and every foll 6th row until there are 105(119) sts. Cont without shaping until work measures 19 in (48 cm) or length required, ending on wrong side. **To make up** follow pressing instructions on yarn label. Join right shoulder seam.

NECK BORDER

With right side facing, No. 12 needles and A yarn, pick up and knit 59(65) sts down left side of neck, place a marker in centre V, pick up and knit 59(65) sts up right side of neck, K across 45(51) sts from back dec 1 st at centre (162(180) sts). **1st row** (wrong side) * K2 P2, rep from * to 3 sts before marker, K2tog, P2 K2tog, work in rib to end, starting with P2. **2nd row** work in P2 K2 rib to 3 sts before marker, P2 tog K2 P2tog, rib to end. Rep 1st and 2nd rows 4 times, then 1st row again, dec once on each side of 2 centre sts on every row, keeping these sts as K sts on right side. Cast off evenly in rib, dec as before.

TO COMPLETE

Join left shoulder seam and neck border. Fold sleeves in half lengthways and mark centre of cast-off edge. Place sleeve along armhole edge, with marker at shoulder and sew in position. Join side and sleeve seams.

High Flyer

Clear the runways for a real flight of fancy – a knitted bomber top that will turn your favourite flyer into the high ace in any pack. Flying jackets always bring out the best in a man – the nonchalant raglan shoulders, deep collar and jaunty pockets have a built-in swagger that's dashing, daring and devil-may-care. The original jackets were grounded along with the Tiger Moths and Sopwith Camels but you can't keep a good design down, particularly when it adapts so well to knitting. Fully lined in creamy sheepskin-look yarn, *High Flyer* has all the warmth and half the weight of the leather versions, so he'll feel as good as he looks all the year through.

MATERIALS **Yarn:** Sirdar Pullman Chunky Knitting, 16(17:18) 50 gm balls in main colour. Sirdar Palamino, 14(15:16) 50 gm balls in contrast colour. **Needles:** 1 pair each No. 3 British (No. 10 American, 6½ mm), No. 4 British (No. 9 American, 6 mm), No. 5 British (No. 8 American, 5½ mm). **Trimming:** 1 24 in (61 cm) open-end zip fastener.

MEASUREMENTS **To fit:** 38(40:42) in (97(102:107) cm) chest. **Actual size:** 42(44:46) in (107(112:117) cm). **Full length:** 27½(28:28½) in (70(71:72) cm). **Sleeve length:** 19 in (48 cm) for all sizes.

TENSION 7 sts to 2 in (5 cm) on No. 3 needles for Pullman, 7 sts to 2 in (5 cm) on No. 4 needles for Palamino.

ABBREVIATIONS **See** page 176. **Colours:** MC=main colour, Pullman Burnt Almond, CC=contrast colour, Palamino.

BACK
Using No. 5 needles and MC, cast on 75(79:83) sts. **1st row** sl.1 K1 * P1 K1, rep from * to last st, K1. **2nd row** sl.1 * P1 K1, rep from * to end. Rep 1st and 2nd rows 7 times more. Change to No. 3 needles and proceed in st.st until work measures 17½ in (44 cm), ending with a purl row. **Shape raglan armholes:** cast off 3 sts at beg of next 2 rows, then dec as follows. **1st row** K2 sl.1 K1 PSSO, knit to last 4 sts, K2tog, K2. **2nd row** P to end. Rep 1st and 2nd rows until 21(23:25) sts rem. Work 1 row. Cast off.

LEFT FRONT
Using No. 5 needles and MC cast on 38(40:42) sts. **1st row** sl.1 * K1 P1, rep from * to last st, K1. Rep 1st row 15 times more. Change to No. 3 needles and proceed as follows. **1st row** K to end. **2nd row** sl.1 K1, P to end. Rep 1st and 2nd rows until work measures 17½ in (44 cm), ending with a purl row. **Shape raglan armholes:** cast off 3 sts at beg of next row, K to end. Work 1 row, then dec as follows. **1st row** K2 sl.1 K1 PSSO, K to end. **2nd row** sl.1 K1, P to end. Rep 1st and 2nd rows until 20(21:22) sts rem, ending with a right side row. **Shape neck: next row** cast off 4(5:6) sts, P to end. Cont to dec armhole edge as before **at the same time** dec 1st at neck edge on next and every alt row until 6 sts rem. Now dec armhole edge only until 3 sts rem. **Next row** P3. **Next row** K1 sl.1 K1 PSSO. **Next row** P2. **Next row** sl.1 K1 PSSO. Break yarn and fasten off.

RIGHT FRONT
Using No. 5 needles and MC, cast on 38(40:42) sts. **1st row** sl.1 * P1 K1, rep from * to last st, K1. Rep 1st row 15 times more. Change to No. 3 needles and proceed as follows. **1st row** sl.1 K to end. **2nd row** P to last 2 sts, K2. Rep 1st and 2nd rows and complete to correspond with left front, reversing all shapings and dec by K2tog instead of sl.1 K1 PSSO.

SLEEVES (Alike)
Using No. 5 needles and MC, cast on 33(37:41) sts. **1st row** sl.1 K1 * P1 K1, rep from * to last st, K1. **2nd row** sl.1 * P1 K1, rep from * to end. Rep 1st and 2nd rows 7 times more. Change to No. 3 needles and proceed in st.st inc 1 st each end of 5th and every foll 6th row to 55(59:63) sts. Cont straight until work measures 19 in (48 cm), ending with a purl row. **Shape top:** cast off 3 sts at beg of next 2 rows, then dec as follows. **1st row** K2 sl.1 K1 PSSO, K to last 4 sts, K2tog K2. **2nd row** P to end. **3rd row** K to end. **4th row** P to end. Rep 1st to 4th rows 2(1:0) times more, then rep 1st and 2nd rows only until 7 sts rem. Work 1 row. Cast off.

LINING BACK
Using No. 4 needles and CC cast on 71(75:79) sts and proceed in reversed st.st commencing with a purl row until work measures 14½ in (37 cm), ending with a knit row. **Shape raglan armholes:** cast off 3 sts at beg of next 2 rows, then dec as follows. **1st row** P2 sl.1 P1 PSSO, P to last 4 sts, P2tog P2. **2nd row** K to end. Rep 1st and 2nd rows until 17(19:21) sts rem. Work 1 row. Cast off.

LINING LEFT FRONT

Using No. 4 needles and CC, cast on 36(38:40) sts and proceed in reverse st.st commencing with a purl row until work measures 14½ in (37 cm), ending with a knit row. **Shape raglan armhole:** cast off 3 sts at beg of next row, P to end. Work 1 row, then dec as follows. **1st row** P2 sl.1 P1 PSSO, P to end. **2nd row** K to end. Rep 1st and 2nd rows until 18(19:20) sts rem, ending with a purl row. **Shape neck: next row** cast off 2(3:4) sts, K to end. Cont to dec armhole edge as before **at the same time** dec 1 st at neck edge on next and every alt row until 6 sts rem. Now dec armhole edge only until 3 sts rem. **Next row** K3. **Next row** P1 sl.1 P1 PSSO. **Next row** K2. **Next row** sl.1 P1 PSSO. Break yarn and fasten off. **Right Front:** work as given for the lining left front reversing all shaping and dec by P2tog instead of sl.1 P1 PSSO.

LINING SLEEVES (Alike)

Using No. 4 needles and CC cast on 29(33:37) sts and proceed in reversed st.st commencing with a purl row and inc 1 st each end of 5th and every foll 6th row to 51(55:59) sts. Cont straight until work measures 16 in (41 cm), ending with a knit row. **Shape top:** cast off 3 sts at beg of next 2 rows, then dec thus. **1st row** P2 sl.1 P1 PSSO, P to last 4 sts, P2tog P2. **2nd row** K to end. **3rd row** P to end. **4th row** K to end. Rep 1st to 4th rows 3(2:1) times more, then rep 1st and 2nd rows only, until 5 sts rem. Work 1 row. Cast off.

POCKETS

Left pocket: using No. 3 needles and MC, cast on 22 sts. **1st row** sl.1 K to end. **2nd row** sl.1 P to last 3 sts, K3. Rep 1st and 2nd rows 15 times more, then 1st row again. Cast off. **Right pocket:** using No. 3 needles and MC, cast on 22 sts. **1st row** sl.1 K to end. **2nd row** sl.1 K2, P to last st, K1. Rep 1st and 2nd rows 15 times more, then 1st row again. Cast off.

COLLAR

Using No. 4 needles and 2 strands of CC yarn, cast on 53 sts and proceed as follows noting that yarn is used double throughout. **1st row** sl.1 K1 * P1 K1, rep from * to last st, K1. **2nd row** sl.1 * P1 K1, rep from * to end. Rep 1st and 2nd rows once. **5th row** sl.1 K1 ML * P1 K1, rep from * to last 3 sts, P1 ML K2. **6th row** sl.1 K1 * P1 K1, rep from * to last 3 sts, P1 K2. **7th row** sl.1 K1 * K1 P1, rep from * to last 3 sts, K3. **8th row** as 6th row. **9th row** sl.1 K1 ML * K1 P1, rep from * to last 3 sts, K1 ML K2. **10th row** as 2nd row. **11th row** as 1st row. **12th row** as 2nd row. Rep 5th to 12th rows once more (61 sts). Cast off loosely in rib.

TO MAKE UP

Sew raglans, side and sleeve seams of jacket. Sew raglans, side and sleeve seams of lining **noting** that the reverse st.st side is the right side of work for the lining. Place lining inside jacket and slip stitch above rib at lower edge and above rib of sleeves. Insert zip fastener to front opening, then slip stitch lining to front edges. Sew cast-on edge of collar to neck edge, then slip stitch lining to neck edge. Sew pockets above rib and 9 sts in from front opening **noting** opening of pocket is vertical. Press seams.

Gleneagles

You'd have to go a fair way to beat the versatility of this classic weekend slipover – an ideal solution when uncertain weather and swift changes of scene threaten to put a cramp in his style. Slipovers are far more comfortable to wear during the sun-and-shower season than the usual lambswool sports sweaters, and the Fair Isle stitches make the fabric more substantial, so the slipover doesn't have a saggy, baggy locker-room look. The colours leave plenty of room for interesting shirt, tie, cravat and polo neck combinations, and Suzanne Barnacle's light and airy geometric pattern goes well with plain, flecked or checked trousers, jeans or pinwhale cords. It even looks good with Plus Twos and Plus Fours, so we've included matching legwarmers. If he won't wear them, knit a slipover for yourself and wear it long and loose as the French do, with a man-tailored shirt. Slip the legwarmers on over your tights or trousers – and you'll be a very smart birdie indeed.

DETAILS Slipover shown with tights by Aristoc, skirt by Cavalier from Top Shop and shirt by Willi Smith from Barney's, New York. Photographed at the Hyde Park Hotel, London.

MATERIAL **Yarn** for slipover: Sirdar Superwash 4-ply, 11(12:13:14) 25 gm balls in main colour A, 2(2:2:2) 25 gm balls in first contrast colour B, 1(1:2:2) 25 gm ball in colour C, 3(4:4:5) 25 gm balls in colour D, 2(2:2:3) 25 gm balls in colour E, 1(1:2:2) 25 gm ball in colour F, 2(2:2:3) 25 gm balls in colour G. **Yarn** for legwarmers: Sirdar Superwash 4-ply, 5(6) 25 gm balls in main colour A, 1 25 gm ball each in colours B, C, E, F, G, 2 25 gm balls in colour D. **Needles:** one pair each of No. 12 British (No. 1 American, 2¾ mm), No. 10 British (No. 3 American, 3¼ mm), set of four double pointed needles size No. 12 British (No. 1 American, 2¾ mm). **Also:** stitch holder.

MEASUREMENTS **To fit:** bust/chest 34(36:38:40) in (86(91:97:102) cm). **All round:** 36(38:40:42) in (91(97:102:107) cm). **Length:** 23½(24½:25:26) in (60(62:63:66) cm). **Legwarmers:** two sizes, length 23½ (24½) in (60 (62) cm) measure top and bottom cuffs folded in half.

TENSION 30 sts to 4 in (10 cm) over Fair Isle on No. 10 needles, and approximately 30 rows to 4 in (10 cm).

ABBREVIATIONS See page 176. **Colours:** A=Camel, B=Light Navy, C=Festive Scarlet, D=Ivory, E=Wine, F=Peppermint, G=Triton Green.

BACK

With No. 12 needles and A yarn cast on 129(145:145:161) sts. **1st row** K1 * P1 K1, rep from * to end. **2nd row** P1 * K1 P1, rep from * to end. Rep these 2 rows for 3 in (7 cm), ending

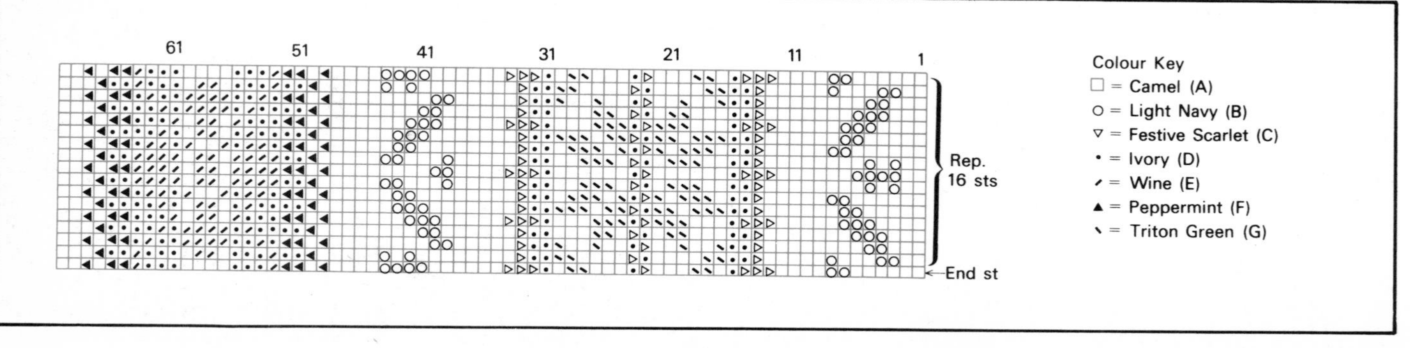

with a 2nd row. Change to No. 10 needles and beg with a K row cont in st.st, working in patt from chart, until work measures 16 in (41 cm) from beg, or length required, ending with a P row. **Armhole shaping:** cast off 4(9:6:11) sts at beg of next 2 rows, then 2 sts at beg of next 4 rows (113(119:125:131) sts). Dec one st at each end of next and foll 7(8:9:10) alt rows (97(101:105:109) sts). Cont without shaping until armholes measure 7½(8½:9:10) in (19(21:23:25) cm), ending with a P row. **Shoulder shaping:** cast off 9(9:10:10) sts at beg of next 4 rows, then 8(9:8:9) sts at beg of next two rows. Leave rem 45(47:49:51) sts on holder.

FRONT

With No. 12 needles and A yarn cast on 145(145:161:161) sts and work as given for Back until 16(10:10:4) rows less than back to armholes, ending with a P row. **Divide for neck: next row** K72(72:80:80), turn and leave rem sts on a spare needle. Dec one st at neck edge on next row, then work 2 rows. Rep the last 3 rows 4(2:2:0) times more, thus ending at side edge. **Armhole shaping: next row** cast off 12(9:14:11), K to last 2 sts, K2tog. At armhole edge, cast off 2 sts at beg of foll 2 alt rows, then dec one st at beg of foll 8(9:10:11) alt rows and **at the same time** at neck edge cont to dec one st on every 3rd row as before until 26(27:28:29) sts rem. Cont without shaping until armhole measures the same as on Back, ending at armhole edge. **Shoulder shaping:** cast off 9(9:10:10) sts at beg of next and foll alt row. Work 1 row, then cast off rem 8(9:8:9) sts. Return to the sts on spare needle, slip 1st st on to safety-pin for front neck, rejoin yarn and K to end. Cont to match first side, reversing all shaping.

NECK BAND

Join shoulder seams. With set of four No. 12 needles pointed at both ends and A yarn and right side facing, K back neck sts, pick up and K60(64:68:72) sts down left front neck, K centre front st from safety-pin, then pick up and K60(64:68:72) sts up right front neck. 166(176:186:196) sts. **Next round** work in K1 P1 rib to 2 sts before centre front st, P2tog, K1 P2tog tbl rib to end. Rep this round 9 times more. Cast off in rib, still dec at centre front.

ARMHOLE BORDERS

With No. 12 needles and A yarn and right side facing, pick up and K129(137:145:153) sts round armhole. Beg with a 2nd rib row, work 10 rows in rib as on back. Cast off in rib. **To make up:** press work with a warm iron over a damp cloth. Join side seams. Press seams.

LEGWARMERS

With No. 12 needles and A yarn cast on 65(81) sts and work in rib as on slipover for 3(4) in, 7(10) cm, ending with a 2nd row. Change to No. 10 needles and beg with a K row cont in st.st, working in patt from chart, inc one st at each end of every 4th row until there are 113(129) sts, working the extra sts into patt, then cont without shaping until work measures 23(24) in or 58(61) cm from beg, ending with a P row. Change to No. 12 needles and cont in rib for 4(5) in or 10(13) cm. Cast off **loosely** in rib. **To make up:** press work with a warm iron over a damp cloth. Join seam, reversing seam on half of the rib at top and bottom for turn-over. Press seam.

Handknitting Abbreviations

alt	alternate
beg	beginning
cm	centimetre
cont	continue
dec	decrease by working 2 stitches together
foll	following
g.st	garter stitch (every row K)
in	inch
inc	increase by working into front and then back of stitch
K	knit
KC	knit cross (lift loop between stitches and knit into back)
K2tog	knit 2 together
K2OA	K20 using A single and so on for all colours
MC	main colour
ML	make 1 stitch by picking up horizontal loop lying before next stitch and working into back of it
mm	millimetre
0	0times
patt	pattern
P1AH	P1 using A and H tog and so on for all colours
PSSO	pass slipped stitch over
P	purl
P2tog	purl 2 together
RC	right cross
rem	remaining
rep	repeat
reqd	required
reverse st.st	reverse stocking stitch (1 row purl, 1 row knit)
sl.1	slip 1 stitch knitways
sl.1p	slip 1 stitch purlways
sl.1K1PSSO	slip 1 knit 1 pass slipped stitch over
sp	space
ss	slip stitch
st	stitch
st.st	stocking stitch (1 row knit, 1 row purl)
tog	together
tbl	through back of loops
yfwd	yarn forward
yrn	yarn round needle
ybk	yarn to back of work

Notes on measurements Measurements have been given in inches and centimetres. Use only one set of measurements; do not mix the two.

Notes on charts and graphs One square represents one stitch of knitting, and each horizontal row of squares represents one row of knitting. For flat knitting, the first and all following odd-numbered rows should be read from right to left, and the second and all following even-numbered rows should be read from left to right. For knitting in rounds, all rows should be read from right to left.

Technological Chic

Introduction

Colour is a designer's best friend – the more you do with it, the more it does for you. Some designers prefer to work with a piquant palette of contrast colours, others contrive whole collections around a few subtle shades. Whatever their differences, all top design collections have this in common: their colours are '*true*' – linking outfit to outfit, texture to texture and fabric to fabric. Whether it's a straight colour match or a tonal counterpoint, pattern on pattern or a medley of plains, good colours tie pieces together, stretch one design into several and, best of all, give whole collections, top to toe, an effortless gloss of style.

Exclusive yarns and fabrics, usually dyed and woven to order, give the *couture* client a privileged passport to good looks. But things are not so easy on the every-day level of design, as anyone who has tried to find a red sweater to match some red slacks will know only too well. If there are so few primary colours, why are there so many reds around – and none of them right? Finding knitting yarns and fabrics in complementary colours presents just as many problems, particularly in the case of colours outside the classic range. These obstacles can be overcome at a price – but not one that everyone is willing or able to pay. So why isn't there a good basic fibre that translates into fabric and knitwear, in colours that you can really do something with?

It's easy to blame a lack of imagination in the yarn and fabric industry and leave it at that, but there *is* another side to the story. Colour forecasts become available to the trade a year to eighteen months before the season in question. Traditionally, spinners and knitters have had to order their colours – again well in advance of the season – in *minimums* of 5-tonne lots. The quantities involved, and the time factor, represented a risk that few were willing to take. With little incentive to experiment so far as colour was concerned, the industry backed off – and fashion started to fade.

Now, however, a technological innovation offers spinners and knitters an alternative – one that could have a dramatic effect on design. Courtaulds, the international fibre specialists, have developed a new Neochrome dyeing process for their Courtelle acrylic fibre. As a result, industrial users of Courtelle can now buy dyed fibre from a carefully researched spectrum of fashion and classic colours all through the season, in quantities as small as sampling weights. This takes the risk away from the spinners, so they can afford to be creative with colour.

Acrylic is an extremely versatile fibre, lending itself equally well to jersey fabrics, machine knitting and handknits. The design potential in a colour range linking these three key areas of popular priced fashion is tremendous. Victor Herbert was commissioned to devise a fashion collection with a difference – one that would give the spinners and knitters an imaginative look at what can be done with colour, and set them thinking about how they can make colour work for *you*.

Victor Herbert

'The New Technology' was the starting point of the collection – the object was to show just what it was all about, and how to make the most of it. Obviously, the new dyeing process and coloured fibre stocks have very exciting implications for the industry. It makes working with colour much more immediate and viable, and it ought to involve the industry much more in fashion and design. So that was my aim – to present a really lively approach to colour and pattern, something that would show the potential in the new technology, and introduce the spinners to a new way of thinking about colour and using it to put their ranges together.

The new processes are a break with tradition, so I wanted the collection to break away from tradition too. To be appealing, but *different*. One could have approached it in a perfectly conventional way and ended up with tailored shapes and separates, tuck stitch patterns, tight textures – everything you find in a normal commercial range. For me, that treatment was never a possibility. You don't need a fashion designer for that. You need a fashion designer to give you something *new*, show you there are *other* ways of doing things.

To begin, I concentrated on what I wanted to achieve, and then how best to put it across within the technical limits of those parts of the industry directly affected – the producers of jersey fabrics, industrial knits and handknitting yarns. You have to be positive, and take advantage of what the machinery can do *best*. The important thing about knitting is that, unlike other fabrics, the pattern can be fairly direct. You can program a computer or knitting machine to knit a pattern – a stripe or zigzag say – about as easily as you could paint it on paper with a brush. It's exactly the same for handknits. The objective was to put colour across very strongly and the affinity of knitting for strong patterning was the ideal means. So there was the basis of the collection.

I decided to use just six of the new fashion colours that would be available to the spinners – Japonica pink, Wisteria purple, Kingfisher blue, Chilli rust, Ochre yellow and Courgette green – and use them to 'paint' geometric patterns onto the fabric. There was to be a variable scale for the patterns – small for jersey, medium for industrial knitting and large for handknitting. I put a team together, including Anna Watson to do the jersey and my partner Christine Ruffhead to do the handknits, and work got under way.

Colour was the most important thing, so in order not to detract from it or complicate the message, all the knits were kept flat rather than textured. It was essential to show how much could be achieved by working with colour alone. I also wanted to demonstrate to the spinners and knitters that they could do a lot with colour just by using in a new way the resources they already had. So we went round the country, looking in the archives of the factories that were producing the fabrics for the collection. Most factories keep cuttings of the fabrics they have produced along with the original pattern or tape, and in some cases the archives go all the way back to the 1920s. We'd go through the samples and pick out geometrics we liked – diamonds, spots, triangles, stripes and so on. Some we just re-coloured, in other cases we asked for minor variations – could they make the triangles larger, or do the diamonds in three colours instead of one? No problem. Or we might ask them to run a diagonal stripe across what was already a good basic design. All they had to do was take the pattern out of the archives, make the adjustments, get it onto the machine and put it into production. We managed to get most of our fabrics that way. Which shows what you can do with what you've got. It also represents a great potential saving of time and money.

As a general theme, the combination of technology and colour made me think of bright fabrics against sharp, shiny metal – and of heralds, jesters and knights in armour. While a medieval theme wouldn't have been appropriate in the context of fibres and fashions for the future, the idea led on to the themes of circuses, masquerades and Harlequins – all the striking colours, patterns and shapes that you see on fairground and pantomime posters. It seemed an ideal way to present colour and pattern very strongly, and to highlight the further potential in using them in combinations. In order to do that, we had to educate ourselves first. It was like learning to use a new palette of make-up – we went over the colours again and again, using them like building blocks, seeing what effect they had on each other. You can make a colour look completely different by combining it with others, and you can change the look of a pattern by putting it into different colourways. Once you start, you can go on forever.

The result of all this was over a hundred fabrics in different patterns and colourways, and we set about putting them together. From the very beginning, I had intended to use easy shapes. For one thing, I didn't want the cut to interfere with the patterns and colours. But equally, I wanted the collection to reflect the trend towards young, easy-wear clothes in fashion – things like track suits and T-shirts. Almost all of the final machine-knit and jersey designs involved combinations of fabrics, often in a classic quartering placement, like the Masquerade track suit. There were dresses that mixed diagonal stripes with penny spots, skirts in diamonds and chevrons, coats that mixed circles with triangles and squares, sometimes in the same colours but more often in several, all trimmed with matching or contrast striped ribbing.

The third part of the collection was the handknits designed by my partner Christine Ruffhead. The visual effect had to be very strong and striking, the treatment completely new. It was particularly important to break with tradition in this most traditional area of design.'

Christine Ruffhead

'The only difference between the handknits and the rest of the collection is the fact that the fabric is produced by hand, not by machine. By treating them as 'shapes' rather than as 'sweaters', it was possible to approach the designs in a completely different way.

All the handknits were one-size, requiring little or no shaping, leaving me free to concentrate on colour and pattern. Again, it was necessary to look closely at the strengths and limitations of the medium. Handknitting gives you two basic options – textured knitting and flat knitting – and I decided to concentrate on flat colour knitting. Handknitting also provides you with two basic stitch configurations – horizontal and vertical. Conventionally, you knit your pattern from the hem going up. But logically it's just as easy to turn the pattern round and knit from side seam to side seam. In practice it isn't often done because of the shaping involved. But if you minimize the shaping, or do away with it completely – what then?

The *Harlequin* design developed out of this approach. It's knitted in four panels, two plain and two patterned, with ribbing and welt added at the end. The Fair Isle panels use all six of the fashion colours – one is knitted from the bottom in the usual way. The other panel is knitted by rotating the design and starting from the line that will become the centre back when the pieces are put together. It's a great way to get lots of colour and variety of pattern onto the surface in the easiest possible way.

Harlequin shows that designs don't have to be conceived and constructed in the conventional handknit mode to be effective. In the course of the collection, many of our samples often led us to devise unusual methods of production. For example, diamond motley patterns that were, basically, patchwork. Just knitting squares, turning them round to make diamonds, and crocheting them together. Knitting a classic diamond pattern fabric is very demanding – this way it couldn't have been simpler.

Not all the designs were asymmetric to begin with, but many developed in that direction because of the geometric nature of the colour patterns. In *Jester* you get a pattern break at the shoulder, so the two different main colours just carried on naturally onto the sleeves. One of the sleeves was striped to link up to the third colour on the design, and also to the striped ribbing on the rest of the collection.

As you can see, the pattern instructions for these designs aren't completely conventional – if anything, they're much simpler. Things that are different don't have to be difficult – all they require is ingenuity and an open mind.'

Harlequin Instructions

MATERIALS **Yarn:** Courtelle Double Knitting, 6 25 gram balls in colours 1 and 2; 3 25 gram balls in colours 3, 4, 5, 6. **Needles:** No. 7 British (No. 6 American, 4½ mm). No 9 British (No 4 American, 3¾ mm)

MEASUREMENTS One size, to fit **bust** 34–38 in (86–97 cm).

COLOURS Col 1, Japonica pink. Col 2, Wisteria purple. Col 3, Ochre yellow. Col 4, Kingfisher blue. Col 5, Chilli rust. Col 6, Courgette green.

NOTE The sweater is knitted in 4 panels. The right front and left back are knitted in plain stocking stitch, the right front in colour 2, the left back in colour 1. The left front and right back are knitted in different Fair Isle patterns. The left front is knitted from the bottom upwards, the right back is knitted from the centre outwards. The ribbing for sleeves, neck and welt are picked up and knitted later.

ABBREVIATIONS: See page 176.

PLAIN PANELS (Right Front and Left Back)
With No. 7 needles and col 1, cast on 63 sts and work into the backs of sts on first row. Now continue in st.st until work measures 14 in (35.5 cm). **Next row** Cast on 32 sts at the end of the row. Cont in st.st for another 8 in (20 cm). Cast off. Rep for second panel in col 2. See Chart A.

LEFT FRONT
With No. 7 needles and col 1, cast on 61 sts, 1 edge stitch included. K 1 row, knitting into backs of sts. P 1 row. Now work in Fair Isle pattern from graph, changing both colours after every 12 rows. If you look at the decreasing in Chart A you will see that you are losing 1 st on one side of the triangle as you go along the K row, then 1 st on the other side of the triangle as you come back on the P row. After 7 colour changes (approximately 14 in or 35.5 cm) cast on 25 sts including 1 edge stitch, and continue in pattern for another 4 colour changes (approximately 8 in or 20 cm). Cast off.

RIGHT BACK
With No. 7 needles and col 1, cast on (for centre back) 134 sts, including 2 edge stitches. Work first row into backs of sts, then P 1 row before beginning pattern. Now work in Fair Isle pattern from graph, keeping 1 edge stitch at either end. After 5 colour changes (approximately 10 in or 25.5 cm) cast off 85 sts at beg of next row. Continue in pattern with only 1 edge stitch now, at shoulder edge. Work for another 3 colour changes (approximately 6 in or 15 cm) and cast off.

JOIN PANELS
Line up front panels and oversew together neatly along the centre seam. Repeat for back panels.

NECK (Front and Back alike)
Working along the top of the joined panels, measure and mark the points 6 in (15 cm) to the right and left of the centre seams. Using No. 9 needles and col 4, pick up and K 52 sts across the front. Work in K1 P1 rib for 2 rows of each colour in the following order: cols 4, 5, 6, 1, 3, 2. Cast off neatly in rib. Repeat for back.

SLEEVES
Sew shoulder seams together including neck ribbing. With No. 9 needles now pick up 52 sts along sleeve edge and work in K1 P1 rib for 10 in (25.5 cm). Repeat for second sleeve.

WELT (Front and Back alike)
With No. 9 needles, pick up and K 78 sts along lower edge of front. Work in K1 P1 rib in 2 rows of each colour for 26 rows. Cast off neatly in rib. Repeat for back. Join side seams.

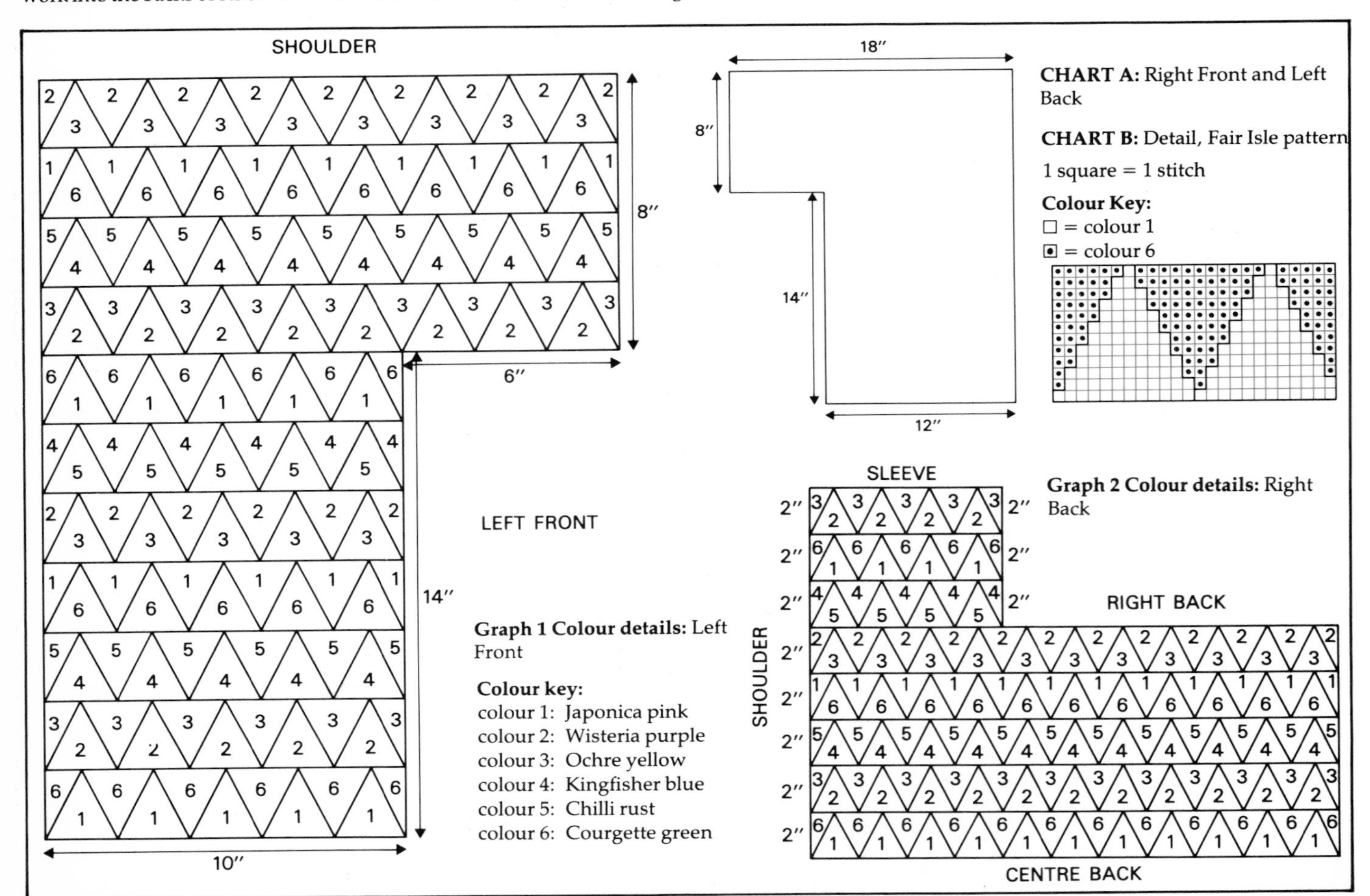

CHART A: Right Front and Left Back

CHART B: Detail, Fair Isle pattern

1 square = 1 stitch

Colour Key:
□ = colour 1
▣ = colour 6

Graph 1 Colour details: Left Front

Colour key:
colour 1: Japonica pink
colour 2: Wisteria purple
colour 3: Ochre yellow
colour 4: Kingfisher blue
colour 5: Chilli rust
colour 6: Courgette green

Graph 2 Colour details: Right Back

Jester Sweater Dress Instructions

DETAILS shown with bracelet, earrings and hair comb from Detail.

MATERIALS **Yarn: Courtelle Double Knitting**, 12 25 gram balls in colour 1, 2 25 gram balls in colour 2, 12 25 gram balls in colour 3. **Needles** No. 7 British (No. 6 American, 4½ mm), No. 9 British (No. 4 American, 3¾ mm).

MEASUREMENTS One size, to fit **bust** 34–38 in (86–97 cm).

COLOURS Col 1, Chilli rust. Col 2, Kingfisher blue. Col 3, Ochre yellow.

ABBREVIATIONS: See page 176.

NOTE Work one sleeve in colour 3 and one sleeve in stripes as follows: work 1½ in (4 cm) in colour 1 and ½ in (1.5 cm) in colour 2. For the sweater length, knit three panels only.

FRONT

Using No. 9 needles and col 1, cast on 104 sts and work in K1 P1 rib for 4 in (10 cm). Change to No. 7 needles and stocking stitch. Now work from Graph 1. **Beginning at point A****K to last 4 sts, turn and P to end, **slipping first stitch**. Repeat from ** until all the sts are held. **Next row:** using col 1, K across all sts. **Next row:** change to col 2 and work 4 rows stocking stitch. **Next row: Beginning at point B** change to col 3. *** P4, turn and K to end, **slipping first stitch**. **Next row:** P8, turn and K to end, slipping first stitch. **Next row:** Continue as from ***, but working in an extra 4 sts each P row. This should bring you to a level row, line C to D. Change to col 2. Beginning with a P row, work 3 rows straight. **Repeat whole pattern twice more from ** to **. Now repeat the pattern a fourth time, but do not work the last 4 rows in col 2. Work an extra row in col 3 and cast off.

BACK

See Graph 2. Work exactly as for front, but work 2 in (5 cm) extra in col 3 before casting off.

SLEEVES

See Graph 3. With No. 9 needles cast on 48 sts and work in K1 P1 rib for 10 in (25.5 cm). **Next row:** K across row, knitting twice into every stitch (96 sts). Now change to No. 7 needles and stocking stitch and continue until work measures 20 in (51 cm) from the beginning. Cast off.

NECK

Sew up side seams. When sewing up, leave a gap of 12 in (30.5 cm) across back and front for neck. With No. 9 needles pick up 52 sts across front and work in K1 P1 rib for 2 rows of each colour. Cast off neatly in rib. Repeat for back. To finish, join neck rib seams.

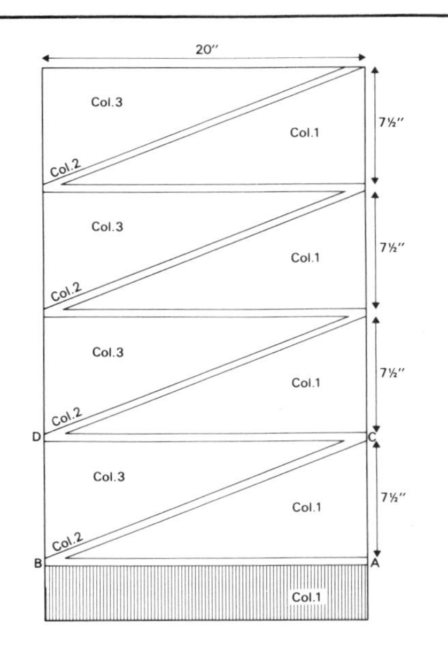

Graph 1: Front

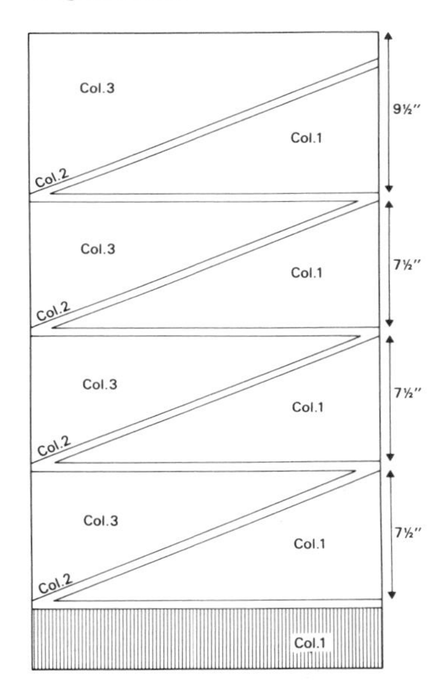

Graph 2: Back

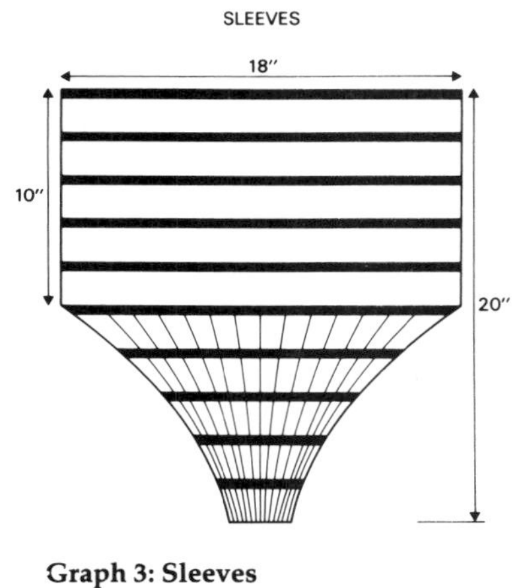

Graph 3: Sleeves

Masquerade Track Suit Instructions

DETAILS Shown with mask by Richard Glassborow, earrings and bracelet from Detail and hair combs by Tailpieces. Fabric by Anna Watson.

MATERIALS **Fabric required for suit in one colour:** 4 yd 28 in (4.38 m) of 60 in (152 cm) wide fabric. **Fabric required for suit in two colours:** 2 yd 14 in (2.19 m) of 36 in (90 cm) wide fabric in **each** colour. **Ribbing:** 25 in (64 cm) of 36 in (90 cm) wide one-way stretch ribbing or one-way stretch fabric. Alternately, ribbing pieces can be handknitted. **Recommended fabrics:** suitable for stretch dress-weight fabrics such as jersey and jacquards. Avoid fabrics that are stiff, thick or heavy. **Made here:** in Courtelle jacquard. **Trimming:** 1 piece of ⅜in (1 cm) wide elastic for waist of trousers. **Also:** a fine ball-point machine needle if making up jersey or acrylic fabric, and stretch sewing thread.

MEASUREMENTS One size, to fit **bust** 32–36 in (81–91 cm), **hips** 33–37 in (84–94 cm). **Back length** of top: 28 in (71 cm) to bottom of ribbing. **Trouser length:** 40½ in (103 cm) to bottom of ribbing.

STYLING The track suit can be made in one colour, but it provides an excellent opportunity for interesting colour combinations. You can quarter two fabrics as shown here or divide them in half, matching panels at the side seams so that the right front and right back are in one fabric, the left front and back in another. You can alternate plain and patterned panels, or use a different fabric for every piece if you can find suitable short lengths at a good price. Do not mix fabrics of different fibres and weights. In all cases, go over the cutting layout carefully and adjust it to suit your requirements.

NOTE All seam allowances are ⅜ in (1 cm) unless otherwise stated.

TOP

PREPARATION **(1)** Cut out pattern pieces following cutting layout. **(2)** Neaten seam allowances by using a zigzag stitch. **ASSEMBLE TOP (1)** With right sides together, join back pieces at centre back seam and front pieces at centre front seam. **(2)** Lay the completed body parts together with right sides inside and stitch both side and shoulder seams. **ASSEMBLE COLLAR (1)** Prepare collar by folding at centre with the right side outside and zigzag together as shown in Figure 1. **(2)** Keeping centre notches together, fold the collar to form the halter shape as shown in Figure 2. **(3)** Machine stitch the 4 layers of fabric into position across the bottom of the halter shape A-B. The stitch line should be ⅛ in (3 mm) from edge.

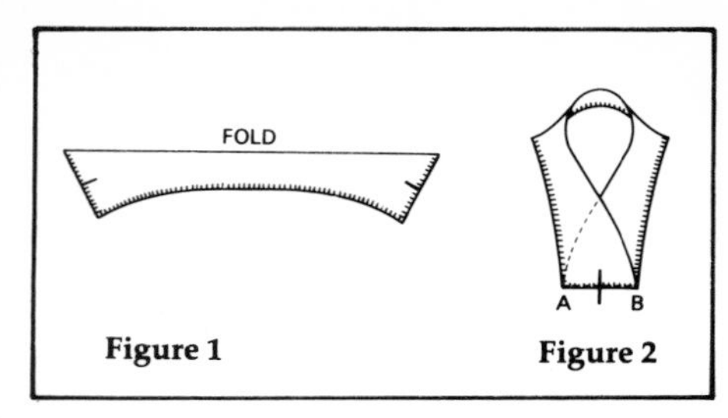

Figure 1 Figure 2

ATTACH COLLAR TO BODY (1) Neaten neck edge on the body part by zigzag all round. **(2)** With right sides together, stitch line A-B on collar to A-B at bottom of neck opening. Stitch firmly **only** up to the exact point, leaving the corner seam allowance free. **(3)** Nip the corners on the body part and complete collar assembly by stitching from A to the back neck, then to corner B. **(4)** Press seams back toward body part lightly. **ATTACH RIBBING TO WELT (1)** Neaten hem by zigzag. **(2)** Fold the hem ribbing in half with right side outside and join ends to make a circle, locking in the seam allowance. **(3)** Zigzag the raw edges together. **(4)** Place the joining seam of the ribbing to one side seam and ease the body part to the hem ribbing evenly all round. Machine stitch into position. **(5)** Press the seam allowance upwards lightly. **TO FINISH (1)** Turn up the sleeve opening allowance and stitch cuffs into position by hand or machine. **(2)** Press lightly.

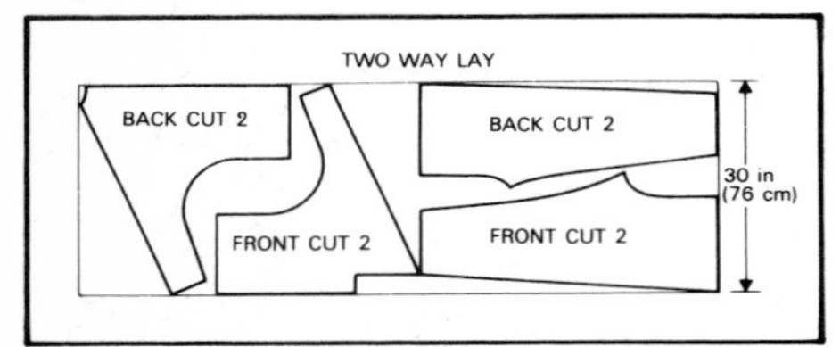

CUTTING LAYOUT (not drawn to scale) for track suit in one colour
Fabric width: 60 in (152 cm)

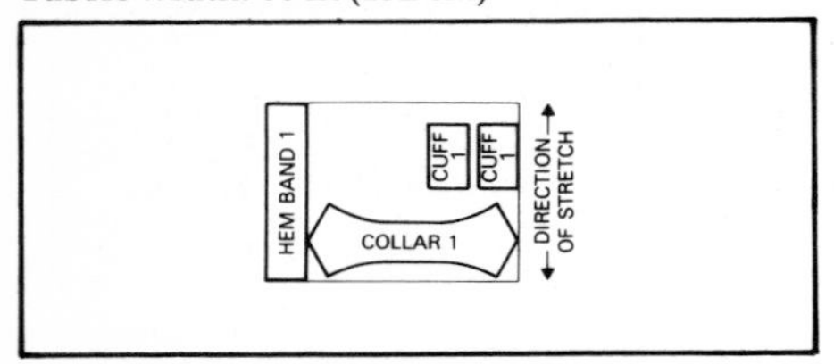

CUTTING LAYOUT (not drawn to scale) for single width stretch ribbing
Fabric width: 36 in (90 cm)

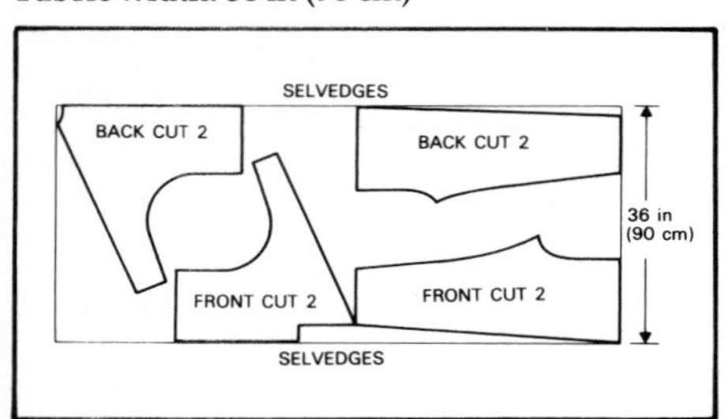

CUTTING LAYOUT (not drawn to scale) for track suit in two colours as in photograph
Fabric width: 36 in (90 cm); 2 lengths

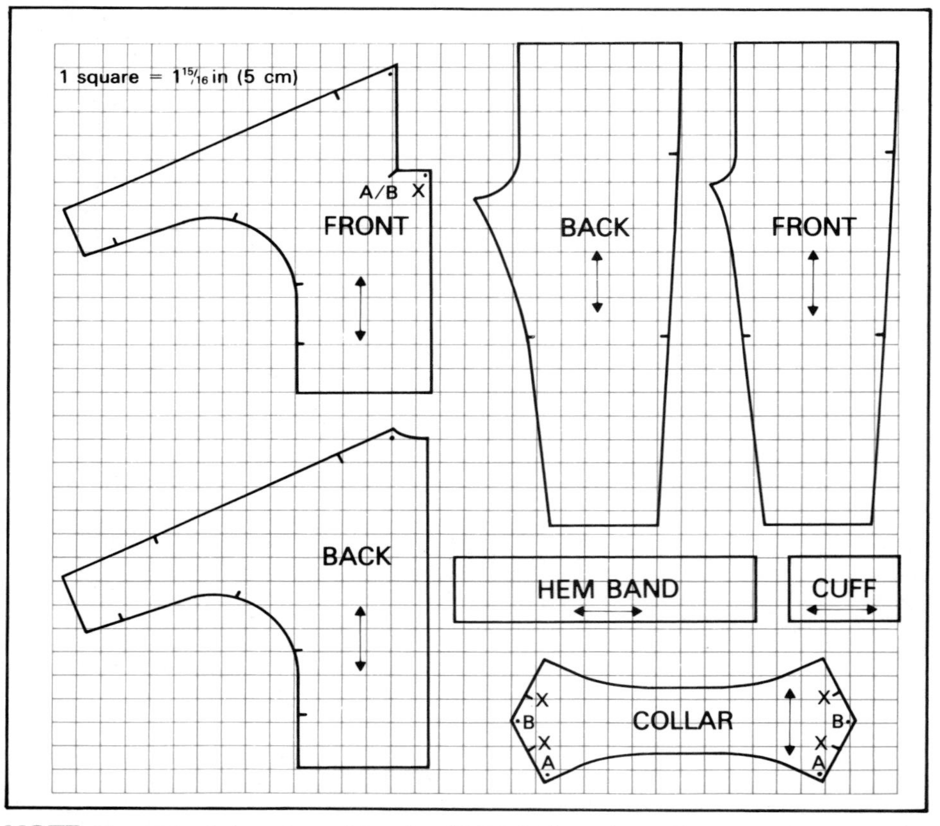

NOTE The cutting layout shows 2 lengths of fabric laid together with wrong sides inside. Following this lay, the right front half of the top will be in the top colour, the left front in the bottom colour. The left front leg will be in the top colour, the right front leg will be in the bottom colour. The backs will be the exact reverse of the fronts.

TROUSERS

ASSEMBLE BODY **(1)** Lay 1 back and 1 front leg piece together with right sides together, matching all notches. **(2)** Stitch the side and leg seams. **(3)** Repeat for second leg. **JOIN LEGS (1)** Lay the finished leg pieces together and pin into position from centre front to centre back seam, matching up the inside leg seams at crotch. **(2)** Stitch twice, for extra strength. **COMPLETE WAISTBAND (1)** Neaten the top edge of the waistline by zigzag. **(2)** Turn the waistband channel allowance down to inside and stitch into position ¾ in (2 cm) from top edge, leaving a small opening at centre back for the elastic. **(3)** Insert elastic through opening and thread through waist channel. **(4)** Fasten off by hand or machine. **TO FINISH (1)** Join the ribbing for ankle cuffs and proceed exactly as for the hem ribbing of top. **(2)** Press lightly.

The Wedding

Introduction

Couture collections traditionally end with a wedding dress, and ours is a very special dress indeed. Every bride should look and feel a princess on her wedding day. Maureen Baker, designer of Princess Anne's wedding dress, talks about brides, beautiful weddings and the dress she created for us.

Maureen Baker

For me, a wedding dress must always be a *wedding dress*. I have many clients who wear their wedding dress just the one day, and put it away in lavender forever. I have many other clients who want to be able to wear their dress more than once. This wedding dress will suit both points of view, but it is still a *wedding* dress, and I don't like to see a bride married in anything less.

This dress has the soft, feminine look and fluid lines that have always been the basis of my designs. Made in silk and lace, it is typically a wedding dress, but you'll find it very adaptable. To begin with, you can make the dress as it's shown here, or inset the trimming on the sleeves so you can see the arms through the lace. After the wedding, you can remove the bottom tier and straightaway you have a very pretty summer cocktail dress. If you made up that model in black trimmed with black lace – the sleeve lace really should be inset – you'd have a cocktail dress to wear all year round. Going back to the wedding dress, you could remove the sleeves as well as the bottom tier and wear it as a summer day dress. Or you could leave on all the tiers and just remove the sleeves – and there's a long summer evening dress. A lot of women insist on having sleeves, but a young bride isn't likely to have to worry about her arms.

The fact that the lines of the dress are very flattering contributes a lot to its versatility. And I must say that I don't think huge, tent-like wedding dresses *are* flattering. A girl's figure is important, and I think a wedding dress should have a proper shape. Even a big girl would look very romantic in this dress. It wouldn't show if she had a thick waist because the skirt starts above the natural waistline. The skirt also makes one look taller, and helps to round out a figure that is on the thin or small side. A wedding dress needn't fit like a glove, but there will be plenty of time for tents later on.

When a bride-to-be comes to me, I think first about what will suit her face and figure. Then I chat to her, to discover the things *she's* fond of. Because it's *her* day, and the dress must bring out everything that she feels. Obviously, some girls prefer very plain dresses and others like very fanciful ones. It can look very cute and pretty, but very often all that froth doesn't do a thing for anyone. All one has eyes for is the dress, not the bride. A wedding dress must never overwhelm the girl who wears it. On the day, it's the bride who must look magnificent – not just her dress.

Should a wedding dress be cream or white? I find that most brides prefer pure white, but it's best to choose the colour that suits your complexion and colouring. White suits very pale complexions, while creamy shades with the slightest touch of pink are more suitable for brunettes and brides with tawny skin tones. But of course it isn't as simple as light or dark, because there are so many shades in between. That's why I love to work with silk for wedding dresses. Silk has a natural sheen, and every silk has the subtlest *hint* of colour, just like pearls do, so you can always find just the right shade to set off a complexion perfectly.

Just as the fabric must suit the complexion, the trimmings must suit the fabric. Chantilly lace is my favourite, because of its pretty femininity and silky look. The finest Chantilly lace is made of silk or silk mixtures. It also comes in cotton, but cotton lace always looks hard against silk – it hasn't got the same depth and subtlety. If I wanted to use a cotton Chantilly, I'd make the wedding dress in a white lawn or man-made chiffon.

I like white and cream equally well, but I *don't* like a coloured wedding dress. A girl can wear any colour she likes every other day of her life. If she's been a bridesmaid, she'll have had a chance to wear a coloured dress in a wedding context, so why not save the pure whites and creams for the one special day? I feel the same way about coloured trimmings, but when it comes to the bride's bouquet then, yes, I do quite like to see some soft flower colours.

What I like to see best is a colour theme carried through an entire wedding. If it's a church wedding, I like to see the flowers in the church, the outfits worn by the mothers of the bride and groom, the dresses worn by the bridesmaids, the costumes worn by the children, and the bouquet carried by the bride all done in a colour theme that *enhances the bride*.

For example, I did a Christmas wedding where the bridesmaids were dressed in pinafores of dusky rose velvet, worn over blouses of soft cream silk. The page boy was dressed in winy maroon velvet, the mother of the bride in a soft browny beige ensemble, the mother of the groom in lighter shades of the same colours, and all the flowers in the church were in shades of beige, cream, brown and yellow. An unusual colour scheme, but one that blended superbly, a perfect setting for the bride in her gown of ivory silk. Naturally it often happens that one member of the wedding party may not want to dress in the theme colours. This will not diminish the beauty of the colour scheme as long as the colours of the attendants' dresses tone with the colours in the bride's bouquet and, if at all possible, the colours of the flowers in the church. I've spoken only about colours, but I should say that pretty, soft prints are perfectly suitable for attendants' dresses. When the print reflects the colours – and better still, the motifs – of the flowers in the bride's bouquet, you couldn't wish for a lovelier summer wedding theme.

A wedding dress is unusual in that the back is what is seen most of the time. So a wedding dress must always have a very pretty back and should, I feel, always be worn with a lovely veil. Traditionally the bridal veil was only lifted after the wedding ring had been placed upon her finger, but today it is customary for the veil to be worn back throughout the ceremony. Veils can be attached to little caps or to a flat headband ending in clusters of flowers, but I much prefer to see them attached to a pretty coronet, as shown here. Veils can be made of net or lace, but for me the loveliest veils of all will always be made of fine silk tulle, soft and light as a cloud.

I don't encourage a bride to wear lots of jewellery. Little pearl earrings are always very pretty, tiny gold earrings are nice as well, and for the neck there should be nothing more elaborate than a single string of pearls or a fine gold chain. Her greatest jewel is her beauty, for every bride is beautiful on her wedding day.

She needs nothing more except, of course, something old, something new, something borrowed and something blue. The wedding dress is new, and brides always have a little something that's old and that's borrowed. The last is my little present to the bride. I always stitch a small blue bow onto the inside of the petticoats of all my special wedding dresses. Yes – even Princess Anne's.

Wedding Dress Instructions

DETAILS Shown with shoes by Charles Jourdan and garland made with fabric flowers from Beale & Inman Ladies Department.

MATERIALS **Fabric required for dress:** 7⅛ yd (6.55 m) of 44–45 in (115 cm) wide fabric; **for underskirt and lining,** 4¾ yd (4.35 m) of 36 in (90 cm) wide fabric; **also for underskirt,** 7⅝ yd (7 m) of 44–45 in (115 cm) wide net for flounce. **Recommended fabrics for dress:** pure silk crepe de Chine, satin crepe, polyester satin, cotton lawn; **for underskirt and lining:** organdie or organza. **Also:** 20 yd (18.3 m) of 2 in (5 cm) wide Chantilly lace; 1 22 in (56 cm) long zip fastener; 4 sets of hooks and eyes. **Made here:** in double weight pure silk crepe de Chine.

MEASUREMENTS One size, to fit **bust** 36 in (91.5 cm), **waist** 26 in (66 cm). **Back length** from shoulder seam to finished hem, 60½ in (153 cm). Length can be adjusted to suit personal requirements. **Note** that the waist of this dress sits slightly above the natural waistline; check measurements carefully before you begin to cut out.

CONSTRUCTION The instructions are given in two stages. Stage 1: bodice, sleeves and the first 2 skirt tiers with lace trimming. Stage 2: the second 2 skirt tiers with lace trimming except for the hem line. **This sequence is recommended for ease of making.**

NOTE **Seam allowances** for side seams, centre back bodice and zip seam on first skirt tier, 1 in (2.5 cm); for neck and cuff rouleau bands, ¼ in (6 mm). All other seam allowances, ½ in (1.2 cm).

PREPARATION **(1)** Cut out pattern pieces following cutting layout. **(2)** Use bodice pattern pieces to cut out bodice linings. **(3)** Tack top and lining layers together and treat as one. **ASSEMBLE BODICE (1)** With right sides together and notches matching, stitch the front side bodice pieces to the front bodice panel. Press seams open and neaten by zigzag or hand. **(2)** Stitch the back bodice side panels to the back bodice pieces. Press seams open, and neaten. **ASSEMBLE SLEEVES (1)** Take 1 sleeve and 1 length of lace trim. Pin the lace to the centre of the sleeve on the right side, as marked on the pattern. **(2)** Machine stitch the lace to the sleeve just inside the edges of the lace. **(3)** Repeat for second sleeve. **(4)** With right sides together, stitch the cuff vent facing to cuff vent, turn facing to the inside and finish by hand. Repeat for second cuff. **(5)** With right sides together, stitch the sleeve seams, neaten the edges by zigzag, and press seams open. **GATHER CUFFS AND ATTACH ROULEAU BANDS (1)** Working with a loose stitch just inside the raw edge, draw up the sleeve cuff edges to measure 8¼ in (21 cm) (5⅝ in or 14.3 cm round the front and 2⅝ in or 6.7 cm round the back) as indicated on pattern. **(2)** With right sides together, stitch the cuff rouleau bindings into position on the cuff edge. **(3)** Turn in the rouleau ends ¼ in (6 mm), fold the rouleau bindings over to the wrong side, turn in the ¼ in (6 mm) seam allowance and finish by hand. **ATTACH SLEEVES TO BODICE (1)** Working with a loose stitch just inside the raw edge, draw up the sleeve head edges between front and back notches to measure 8⅛ in (20.6 cm) as indicated on pattern. **(2)** Carefully tack the completed sleeves into the bodice armholes, matching up the underarm seams and notches. **(3)** Stitch sleeves into position. **(4)** Neaten seams together by zigzag and press lightly. **ASSEMBLE AND JOIN FIRST AND SECOND SKIRT TIERS (1)** Draw up the waist edge of the first tier to measure 26 in (66 cm) or waist size, leaving the zip seam allowance of 1 in (2.5 cm) at each end. **(2)** Make up the second tier by joining the two pieces at the selvedges, as indicated on pattern. Do not join to make a circle. **(3)** Draw up the **top** edge of the second tier to fit the **bottom** edge of the first tier. **(4)** With right sides together and with fullness evenly distributed, stitch the top of the second tier to the bottom of the first tier. Neaten edges together by zigzag. **JOIN SKIRT TIERS TO BODICE (1)** With right sides together and with the fullness evenly distributed, stitch the waist edge of the first skirt tier to the bodice. Neaten the edges by zigzag. **ATTACH LACE TRIMMING (1)** Lay the assembled bodice and 2 skirt tiers flat with right sides up. **(2)** Cut 1 length of lace trim to the waist size **plus** the 2 in (5 cm) back seam allowances. **(3)** Carefully tack the top edge of the lace to the bodice so the bottom edge of the lace covers the waist seam line by ⅜ in (1 cm). **(4)** With a single row machine stitching placed as near to the top edge of the lace as possible, stitch lace to the bodice **and** back seam allowance. **Do not stitch the bottom edge of the lace:** it should fall free over the seam line. **(5)** Attach lace to cover the top seam line of the second tier in exactly the same way. **(6)** Repeat for the bottom of the second tier making sure that the free edge of the lace trim will cover the seam join when the remaining tiers are added. **FINISH CENTRE BACK SEAM (1)** Pin or tack the back seam into position for stitching, matching up the seams and lace trim. **(2)** Stitch the seam from the bottom of the first tier to approximately 3 in (7.5 cm) from the bottom of the second tier. This opening will make it easier to attach the remaining skirt tiers later on. **(3)** Turn back the 1 in (2.5 cm) zip seam allowance and insert the zip fastener. **(4)** Attach and complete the neck rouleau binding exactly as for the cuffs. **ASSEMBLE REMAINING SKIRT TIERS (1)** With right sides together, join the 7 panel pieces that make up the bottom tier,

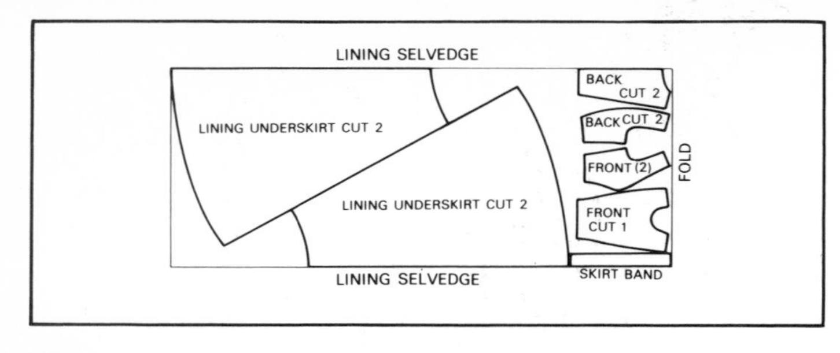

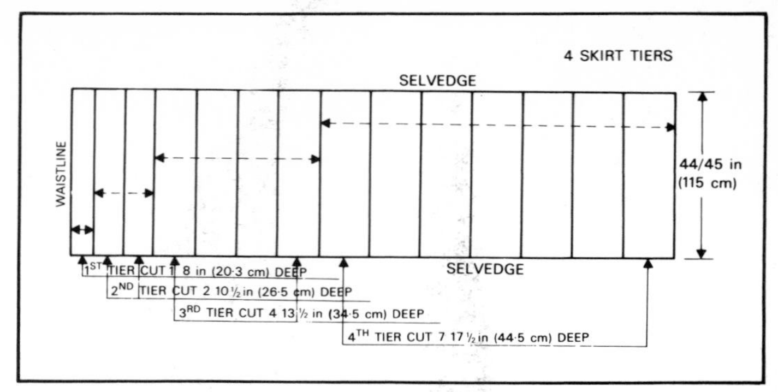

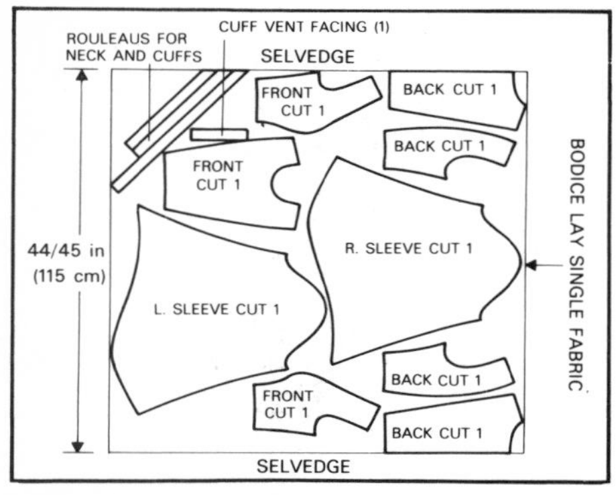

matching up the selvedges. Do not stitch the final seam to make a circle: allow the piece to lie flat. **(2)** Join the 4 panel pieces of the third tier in exactly the same way. **(3)** Draw up the top edge of the bottom tier to fit the bottom of the third tier. **(4)** With right sides together and the fullness evenly distributed, switch into position and neaten as before. **(5)** Attach the lace trim to cover the seam as before. **JOIN TIERS (1)** Draw up the top edge of the third tier to fit the bottom edge of the second tier. **(2)** Stitch together and neaten as before, taking care not to catch the lace trim into the seam line. **(3)** Complete back seam matching seams and lace trim, and press seam open. **TO FINISH (1)** Adjust hem to suit requirements, then finish hem line by turning up twice and stitching into position. **(2)** Attach lace trim as before. **(3)** Finish the neck and cuffs with small hooks and eyes.

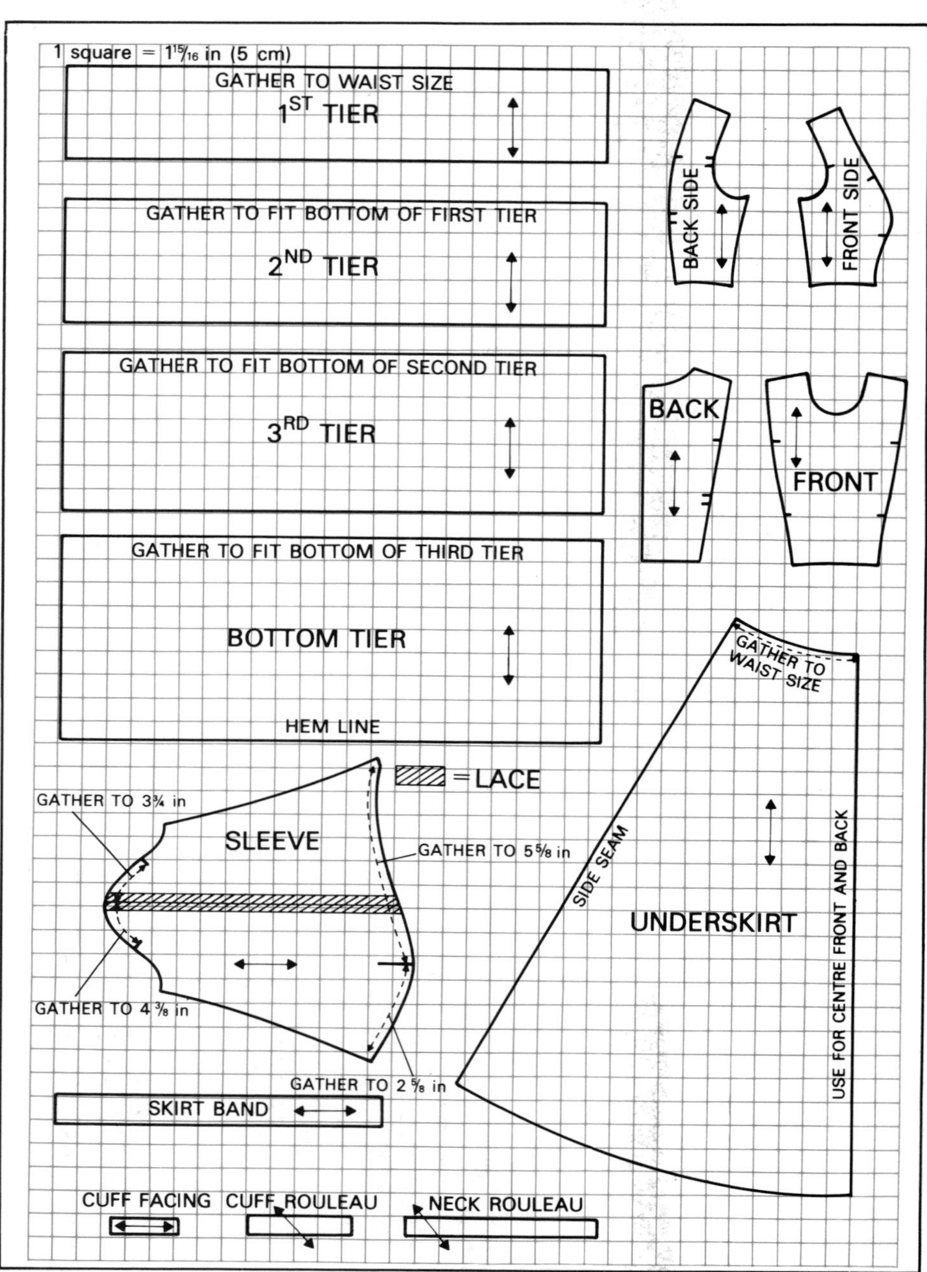

Underskirt Instructions

NOTE Seam allowances for centre back, centre front and side seams: ¾ in (2 cm): for waist edge and waistband, ¼ in (6 mm). Neaten seams by turning in ⅛ in (3 mm) and stitching flat.

PREPARATION (1) Cut out pattern pieces following cutting layout. **ASSEMBLE SKIRT (1)** With right sides together, stitch the centre front and centre back seams. **(2)** Neaten edges and press seams open. **(3)** Pin or tack the front to the back and stitch the side seams, leaving an opening of approximately 7 in (18 cm) at the top of the left side seam. **(4)** Neaten the edges and press seams open. **(5)** With a loose stitch, tack round the waist just inside the raw edge and draw up to 26 in (66 cm) or required waist size. **ATTACH WAISTBAND (1)** Fold the waistband lengthwise with right side outside, and press. **(2)** With right sides together, tack the single edge of the waistband to the skirt top, ensuring that the fullness is evenly distributed. **(3)** Stitch the waistband to the skirt, and turn the band to the inside. **(4)** Turn in the ends and turn under the inside seam allowance ¼ in (6 mm). Finish by hand or machine. **(5)** Stitch hook and eye on to waistband. **ATTACH NET FLOUNCE (1)** Adjust hem to suit requirements, turn up twice and machine stitch. Press. **(2)** Fold net in half along its length and gather it up along the folded edge to make a flounce 22 in (56 cm) deep and 126 in (319 cm) long. **(3)** Pin into place along the outside of the underskirt, keeping hems level. **(4)** Stitch into place, and trim net at hem if necessary.